D0914232

Developments in Geotechnical Engineering 16A

APPLIED SALT-ROCK MECHANICS, 1

Further titles in this series:

1. *G. SANGLERAT*
THE PENETROMETER AND SOIL EXPLORATION

2. *Q. ZARUBA AND V. MENCL*
LANDSLIDES AND THEIR CONTROL

3. *E.E. WAHLSTROM*
TUNNELING IN ROCK

4. *R. SILVESTER*
COASTAL ENGINEERING, I and II

5. *R.N. YOUNG AND B.P. WARKENTIN*
SOIL PROPERTIES AND BEHAVIOUR

6. *E.E. WAHLSTROM*
DAMS, DAM FOUNDATIONS, AND RESERVOIR SITES

7. *W.F. CHEN*
LIMIT ANALYSIS AND SOIL PLASTICITY

8. *L.N. PERSEN*
ROCK DYNAMICS AND GEOPHYSICAL EXPLORATION
Introduction to Stress Waves in Rocks

9. *M.D. GIDIGASU*
LATERITE SOIL ENGINEERING

10. *Q. ZARUBA AND V. MENCL*
ENGINEERING GEOLOGY

11. *H.K. GUPTA AND B.K. RASTOGI*
DAMS AND EARTHQUAKES

12. *F.H. CHEN*
FOUNDATIONS ON EXPANSIVE SOILS

13. *L. HOBST AND J. ZAJIC*
ANCHORING IN ROCK FORMATIONS

14. *B. VOIGT (Editor)*
ROCKSLIDES AND AVALANCHES, 1 and 2

15. *C. LOMNITZ AND E. ROSENBLUETH*
SEISMIC RISK AND ENGINEERING DECISIONS

Developments in Geotechnical Engineering 16A

APPLIED SALT-ROCK MECHANICS 1

The in-situ behavior of salt rocks

by

C.A. BAAR

ELSEVIER SCIENTIFIC PUBLISHING COMPANY

Amsterdam — Oxford — New York — 1977

ELSEVIER SCIENTIFIC PUBLISHING COMPANY
335 Jan van Galenstraat
P.O. Box 211, Amsterdam, The Netherlands

Distributors for The United States and Canada:

ELSEVIER/NORTH-HOLLAND INC.
52, Vanderbilt Avenue
New York, N.Y. 10017

Library of Congress Cataloging in Publication Data

Baar, C A
Applied salt-rock mechanics.

(Developments in geotechnical engineering ; 16A)
Bibliography: p.
Includes index.
CONTENTS: v. 1. The in-situ behavior of salt rocks.
1. Rock mechanics. 2. Salt deposits. I. Title.
II. Series.
TA706.B3 624'.1513 76-46495
ISBN 0-444-41500-9

Printed in Gt. Britain

CONTENTS

Chapter 1

INTRODUCTION

Ever since Man lived in caves he has been exploiting salt deposits; the fact that humans and animals need salt has been recognized long before any scientific or sophisticated knowledge of nutrition and feeding became available.

Salt mining may have been among Man's first mining activities; salt trading certainly was one of Man's first commercial activities, and wars over the possession of salt sources have been going on over thousands of years — until Man learned that buried salt deposits can be found almost everywhere in sedimentary formations a few hundred metres beneath the earth's surface.

Ever since Man went underground for mining salt deposits he has been facing the specific problems related to the unique properties of salt rocks compared to what is usually called rock: salt rocks are readily dissolved in water, and salt rocks flow slowly, but continuously, like glacier ice. As a matter of fact, salt glaciers exist in some parts of the world, particularly in Iran, where the dissolution of salt by meteoric water cannot keep pace with the flow of salt from great depths: buried salt deposits are squeezed to the surface by natural forces to form mountains of salt from which salt glaciers are flowing over surrounding plains.

There should be no need for proving that a corresponding flow of salt must be expected when openings are excavated in salt deposits at great depths. For many centuries, it has been known for a fact that underground openings in deep salt mines are closed completely by slow, but continuous flow which is usually termed creep. The deeper a salt mine and the higher the extraction ratio, the faster the complete closure of mine openings occurs by creep of the surrounding salt into mined-out openings. Under certain conditions described in publications, it took only a few years until mine openings had disappeared due to creep.

The problems encountered in drilling through deeply buried salt deposits are well known: extra-heavy drilling fluids must be used to prevent excessive borehole closure during the relatively short period of time of drilling through salt; even such measures cannot prevent the rapid closure of boreholes in salt at depths exceeding a few kilometres. It is not known how many boreholes had to be abandoned in salt deposits on account of rapid creep closure.

The difficulties caused by creep closure of openings in conventional salt and potash mines and in solution-mined caverns, washed-out for salt production or for the purpose of underground storage, have been dealt with in numerous publications. As the creep of salt rocks is time-dependent and the use of salt caverns for storage is relatively new, the continuous creep closure of salt caverns has not yet resulted in events as spectacular as caused by continuous creep in conventional salt and potash mines.

Only two recent events which produced headlines in newspapers all over the world are mentioned here to show possible consequences of the application of erroneous hypotheses in the design of underground excavations in salt deposits: on June 23, 1975, the strongest earthquake on record in Germany since 1911 was caused by potash mining; similar events occurred previously; in 1953, a rockburst above a potash mine produced surface fractures 400 m long, 4 m wide, and over 50 m deep. Such earthquakes are caused by inadequate design of potash mines operating at relatively shallow depths of 400—700 m.

The second event was caused by potash mining in a salt dome: in July 1975, a potash mine near Hannover, Germany, fell victim to an unexpected inrush of groundwater; craters and sinkholes formed at the surface and made the temporary evacuation of residential areas necessary as buildings collapsed.

At numerous locations all over the world, the extraction of salt deposits resulted in similar damage. In most cases, the solubility of salt rocks was underestimated, or ignored when mining operations underneath salt deposits resulted in subsidence and made the salt formations permeable. Unexpected creep deformations around mined-out openings in salt deposits are the main cause of groundwater inflow that, sooner or later, results in disastrous surface damages; it goes without saying that a continuous flow of groundwater, which dissolves salt before it is pumped back to the surface, must result in uncontrollable rock-mechanical reactions of overburden formations, such as rockbursts, differential subsidence, and sinkhole formation. Frequently, the distances over which groundwater flow systems are affected by such events are underestimated.

Three examples may be cited here as they show three different possibilities of how salt extraction can result in severe damage at the surface; in each case, large areas of German cities fell victim to the uncontrolled dissolution of salt deposits by groundwater, the dissolution being initiated by mining operations.

(1) The city of Stassfurt, known as the birthplace of potash mining which started about hundred years ago, suffered severe damage from conventional potash mining. Several mines operating within the city limits were lost by flooding during the first decades of potash mining as the strength of mine pillars, and the creep deformation of salt rocks were underestimated. At least one of the large surface craters which formed in residential areas is now in use as a lake resort area. Reports indicate continuing surface subsidence; large areas in the city centre had to be abandoned.

(2) The city of Eisleben, known as Martin Luther's birthplace, suffered

similar damage from copper-shale mining which initiated the dissolution of salt deposits above the mining levels. Groundwater from considerable distances was flowing for decades into the mines, dissolved overlying salt, and was pumped back to the surface to be channelled into major rivers.

(3) The city of Lüneburg, near Hamburg, fell partly victim to salt dissolution by inadequate solution-mining techniques; natural salt dissolution by groundwater may also have played a minor role.

These three examples are well documented in the technical literature. Many similar cases in other parts of the world attracted little scientific interest. This is deplorable since a wealth of information is lost; similar costly errors of the same type are made again and again.

The lack of general interest in applied rock mechanics in salt deposits is understandable, as salt and potash mining represent only a small fraction of the mining industry as a whole. Investments required for the construction of a new salt or potash mine used to be small compared to what the mining industry is spending on bringing other mining ventures into operation. However, with the need for going deeper underground in potash mining, and with the increasing need for large and highly mechanized operations to meet economic requirements, the situation has changed considerably: for example, 200—300 million dollars are required presently for a new potash mining operation in Canada, where the richest potash deposits discovered to date are being mined.

It has been claimed in technical literature that the first Canadian potash mine which went on stream in 1963 is the first known potash mine planned from the beginning on rock-mechanical principles developed in a comprehensive research program. However, these much publicized principles proved to be erroneous: instead of providing a mine that was expected to be stable for thousand years to come, the design of the first mined-out areas results in continuing creep closure of the extracted rooms; only 15 years after mining, the rooms are completely closed by creep of the surrounding salt.

In this particular case, no damage from the rapid, unforeseen creep is expected because the potash bed is only 2.3—2.4 m thick, and the extraction ratio is only about 1/3, with 15.6 m wide pillars left in the first mined-out room-and-pillar panels; the pillars, having a height-to-width ratio of nearly 1/5, do not fail when loaded with the full weight of more than 900 m of overburden. In spite of the unexpected and relatively rapid creep closure, the impermeability of a 30-m rock-salt bed between the mine workings and an aquifer above the salt is not jeopardized, clearly demonstrating very positive aspects of creep deformation of salt rocks: any other composition of the mine roof formation would result in fracturing as the continuing creep of the potash pillars is inevitable; water discharge from the aquifer into the mine would almost certainly have occurred, adding another name to the long list of flooded salt and potash mines. It should be mentioned that during the first 100 years of potash mining in Germany one potash mine shaft per year was lost by flooding.

The abovementioned case of unexpected, although foreseeable, creep in a deep potash mine also unveiled problems which are much more detrimental to salt and potash mining than is the inevitable creep experienced engineers have learned to cope with long ago: the designer, a civil engineer who apparently was not aware of previous experience in deep salt mining operations, based the mine design exclusively on theories of elastic behavior as, unfortunately, salt and potash rocks exhibit near-elastic behavior in the laboratory when subjected to standard strength-testing procedures. It is extremely dangerous to use strength parameters obtained in this way in the design of salt and potash mines because the in-situ conditions are entirely different.

However, some theoreticians, particularly university teachers and their students, apparently have too much time for publicizing of what they consider solutions to problems in salt deposits which, in fact, do not even exist; the real problems encountered in deep salt mining are not mentioned in that type of publication as the writers, having never been underground, failed to consult the comprehensive technical literature dealing with actual problems caused by continuing creep in deep salt mining operations. As a result, a vicious circle of academic discussion in ivory towers has developed; unfortunately, it has some extremely dangerous aspects when the hypotheses evolving from such theoretical discussions are used in the design of mining operations in deep salt deposits: some theoreticians seem not to be bothered with disastrous consequences of the practical application of their hypotheses, but continue to claim successful application, misleading other theoreticians, and luring mining engineers into dangerous practical application of erroneous hypotheses.

When such application results in disaster, some theoreticians seem to count on the tendency of those directly involved to disclose only what cannot be concealed because it is visible at the surface; underground data, particularly data obtained in in-situ creep measurements, are usually published in well-prepared, but arbitrary selection designed to "prove" the validity of certain hypotheses. There are only a few exceptions in publications by writers who are not aware of the real meaning of published original in-situ measurement data.

The case referred to above provides a typical example: over ten years ago, it had become obvious from extensive in-situ measurements that the underground creep rates were much higher than expected and predicted in publications; nevertheless, the designer as well as other theoreticians who promote similar hypotheses continue to claim successful application of their theoretical principles in actual mine design. However, the real problems which resulted from the application of erroneous hypotheses are not mentioned: the standard room-and-pillar design resulted in serious roof failure problems; roof failure occurred in individual rooms; under certain conditions, roof failure occurred during, or immediately after cutting the rooms with huge twin-borers.

The application of elasticity theories in attempts to resolve the roof failure problem led to the conclusion that the room width should be reduced to reduce the assumed stress concentration near the rooms; however, for reasons outlined in detail in Chapter 4, the reduction of the room width intensified the roof failure problem. In 1968, this writer was hired to resolve the difficulties; the solution was too simple to be accepted off-hand by the designer, and by mining engineers who had been misinformed for years: the rooms had just to be excavated wider, about twice as wide as originally designed, i.e., at least to 12 m instead of the design width of 6 m.

It took over a year of re-evaluation of accumulated measurements and of extensive underground testing requested by the designer until he was forced to agree that, under the given geological conditions, wider rooms were safer; the roof failure problem disappeared after mine management had given the green light for wider rooms.

The reaction of the designer is typical of what frequently happens when a theoretician is forced to revise his hypotheses which had been publicized in numerous research reports and other publications: not admitting to error in previous publications, and not understanding why the rooms had to be made wider in the particular case described above, the designer immediately published "proof" of the feasibility of wider rooms in other Canadian potash mines which operate under different geological conditions; these conditions are characterized by clay layers which separate salt and potash beds, and under such conditions, the rooms must not exceed certain maximum widths. As the result of the practical application of these hypothetical recommendations, serious roof failure problems developed in most of the Canadian potash mines which became operational in the late 1960's; large areas under development in the early 1970's had to be abandoned on account of roof failure.

To overcome these difficulties, known principles of the inelastic in-situ behavior of salt rocks had to be applied in what is termed stress-relief methods to be dealt with in detail in Volume 2. The reader should not be surprised at recent claims by the theoretician mentioned above that his hypotheses of elastic behavior led to the development of stress-relief methods for stabilizing openings in deep salt deposits, and that his methods are patented in the U.S.A. and in Canada.

Considering the inundation of the technical and scientific literature with publications in which some multi-writers, and their students, claim the successful practical application of their erroneous hypotheses, this writer finds himself in an awkward situation: obviously, it is insufficient to publish in-situ measurement data which show certain hypotheses wrong; such publications are simply ignored by those who continue publicizing erroneous hypotheses, disregarding the serious and frequently disastrous consequences of previous applications of such hypotheses. It is, therefore, necessary to show in detail why certain hypotheses are erroneous, where and when their practical application resulted in difficulties or even in disaster, and what can be done and should be done to prevent such consequences.

The validity of elasticity theories is not questioned — the point to get across to those engaged in applied rock mechanics in salt deposits is that salt rocks do not meet the prerequisite of elastic behavior in situ, in contrast to their behavior in the laboratory under standard strength-testing procedures. However, this is already a statement which is to be shown valid.

The mass of misleading recent publications dealing with these controversial matters made it impossible to present the available in-situ data, and to compare them with hypothetical postulates which proved erroneous, in one single volume. It was therefore decided to present the comprehensive material in two volumes; the second volume will deal particularly with stress-relief methods which proved feasible in deep salt deposits for the prevention of disasters which, to date, have been haunting the salt and potash mining industry.

The apparent potential for more disaster on account of error in the application of rock-mechanical hypotheses in salt deposits makes it necessary to spell out exactly the reasons why previous disasters occurred. Some theoreticians in teaching positions may dislike outspoken statements; they are invited to read carefully the following (slightly altered) foreword to a published seminar held by this writer early in 1971 at the Saskatchewan Research Council (SRC), University Campus, Saskatoon, for the trouble-plagued potash industry in Canada:

"This seminar is being held for the benefit of those who work underground and for those who are responsible for underground operations in salt deposits. The writer has been frank in criticism of what are believed to be errors in theories being applied to deep mining in salt deposits. This has been done because the writer has witnessed many serious accidents involving the loss of many lives and mines, and hence is very concerned that misleading hypotheses may lead to even more serious accidents. The objective is to contribute to a full discussion of published information which is to be identified by the authors' names — not to imply a personal attack on writers whose theories are criticized. The results and conclusions will be published in a different form for public discussion, and all those identified with certain theories will be given the opportunity to comment on criticism of their publications prior to any publication of such criticism."

This has been done; as a matter of fact, the principal multi-writers in the U.S.A. and in Germany were given the opportunity to defend their hypotheses in front of an international audience during a discussion session specially arranged at the Fourth International Symposium on Salt, Houston, Texas, April 1973. The outcome is summarized in the proceedings — however, recent publications by the same theoreticians indicate the failure of this approach as these writers resorted to more manipulation of data for alleged proof of successful application of their hypotheses in various salt deposits.

This attitude, and the recent disasters referred to above, no longer allow restraint in criticism of hypotheses which proved extremely dangerous when applied to the design of underground openings in salt deposits — particularly in view of the increasing use of salt deposits for underground storage, and for disposal of dangerous industrial wastes, including radioactive waste.

Chapter 2

GEOLOGY OF EVAPORITE DEPOSITS

2.1 OCEANIC ORIGIN OF EVAPORITE DEPOSITS

2.1.1 Introduction

This world's oceans are the source of the evaporite deposits dealt with in this text. Most major deposits were directly precipitated from concentrated sea water. Such primary deposits may have been dissolved and re-deposited as "second-cycle" deposits which are usually much smaller and thinner than the original marine deposits.

Second-cycle deposits are presently forming on all continents. These processes have been studied extensively. It should be borne in mind that such present-day salt deposition in shallow intracontinental depressions differs in many regards from the processes which governed the formation of large marine evaporite deposits in the geological past.

There are a few cases of marine evaporites forming presently under exceptional conditions along the coast of the oceans; in addition, Man had learned long ago how to create the conditions for the solar evaporation of sea water to produce salt in relatively small ponds. Again, caution is indicated in comparing such present-day salt deposition with the geological processes involved in the formation of large oceanic salt deposits.

The average salt content of ocean water amounts to 3.5 wt.%. The main components of this salt content of sea water are listed in Table 2-I. Several lines of evidence show the composition of sea water virtually unchanged since Precambrian times (Valyashko, 1956; Lotze, 1957; Holser, 1966; Braitsch, 1971; many other authors).

Since Precambrian times, huge evaporite deposits were formed and buried in sedimentary formations. Although these evaporite deposits originated from the same source, the chemical composition of which has not changed, the evaporites exhibit a variability in their present composition unknown in other sediments; this is mainly because of the different solubilities of the original components of sea water: upon evaporation of water, highly soluble components remained in the concentrating solution while less soluble components precipitated.

Other reasons why evaporite deposits exhibit extremely variable compositions are: differences in the environmental conditions of sea water evaporation, post-depositional changes to the original composition of precipitates,

TABLE 2-I

Main components of ocean waters (after Braitsch 1962; 1971, table 1)

Ions	Weight %	Mol. in 1000 mol. H_2O	Fictive compounds weight % of sum of components
Na^+	1.056	8.567	78.03 NaCl
Mg^{2+}	0.127	0.976	9.21 $MgCl_2$
Ca^{2+}	0.040	0.186	3.48 $CaSO_4$
K^+	0.038	0.181	2.11 KCl
Cl^-	1.898	9.988	
SO_4^{2-}	0.265	0.514	6.53 $MgSO_4$
HCO_3^-	0.014	0.043	0.33 $CaCO_3$

etc.; as a matter of fact, the variability in the composition of evaporites, particularly of salt and potash deposits, is virtually unlimited. Detailed studies of newly discovered deposits often reveal that the actual processes of deposition have little in common with simplified theoretical models of sea water evaporation. The potash deposits encountered in the Prairie Evaporites of Canada represent one example (Baar, 1974a).

Many long-lasting controversies between various schools of thought developed around questions of how particular salt deposits were formed and transformed into their present conditions; some of these scholarly controversies are still going on. As D'Ans (1947) put it: "the task facing the geologists in giving an explanation of salt deposition is of a special kind as salt deposits consist of substances which are readily dissolved in water"; this statement applies particularly to potash salts which exhibit high solubilities and highly complex physico-chemical relationships.

The basic solution equilibria of various systems of oceanic salts at different temperatures were determined between 1896 and 1906 by Van't Hoff and numerous co-workers (D'Ans, 1947). These relationships are of particular interest in the production of various commercial potash salts and their by-products; for this reason, research institutions in some countries — U.S.S.R., Germany — have been working continuously to further improve the knowledge of the solution equilibria of salt systems under various conditions (D'Ans, 1933; Lepeschkow, 1958; Serowy, 1958; Wjasowow, 1958, Autenrieth, 1970). For English summaries, the reader is referred to Borchert and Muir (1964), and Braitsch (1971).

Disregard of long-established physico-chemical principles by geologists who dealt with the deposition of evaporites, has repeatedly drawn criticism from physico-chemists, e.g., D'Ans (1947, and various subsequent publications). However, it appears that scholar oversimplification, and in some cases overemphasis on local conditions, prevented to date any significant narrowing of the gap between some geological schools of thought and those scientists who postulate that well-established physico-chemical principles must not be neglected in evaporite geology.

In view of the many hundreds of geological publications on evaporite deposits — Lotze's (1957) comprehensive textbook on salt deposits is based on reference publications listed on 75 pages! — Holser's foreword to Braitsch (1971) may be quoted: (Braitsch's) "most complete and authoritative treatment, in any language, of the geochemistry of evaporites... can no longer be ignored by any (geologist)... Today, we must demand of geological investigations insofar as they have pretensions to pose and answer questions concerning the genesis of salt deposits, adequate documentation of their qualitative and quantitative composition, and over and above this, an understanding of the underlying physico-chemical principles."

2.1.2 Conditions of evaporite formation

2.1.2.1 Physico-chemical principles of evaporite deposition. Considering the complexity of the physico-chemical relationships between sea water and solids which precipitate upon removal of water by evaporation, only a few points of major significance can be outlined here; for details, the reader is referred to the listed reference publications.

Borchert and Muir (1964) emphasized the differences between isothermal and dynamo-polythermal evaporation of sea water. Isothermal evaporation does not occur under natural conditions; however, isothermal experimentation has been extremely important in establishing the solution equilibria of sea water components during processes of concentration and precipitation of solid salts.

Fig. 2-1 is a simple model to demonstrate the effects of isothermal evaporation of sea water in a conical vessel that could be regarded representative of a regular natural basin of evaporation; each circle indicates the remaining surface area after evaporation of water during various phases of precipitation of solids; it is emphasized that this model assumes straightforward evaporation of a given amount of sea water, no water from any source being added during the process of evaporation to complete dryness.

During phase I, only small amounts of carbonates precipitate. As the original sea water contains relatively few HCO_3^- ions (see Table 2-I) only about 1% of the sum of precipitated solids is made up of carbonates.

Phase II is the phase of $CaSO_4$ precipitation. It depends mainly on the temperature whether gypsum or anhydrite crystallize; this question is still debated. Calcium sulphates precipitating during phase II amount to about 3%.

In phase III, the concentration increases from 12.1 times to 63.6 times the original volume. Common rock salt (halite) crystallizes along with very small amounts of calcium sulphates, representing no less than 69% of the sum of precipitated solid salts.

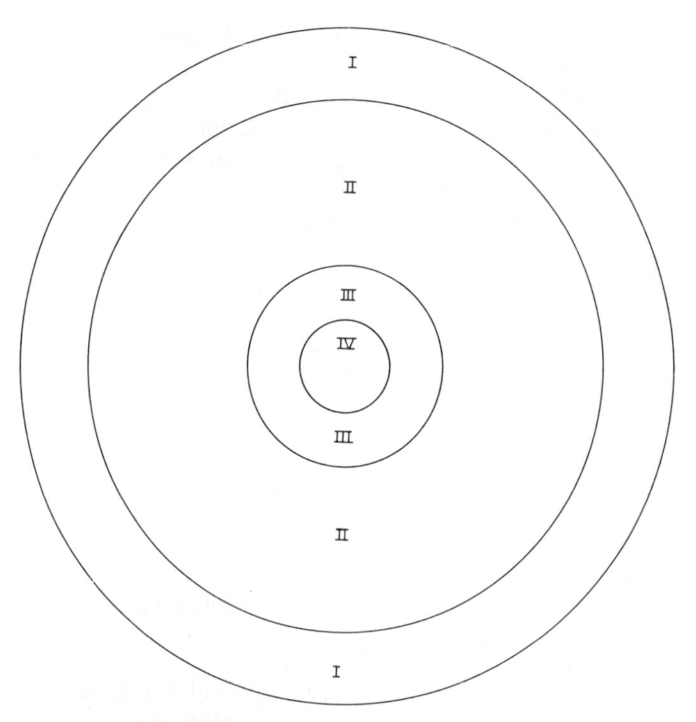

Fig. 2-1. Successive phases in the progressive evaporation of sea water. The areas of the circles represent the volume of brine at the beginning of various phases of precipitation. (After Borchert and Muir, 1964).

I = Ca—Mg carbonates; *II* = Ca sulphates; *III* = Na chloride (common rock salt); *IV* = Na—K—Mg sulphates and chlorides (potash salts and bischofite).

During phase IV, various Na—K—Mg sulphates and chlorides crystallize. As magnesiumchloride possesses the highest solubility among the main original components of sea water, the last mineral to crystallize out of concentrated sea water is bischofite ($MgCl_2 \cdot 6H_2O$). The salts of phase IV represent the remaining 27% of solids which crystallize when sea water is evaporated to complete dryness.

Borchert and Muir (1964) explicitly stated: "...these proportions would be obtained *only if* a marginal sea evaporated to dryness and was constantly and completely cut off from the open sea; such conditions have probably never been fulfilled". In conclusion, these authors reject the application of this oversimplified model to natural evaporation of sea water; the model had been used many decades ago to explain lateral facies changes in evaporite deposits. (See also Lotze, 1957.)

A similar oversimplified model resulting in the bull's eye pattern of Fig. 2-2, was recently re-introduced in various publications; for references, see Hsü et al. (1973). It is called the "desiccated deep-basin model"; the figures

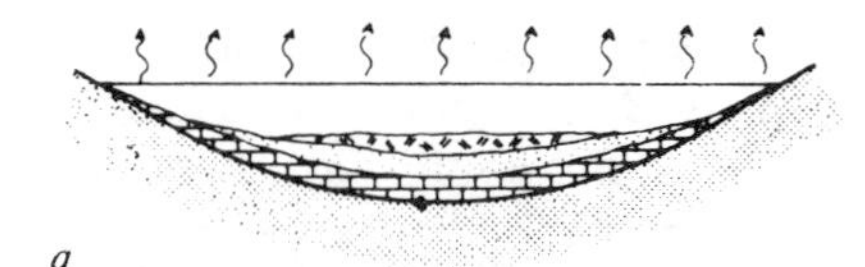

a

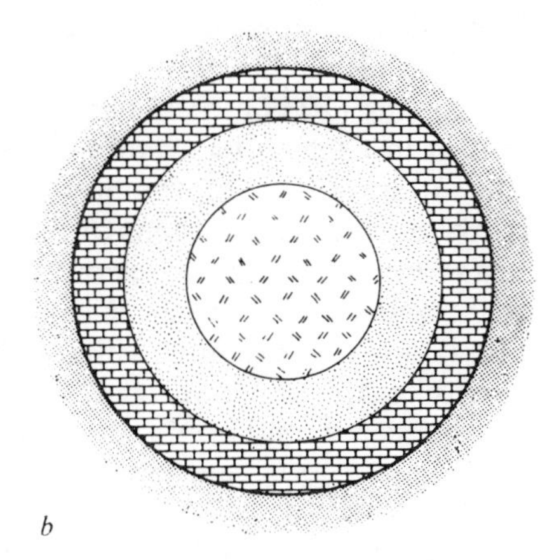

b

Fig. 2-2. Idealized bull's eye pattern of evaporite distribution, typical of isolated basins. (After Hsü et al., 1973.)
a. section; b. map.

TABLE 2-II

Water balance of the Mediterranean Sea, excluding the Black Sea

	1 *	2	3	Unit
Area covered by the Mediterranean Sea	2.5	2.496	2.5	million km^2
Water volume		3.7	3.7	million km^3
Average depth	1431		1500	metres
Annual loss by evaporation	3.5	4.69	4.7	$km^3 \cdot 10^3$
Annual precipitation		1.15	1.2	$km^3 \cdot 10^3$
Annual volume delivered by river influx		0.23	0.2	$km^3 \cdot 10^3$
Annual net loss (evaporation less prec. and river influx)	2.1	3.31	3.3	$km^3 \cdot 10^3$
Annual influx through Strait of Gibraltar	55.2	55.2		$km^3 \cdot 10^3$
Annual reflux through Strait of Gibraltar	53.1	51.89		$km^3 \cdot 10^3$

* Column 1 from d'Ans (1947), column 2 from Lotze (1957), column 3 from Hsü et al. (1973).

compiled in Table 2-II are used in "a detailed computation of the material balance budget" of salt deposition in the Mediterranean basin. The authors conclude that "eight or ten marine invasions could have been sufficient to account for all the salts under the Mediterranean abyssal plains".

Lotze (1957, p. 142) used exactly the same data in his water balance calculations for the present Mediterranean Sea; as none of the authors reveals the source of the data, Lotze's figures were also included in Table 2-II, along with slightly different figures used by D'Ans (1947) in similar calculations. In essence, the data indicate evaporation in excess of precipitation and river influxes; the resulting influx from the Atlantic through the Strait of Gibraltar contains 3.55 wt.% salt. However, the average salt content of the Mediterranean remains constant for there is a reflux to the Atlantic, containing 3.65 wt.% salt (D'Ans, 1947). The reflux current runs at depth; it can be traced for hundreds of miles into the Atlantic (Borchert and Muir, 1964, p. 15).

If the Strait of Gibraltar were closed completely, *and if* all other conditions remained unchanged, the Mediterranean would evaporate to dryness in about 2000 years (D'Ans, 1947) or in about 1000 years according to Hsü et al. (1973). The resulting average thickness of salt would be 24 m (D'Ans, 1947; Borchert and Muir, 1964), based on an average depth of 1431 m.

Hsü et al., although using the slightly greater average depth of 1500 m, arrive at an average thickness of salt of "only about 20 m that could be precipitated from one basinful of Mediterranean waters". "Even if all the salt deposited within restricted basinal areas, covering only one third of the total (according to the bull's eye pattern of Fig. 2-2), the resulting deposit should be only 60 m thick." It is apparent that a great many more than eight or ten marine invasions are required to account for salt thicknesses of 2—3 km assumed to exist in certain areas of the Mediterranean basin (Hsü et al., 1973).

In principle, however, the "desiccated deep-basin model" is not "improbable because none of the desert basins today is comparable in size and depth to a desiccated Mediterranean", as argued by Hsü et al. Comparable conditions controlled the deposition of the Prairie Evaporites of Saskatchewan, Canada, as shown by Baar et al. (1971) and, in more detail, by Baar (1974a): the central areas of bull's eye patterns in several sub-basins are indicated by potash beds; repeated invasions of less concentrated marine brines into partly desiccated sub-basins resulted in peculiar features, see p. 23 and Fig. 2—10.

The actual development of barriers, such as the present Strait of Gibraltar, is discussed in numerous publications, e.g., Borchert and Muir (1964) and Lotze (1957); the latter author emphasized the consequences of tectonic uplift of the Strait of Gibraltar as follows: the reflux of more concentrated brines out into the Atlantic would be cut off; the influx of sea water from the Atlantic would become insufficient to balance the loss by evaporation, resulting in lowering of the level of the Mediterranean, in increasing concentration, and in desiccation of marginal areas. However, with lower levels of the Mediterranean, the velocity of the influx current would increase and result in increased erosion until the balance may be restored. Such a mechanism would provide the possibility of salt deposition in the deepest parts of the basin, particularly in the distant eastern parts.

This appears to represent a plausible combination of the two extreme theoretical models shown in Figs. 2-2 and 2-3: the "desiccated deep-basin model" calls for evaporation to complete dryness, while the opposite extreme model calls for deep, pre-existing interior basins such as the Mediterranean basin, the net loss by evaporation being continuously balanced by sea water influx, but reflux of concentrated brines to the ocean being restricted or non-existent; the level of the evaporating basin is assumed to be at sea level. Such a basin may be filled to the rim with evaporites.

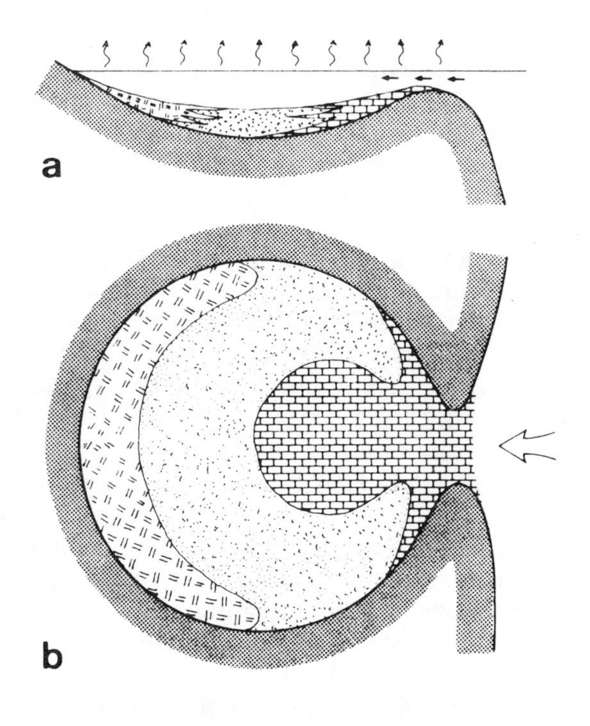

Fig. 2-3. Idealized tear-drop pattern of evaporite distribution, typical of partially restricted basins. (After Hsü et al., 1973.)

a. section; b. map. Note: no precipitation of calcium carbonates and sulphates at the distant end of the basin.

Hypotheses postulate the tear-drop pattern of Fig. 2-3 in the evaporites: the precipitation during phase I of Fig. 2-1 should take place near the inlet; phase II would occur in an intermediate area; salt would be deposited only near the distal end of the basin (Hsü et al., 1973). In other words: the concentration would increase regularly with the distance from the inlet. The effects of vertical movements of concentrating brines are disregarded, see pp. 25—27 for further details.

Other authors, particularly Richter-Bernburg (1955), call for "saturation shelves": sea water travelling to a basin where salt is deposited, is assumed to reach saturation, with regard to phase-I precipitates, prior to entering the salt basin; phase-II precipitates (gypsum, anhydrite) are assumed to accumulate preferably near the inlet, at the slope of the deep salt basin. Reflux of concentrated brines would not occur.

The effects of carbonate banks and reefs on the influx—reflux regime has been given particular attention in some recent publications, e.g., Hite (1970).

Borchert and Muir (1964) discussed extensively models of what is termed dynamo-polythermal progressive evaporation of sea water in series of restricted basins with water levels at or near sea level; before entering the terminal basin of potash deposition (phase IV of Fig. 2-1), progressive concentration would take place in forebasins separated by bar zones; tectonic movements in the bar zones would control the influx—reflux regime. It appears that the latter postulate could be reduced considerably when the effects of carbonate banks and reefs are considered carefully for any particular deposit.

It is possible to construct models in which the phases I—III of Fig. 2-1 would occur in three separate forebasins, while phase IV would result in non-proportional deposition of potash salts in the terminal basin. Evidence of such fractional precipitation has been found repeatedly, e.g., in the Upper Rhine salt and potash deposits (Baar and Kühn, 1961; Braitsch, 1971), and in the Prairie Evaporites of western Canada (Wardlaw, 1963). Potash deposition in the terminal Prairie Evaporite basin of Saskatchewan was controlled by carbonate banks and reefs; see Baar (1974a) for a summary of related publications.

Detailed studies of other large evaporite deposits may reveal similar conditions of evaporite deposition; regarding the Salina basin of Michigan, Matthews and Egleson (1974) emphasized: "The numerous reefs and carbonate banks around the rims had considerable relief, and certainly they contributed to the degree of restriction in the early basin."

It should be apparent from the foregoing that simple theoretical models of isothermal or polythermal evaporation of sea water are of little, if any, practical significance; extensive efforts in calculating theoretical profiles of evaporite deposits from sea water — e.g., Borchert and Muir, 1964, fig. 6.3, with reference to previous attempts by several authors — have turned out only one undisputed result: to date, no evaporite deposit has been found yet to match any preconceived theoretical model of the type shown in Fig.

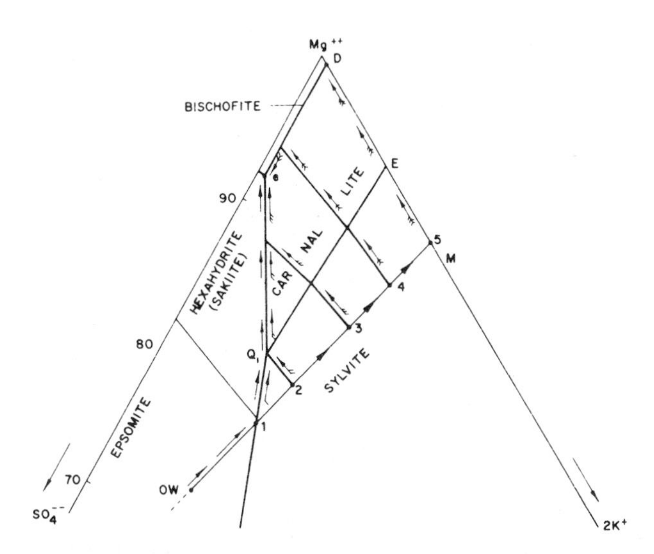

Fig. 2-4. Sequence of crystallization in phase IV of Fig. 2-1. (Baar, 1966a, after Valyashko, 1956.)

OW indicates the composition of ocean water in the case of evaporation according to Fig. 2-1. Points *1—5* indicate various degrees of loss of SO_4^{2-} in concentrating sea water. Arrows indicate the resulting crystallization paths.

2-2 and 2-3. The main reason for this situation appears to be that any natural evaporation process can be halted, even reversed, during any of the four phases shown in Fig. 2-1; only relatively small changes to any of the various environmental conditions which control an evaporation process, may bring about changes to the depositional sequence—not to mention post-depositional changes which affected most evaporite deposits.

Fig. 2-4 may serve to demonstrate possible changes to depositional sequences during phase IV of Fig. 2-1; it shows a Jänecke diagram — see Borchert and Muir (1964) or Braitsch (1971), for detailed explanation — with the fields of crystallization of various Na—K—Mg sulphates and chlorides. Point *OW* represents the composition of sea water; the points *1—4* indicate the composition of sea water which has lost increasing amounts of SO_4^{2-} prior to reaching the terminal basin of evaporation; at point *5*, no sulphate ions are left in the solution.

The theoretical profiles of progressive evaporation according to Fig. 2-4 are shown in Fig. 2-5; with increasing losses of sulphate ions, the theoretical sequences become less complex: $MgSO_4$-salts gradually disappear; in extreme cases of SO_4^{2-} removal, only chlorides are present in the salt sequences, as is the case in the Prairie Evaporites of Canada, the Upper Rhine salt deposits, and in other salt sequences which contain the richest potash beds; apparently, the length of the travel way for the concentrating sea water from the

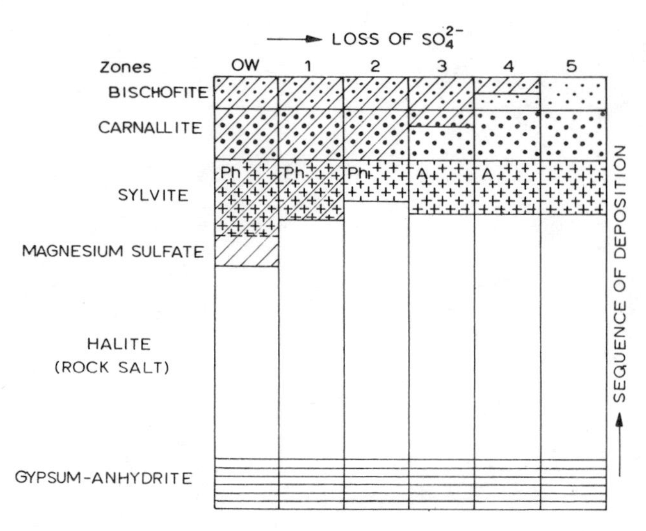

Fig. 2-5. Salt sequences resulting from sea water evaporation. (Baar, 1966a, after Valyashko, 1956.)

Alterations according to Fig. 2-4 assumed. *Ph* (polyhalite) and *A* (anhydrite) occur in small amounts.

open ocean to a terminal basin of potash deposition is an important factor. However, there is no agreement among geochemists as to the chemical processes which may have caused the partial or complete removal of SO_4^{2-} from the concentrating sea water, see Braitsch (1971) for details; apparently, there are several theoretical possibilities. The fact that some potash deposits do not contain any sulphates leaves no doubt of corresponding alterations according to Figs. 2-4 and 2-5.

In cases of complete removal of SO_4^{2-} prior to arrival at a terminal basin of evaporation, the evaporation process can be calculated quantitatively on the basis of known solution equilibria, provided that the composition of a potash bed which was deposited after progressive evaporation, is known, and provided that no post-depositional alterations occurred. This has been shown for the Lower Potash Bed of the Upper Rhine valley; this potash bed has been mined extensively for many decades, so its quantitative composition is well-established. It contains 39.31 wt.% KCl; primary sylvite is the only potash mineral present in the lower two thirds of the potash bed, for which the process of deposition was reconstructed by Baar and Kühn (1961). The quantitative model proposed by these authors results in the theoretical KCl content of 39.13 wt.%; it matches surprisingly well the actual KCl content. For further details, the reader is referred to the original publication, and to Braitsch (1971) who provided additional geochemical evidence of the validity of the depositional model. To date, this is the only known case in which a quantitative model of deposition could be developed to match the actual composition of a well-known deposit, demonstrating the significance of the statement quoted at the end of sub-section 2.1.1.

2.1.2.2 Bromine profiles indicate the course of salt deposition. The processes of sea water evaporation in restricted basins are extremely sensitive to changes in depositional conditions, particularly to changes in the influx—reflux regime between the open sea and the terminal basin of evaporation. As emphasized before, any of the four phases shown in Fig. 2-1 can last for any length of time; in addition, a process of progressive evaporation can be reversed at any time, i.e., at any degree of concentration reached in a terminal basin of evaporation. For these reasons, the relative amounts of precipitates in natural evaporite successions can vary almost infinitely, as stressed by Borchert and Muir (1964) in view of frustrating attempts to develop models of evaporation which could be applied generally.

It has been proposed by Richter-Bernburg (1955) to apply the term "recessive evaporation" in cases where a process of progressive evaporation with increasing concentrations is reversed to the effect that the concentrations gradually decrease, resulting in reversal of the sequence of the four phases of Fig. 2-1. The term "recessive" is used in this text, regardless of the source of the water that causes the dilution of a body of concentrated brines.

In cases where the brine level in a terminal basin had been lowered considerably below sea level during progressive evaporation, recessive processes result in rising water levels, i.e., in transgressions over possibly large marginal areas exposed due to progressive evaporation. At first glance, this terminology may sound confusing; however, it certainly would avoid many misunderstandings if all writers would adhere to it.

Some authors, e.g., Matthews and Egleson (1974), are inclined to consider the gradual reversal of progressive evaporation the normal development. This means that potash precipitation in phase IV of Fig. 2-1 should normally be followed by halite precipitation of phase III; further dilution of the brines contained in a restricted basin would yield calcium sulphates, and, finally, normal marine conditions could be restored, unless the influx—reflux conditions were changed again.

Such gradual reversals are indeed indicated by the bromine profiles of some salt deposits, including the one to which Matthews and Egleson (1974) refer. Other examples are provided by the Zechstein 1 deposits in the Werra potash mining district of Germany (see Fig. 2-9), and by the Zechstein 3 and 4 of northern Germany (Baar, 1963, 1966c).

However, in many other cases, potash deposition was terminated rather abruptly; frequently, potash deposition was followed by sedimentation of non-evaporitic material, grading into another cycle of evaporation with the evaporites following each other in the typical progressive order shown in Figs. 2-1 and 2-5.

The bromine method proved indispensable in modern salt geology to elucidate the course of evaporation processes which can change gradually or abruptly. The bromine content of sea water has not changed since Precambrian times. Only a fraction of the bromine present in concentrated sea water enters crystallizing chlorides, Br^- substituting Cl^- following known physico-chemical principles. For the most common chlorides such as halite, sylvite, and carnallite, the controlling principles were established by Boeke (1908) and confirmed in investigations by D'Ans and Kühn (1940), and Braitsch and Herrmann (1963). Detailed discussions in the English language can be found in publications by Schwerdtner and Wardlaw (1963), Baar (1966b), Braitsch (1966, 1971), Holser (1966), Raup (1966), Kühn (1968).

It goes without saying that only the bromine contents of primary minerals can be used in investigations of the course of evaporation processes. As the bromine contents of chlorides are extremely sensitive to post-depositional influences, e.g., recrystallization, certain precautions are necessary in establishing bromine profiles of a salt sequence. Large numbers of analyses are required to determine the normal bromine profile; any irregular values, which may be higher or lower than the original primary values, must be recognized and eliminated prior to drawing any conclusion regarding the concentration of the brine body from which the chlorides originally crystallized.

Irregular bromine contents are frequently found at the base and at the top of salt sequences, or throughout relatively thin salt deposits; such irregular values were caused by diagenetic recrystallization and must not be considered indicative of the course of concentration or dilution of the whole original brine body. The recent upsurge in the application of the bromine method resulted in many publications which do not meet basic requirements for reliable genetic evaluations; some premature conclusions based on few bromine determinations proved untenable as shown by Baar (1963, 1966c), Baar et al. (1971), Braitsch (1971), and others. On the other hand, the bromine method proved extremely useful for stratigraphic purposes in thick, uniform rock salt sequences such as the Zechstein 2 series underneath the Stassfurt potash bed with thicknesses in excess of 500 m in central areas of the North-European Zechstein basin (Baar 1954b, 1955; Schulze, 1958).

Fig. 2-6 shows the normal, regular bromine profiles of the Zechstein 2 rock salt. In the central basin, the regular profile commences with values of 0.007 wt.% (70 ppm) at the salt basis; the values increase slowly and very regularly through hundreds of metres of rock salt, indicating the slowly increasing concentration of the huge body of brines which covered northwestern Europe from England to Poland. Corresponding profiles in shallower parts of the basin show higher values at the salt basis and much faster increases, indicating brine movements from the shore into the deep basin where halite crystallized due to cooling (Baar, 1954b).

Similar regular bromine profiles were found in other salt sequences (Raup,

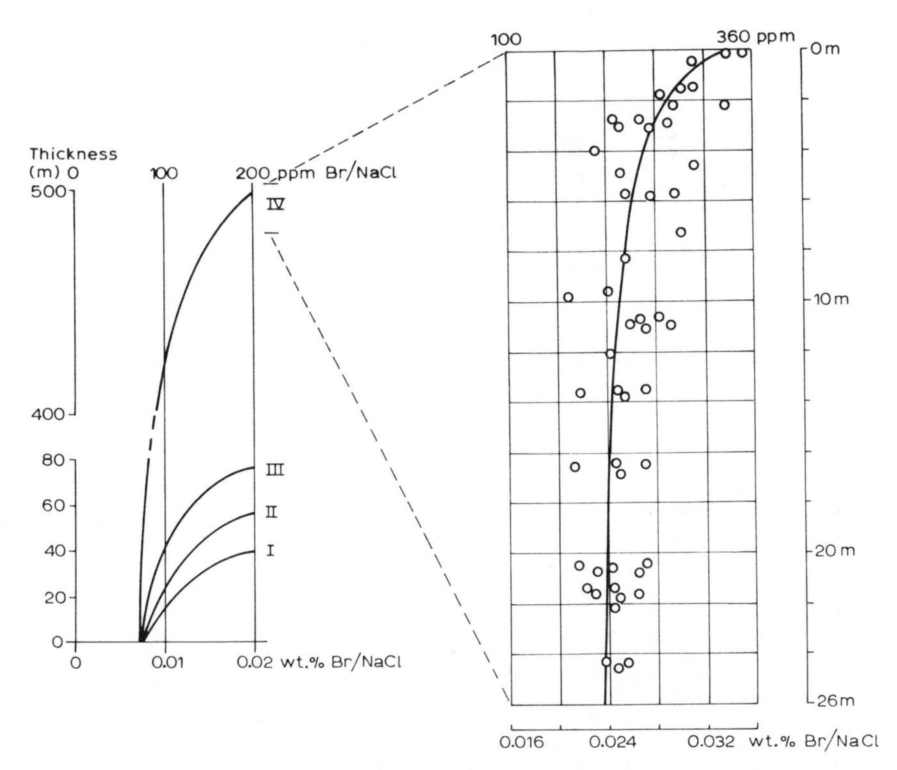

Fig. 2-6. Regular bromine profiles, Zechstein 2 rock salt (Baar, 1954b).
I—III = marginal areas; *IV* = main basin.

Fig. 2-7. Irregular bromine contents in deformed rock salt. (After Simon and Haltenhof, 1970.)
26-m section at the top of profile *IV* of Fig. 2-6.

1966; Raup et al., 1970; Hite, 1974); potash beds at the top may or may not have survived the rather abrupt dilution of the brine bodies.

The normal bromine profiles of rock-salt sequences from regular progressive evaporation reflect the original depths of the basins (Kühn, 1955a,b; Kühn and Baar, 1955; Holser, 1966). It is re-emphasized that only regular bromine profiles must be used in such calculations.

In salt domes and similar, heavily deformed structures, the original bromine contents may have been altered to such an extent that conclusions as to the evaporation processes cannot be drawn; this holds true particularly for salt sequences with potash beds. Extensive recrystallizations in the presence of brines with different bromine levels may have lowered or raised the original bromine contents (Baar, 1963, 1966a). Fig. 2-7 may serve as an example of irregular bromine contents near the top of the Zechstein 2 rock salt; corresponding profiles in less disturbed areas (Fig. 2-6) are reasonably regular

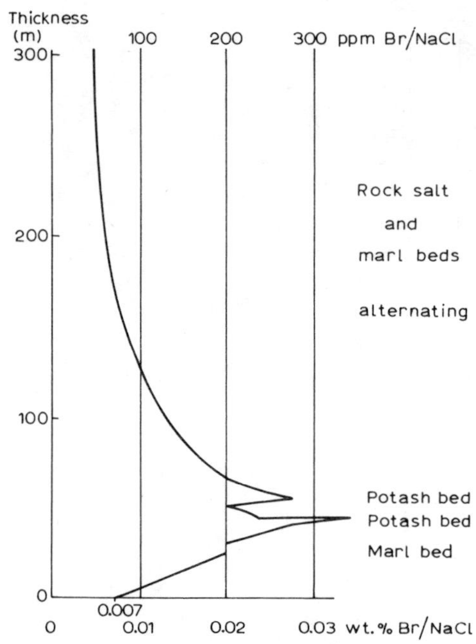

Fig. 2-8. Average bromine contents, rock salt of the Upper Rhine deposits (Baar, 1963). The profiles indicate gradual changes in concentrations.

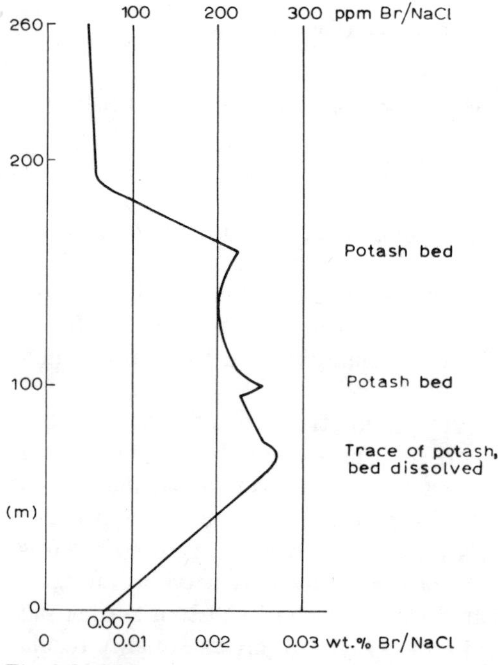

Fig. 2-9. Average bromine contents, Zechstein 1 rock salt, Werra district (Baar, 1963). Gradual changes in concentrations indicated.

(see also Schulze, 1958, with discussions by Baar, Borchert, Haase, Waljaschko).

Very irregular bromine contents may also be found in thick salt sequences, including potash beds, which have suffered no deformation since deposition. The Prairie Evaporites of Saskatchewan provide an example which has been dealt with extensively in recent publications (Schwerdtner, Wardlaw, and students). Irregular bromine contents, and other peculiar features, show that these deposits were severely affected by rainstorms during temporary exposure (Baar et al., 1971; Baar, 1974a). Fig. 2-11 shows anomalies encountered in marginal areas where the first potash bed lacks stratification (Keys and Wright, 1966); shortly after deposition, potash minerals were dissolved and flushed, along with clay minerals, to local sink areas where they seeped into the unconsolidated deposit. The drainage channels and the sink areas show the original potash bed replaced by "dirty" rock salt. Rising brine levels, caused by increased influx of less concentrated sea water, resulted in another type of anomaly — the potash bed was replaced by clean salt.

Fig. 2-10 shows a different type of anomaly frequently encountered in the uppermost potash bed which is extracted in the Saskatoon area; it exhibits regular bedding and several prominent clay layers, indicating that this potash bed was originally deposited under a cover of brines. Apparently, in deeper depressions of the Prairie Evaporite basin, the brine cover never disappeared completely. For further details, the reader is referred to Baar (1974a). It appears that various combinations of the two basic theoretical

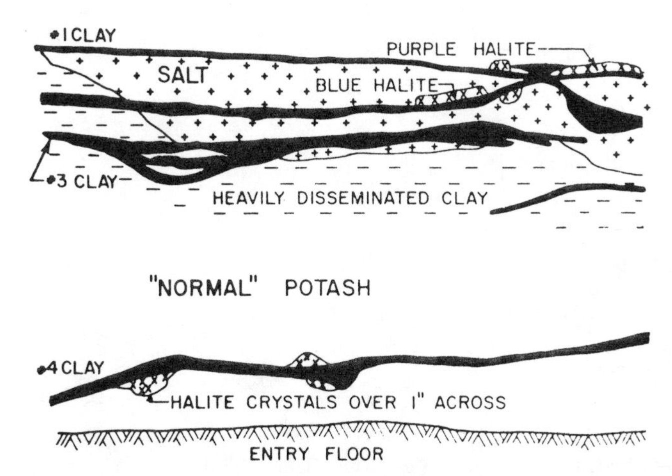

Fig. 2.10. Anomalies in the top potash bed, Prairie Evaporites (Baar, 1974a).

Seasonal bedding; *1—4* "clay" layers are normally up to 10 cm thick; the normal distance between *1* and *4* clays is approximately 3 m.

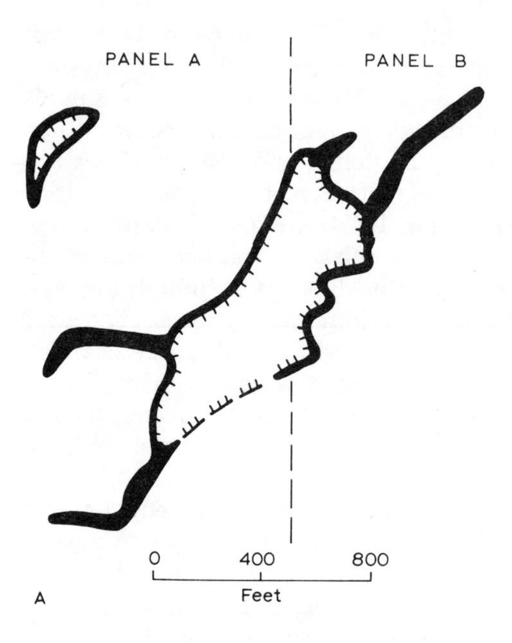

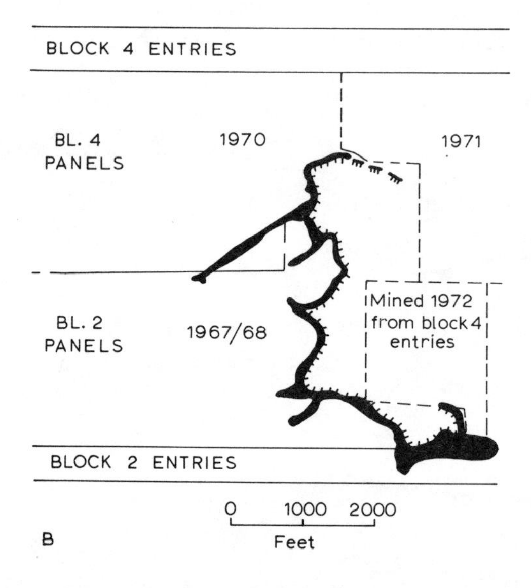

Fig. 2-11. Anomalies in potash bed, IMC-K 1, Saskatchewan (Baar, 1974a).

Marginal area of the Prairie Evaporite basin. Drainage channels (black) filled with salt and insolubles, leading to sink areas.

A. "Small" anomaly left unmined. B. Large anomaly required changes to mining plan.

models shown in Figs. 2-2 and 2-3, resulted in the present irregularities; even the "dry lake" model stressed by Valyashko (1956) — identical with Waljaschko (1958) — appears applicable in explaining some of the observed features. It goes without saying that the rock mechanical behavior is severely affected by such anomalies; more details are shown in subsequent chapters.

2.1.2.3 Effects of brine stratification. Lotze (1957) documented numerous examples of vertical temperature and concentration gradients in concentrated natural brine bodies. The temperatures and the densities increase in diluted surface layers to depths of several metres; with greater depths, the temperatures decrease to average temperatures which are not affected by seasonal climatic conditions. In contrast, the densities continue to increase slightly with increasing depths.

Most authors acknowledge the significance of brine stratification in sea water concentrates which are intermittently covered with dilute influxes. Brongersma-Sanders and Groen (1970) emphasized the influence of trade winds on circulation systems which develop in restricted basins.

Laboratory experiments with various solutions of the major components of sea water are reported by Borchert and Muir (1964) with reference to earlier investigations by the first author; for evaporation and precipitation of solids, the solutions were heated *from below* in long, narrow vessels. The application of the results of these experiments to actual conditions in restricted marine basins has repeatedly drawn heavy criticism (D'Ans, 1947; Braitsch, 1971), mainly because of the fact that the solar heat is applied to the surface of evaporating brine bodies.

Rapid evaporation of diluted surface layers may result in halite crystallization at the surface to produce "hopper crystals" which eventually settle down to the bottom of the evaporating basin (Dellwig, 1953). The chevron patterns of such crystals are produced by daily temperature drops which cause relatively fast growth of the crystals. After having arrived at the bottom of the brine body, the crystals may continue to grow without developing chevron patterns in basins of sufficient depth.

Such conditions have been verified regarding the Upper Rhine evaporites where hopper crystals are abundant in some seasonal layers of the potash beds (Baar and Kühn, 1961, figs. 8—10). The sequence of precipitation of such layers is frequently as follows: thin clay-anhydrite layers reflect the seasonal influx of fresh water; evaporation of the surface layer of diluted brines results in hopper crystals of halite which are embedded in a mass of clear halite; the following layer consists mainly of sylvite crystals. Some thicker beds of clay and similar non-marine material correlate across the whole basin with several sub-basins (Sturmfels, 1943; Wagner, 1955).

These observations, combined with bromine investigations, were the basis for the quantitative calculation of the evaporation process which resulted in the Lower Potash bed (Baar and Kühn, 1961; see sub-section 2.1.2.2). The progressive evaporation of the main brine body was not significantly influenced by seasonal influx of fresh water which carried non-marine material into the basin, spreading like a blanket over the surface of the concentrated brines and subsequently evaporating.

It may be pointed out that the halite hopper crystals which are abundant in the Upper Rhine deposits, are equivalent to so-called chevron crystals referred to in a number of recent publications, e.g., Wardlaw and Schwerdtner (1966); such chevron crystals develop in very shallow brine bodies, the dark crystal layers with abundant inclusions reflecting diurnal cooling periods (Valyashko, 1956; Shearman and Fuller, 1969). Both the hopper and the chevron crystals continue slow growth, which results in clear crystals, after removal from the range of influence of diurnal temperature changes: the hoppers settled to the bottom of the basin where seasonal cooling remained effective (Braitsch, 1971); the chevrons were covered with deeper brine, due to the influx of less-concentrated sea water (Baar, 1974a; see Fig. 2-11, B). The "brine mixing" effect shown by Raup (1970) may also have contributed to clear growth of halite crystals which started out as hoppers or chevrons, respectively.

The blanketing of concentrated brines by intermittent influxes of less-concentrated brines is evidenced by many observations in other salt deposits. One more example may be cited: rock salt beds up to 1/2 m thick are present in the carnallitic Stassfurt potash bed at the top of the Zechstein 2 rock salt of northwest Europe; these halite beds correlate over thousands of square kilometres around the Harz mountains of central Germany, and they serve as marker beds in the potash mines. The highly concentrated brines from which the carnallite crystallized could not have contained sufficient NaCl to account for the rock salt beds; for this reason, blanketing by dilute brines which carried additional NaCl into the Zechstein basin, is clearly indicated, as emphasized by D'Ans (1947); bromine investigations (Baar, 1955) provided additional evidence.

Due to similar persisting stratification, concentrated brines apparently survived marine ingressions into partly desiccated large basins, the concentrated brines being conserved in relatively deep parts of the basins; when the influx—reflux regime changed to result in another cycle of evaporation, the rock salt at the basis of phase III of Fig. 2-1 is characterized by irregularly high bromine contents carried over from the preceding cycle. In such cases, the bromine profile shows decreasing values with progressive evaporation until the excess bromine is incorporated in the first rock salt at the base of the rock salt sequence. Such processes apparently occurred in both the Zechstein basin (Baar, 1966c) and the Paradox basin (Raup, 1966, 1970), as evidenced by bromine profiles. Bromine profiles of the Prairie Evaporites indicate different depositional conditions as outlined in sub-section 2.2.2.

In conclusion, the statement quoted at the end of sub-section 2.1.1 may be modified as follows: answers to questions concerning the genesis of salt deposits require adequate documentation of their composition and an understanding of the physico-chemical principles which govern the precipitation of evaporites from sea water; quantitative calculations must be based on facts which may vary considerably, and, therefore, must be determined individually for any specific deposit. General assumptions, and oversimplifications of processes which are known for being extremely variable, are of little scientific value and of no practical value; to the contrary, they frequently proved misleading when applied to actual geological conditions in evaporite deposits.

2.2 MINERALOGY AND PETROLOGY OF EVAPORITES

2.2.1 Evaporite mineralogy

The mineralogy of evaporites of phase IV (K—Mg salts) of Fig. 2-1 is as complex as is the physico-chemistry of this phase. This is partly due to the fact that many of the rare constituents of sea water remain in solution until the final stage of concentration is reached. Other rare minerals which are occasionally found in evaporite deposits, crystallized from solutions which gained access to the evaporites from other formations.

The following shortened version of the summary by Borchert and Muir (1964, p. 3) is considered sufficient to outline this subject: although almost forty salts have been recorded from evaporite deposits, only a little over half that number are frequently present in more than trace quantities. Some are simple salts; others are double or even more complex salts. Some are anhydrous, others hydrated.

The most commonly occurring of these salts — if we exclude calcite, dolomite and magnesite, which are not commonly regarded as evaporite minerals despite Fig. 2-1 with carbonate precipitation in phase I — are the calcium sulphates, gypsum ($CaSO_4 \cdot 2H_2O$) and anhydrite ($CaSO_4$). Next in order of abundance comes halite (NaCl), often known as rock salt or common salt.

The potash salts are much scarcer. Among this group, sylvite, carnallite, langbeinite and kainite are most important. The potash salts and other salts which are often found in association with potash salts are listed in Table 2-III. Borates are even less common and are occasionally of non-marine origin.

For more details concerning evaporite minerals, the reader is referred to Borchert and Muir (1964), Kühn (1968), Braitsch (1971), or to other pertinent sources.

As it so happened, many of the most soluble potash minerals were discovered in the vicinity of Stassfurt, Germany, where the first potash mines began to operate some 100 years ago. Mineralogists have the undisputed right to give the name of their choice to any mineral they discover and describe for the first time. During the first decades of potash mining, there

TABLE 2-III

Main components of salt and potash deposits in approximate order of abundance (selected from table 3 in Braitsch, 1971)

Mineral	Formula	Color (normal)	Solubility (20°C)	Remarks
Chlorides:				
Halite	$NaCl$	colorless to grey	26.4%	primary or secondary; rare colors: red, blue
Sylvite	KCl	white, reddish	25.6%	primary or secondary
Carnallite	$KMgCl_3 \cdot 6\ H_2O$	red, white	27.3% hygroscopic [1]	primary, secondary in pockets, fissures
Bischofite	$MgCl_2 \cdot 6\ H_2O$	colorless to white	35.1% very hygroscopic [1]	primary terminal mineral secondary in fissures
Tachhydrite	$CaMg_2Cl_6 \cdot 12\ H_2O$	yellow	very hygroscopic [1]	considered secondary
Sulphates:				
Kieserite	$MgSO_4 \cdot H_2O$	colorless to white	hygroscopic [1]	transforms into other $MgSO_4$ hydrates
Langbeinite	$K_2SO_4 \cdot MgSO_4$	colorless	slowly dissolving	usually secondary
Epsomite	$MgSO_4 \cdot 7\ H_2O$	white	26.2%	often forming in mines after kieserite
Polyhalite	$K_2MgCa_2(SO_4)_4 \cdot 2\ H_2O$	red	slowly dissolving	usually secondary
Chloride-sulphate:				
Kainite	$4\ (KClMgSO_4) \cdot 11\ H_2O$	reddish		primary or secondary

[1] Hygroscopic = deliquescent, adsorbing water from moist air.

was an apparent tendency to name newly discovered potash minerals after executives in the industry, resulting in mineral names such as bischofite, carnallite, langbeinite, loeweite; about a dozen names of rare minerals fall into this group. The tendency then changed to preference of names intended to honor scientific merits in fields related to potash; and thus, since the more common potash minerals had been named already, names such as rinneite, vanthoffite, d'ansite, and others are seldom found in the geological literature on evaporites, with the exception of specialized mineralogical and physico-chemical publications, and textbooks such as Borchert and Muir (1964) and Braitsch (1971).

As a matter of fact, the vast majority of all known evaporite minerals have no effect on the rock-mechanical behavior of evaporites, contrasting particularly to the effects of non-marine material present in bedding planes in a great many salt and potash deposits. Regarding the mineralogy of such non-evaporites or semi-evaporites, which are often called "clay" for their very fine grain size, the reader is referred to Kühn (1968) and Braitsch (1971); the composition of clay minerals found in evaporite deposits appears to be even more complex, and controversial in some cases, than the composition of some rare evaporite minerals.

2.2.2 Petrology of evaporites

Thick beds of limestone, dolomite, and anhydrite occur in most evaporite deposits, making up particularly the floor and roof formations of salt sequences. In most cases, the rock-mechanical behavior of the salt rocks around mine openings and solution-mined cavities is not directly affected by the different behavior of carbonate and anhydrite beds, which are usually considered "competent" regardless of their petrology — unless they have been affected by groundwater action that caused karstification. It must be borne in mind that karstification may have taken place at any time since deposition. Therefore, the present depth of burial is of little significance in this regard. On the other hand, it goes without saying that karstification is terminated when there is no more groundwater movement through deeply buried formations.

The petrology of carbonates and calcium sulphates is another subject of controversy between petroleum geologists and petrographers, revolving around the question whether the modern sabkha tidal flats represent the only valid model for the explanation of certain petrographic features observed in anhydrites and dolomites. As Coogan (1974) put it in his recent evaluation of papers presented at the Fourth International Symposium on Salt: "The 'deep versus shallow origin'-controversy continues." For information on recent developments in this much publicized special field, the reader is referred to Coogan's reference papers.

Regarding the petrology of salt rocks, nature prevented most of the potential similar controversies in a very special way: when salt rocks recrystallize — either during early diagenesis or on occasion of large-scale deformations in salt diapirs — the new crystals tend to reject the inclusion of material other than that of which any individual crystal consists. As a result, most recrystallized salt rocks consist of greatly purified crystals of halite, plus sylvite or carnallite crystals in most potash ores. Other components, particularly clay minerals, are accumulated between the salt.

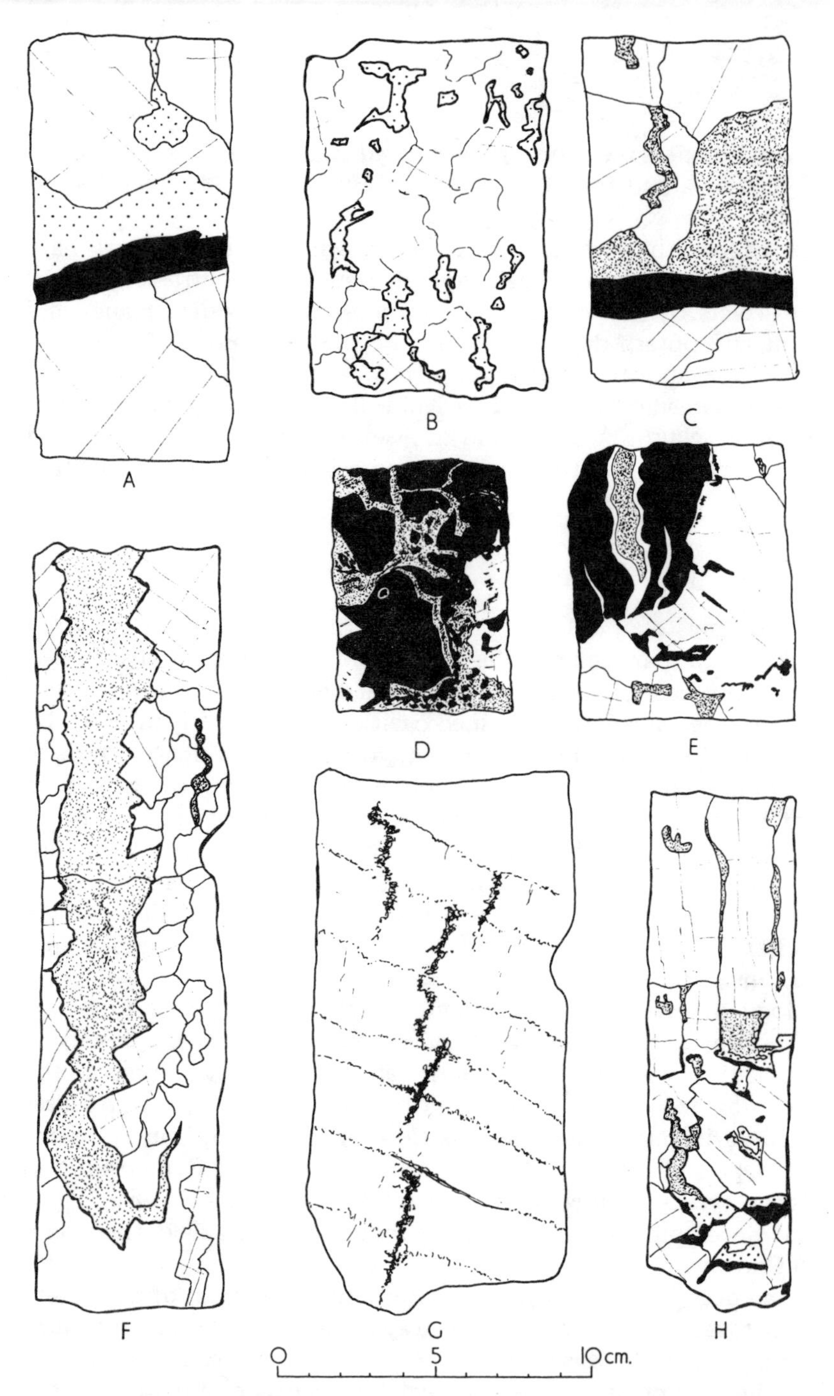

Fig. 2-12. Petrographic structures in Prairie Evaporite salts. (After Wardlaw, 1968.)

Mineral identification: halite (white), sylvite (coarse stipple), carnallite (fine stipple), insolubles (black). Note: in cores *A* and *C*, a clay layer apparently terminated the seepage of brines from which sylvite and carnallite crystallized; in core *H*, insolubles, sylvite, and carnallite, in upward sequence, are occupying irregular pockets and vertical cleavage planes of large halite crystals.

Fig. 2-12 (after Wardlaw, 1968) reveals the sequence of such events in the Prairie Evaporites of Saskatchewan: highly concentrated brines of phase IV of Fig. 2-1 seeped into an apparently unconsolidated mass of halite crystals, replacing the original brine; apparently, the infiltrating brines from the surface of the deposit — see sub-section 2.1.2.2 for description of the depositional conditions — precipitated firstly sylvite, secondly carnallite, due to cooling. In addition, the original chevron patterns of halite crystals disappeared partly or completely due to recrystallization, as evidenced by microphotographs (Valyashko, 1956; Wardlaw and Schwerdtner, 1966). The clay particles accumulated at the bottom of intercrystalline cavities. In this way, all existing cavities were filled, rendering the deposit practically impermeable.

Fig. 2-13 (location map in Fig. 2-14) shows the deposit affected to vary

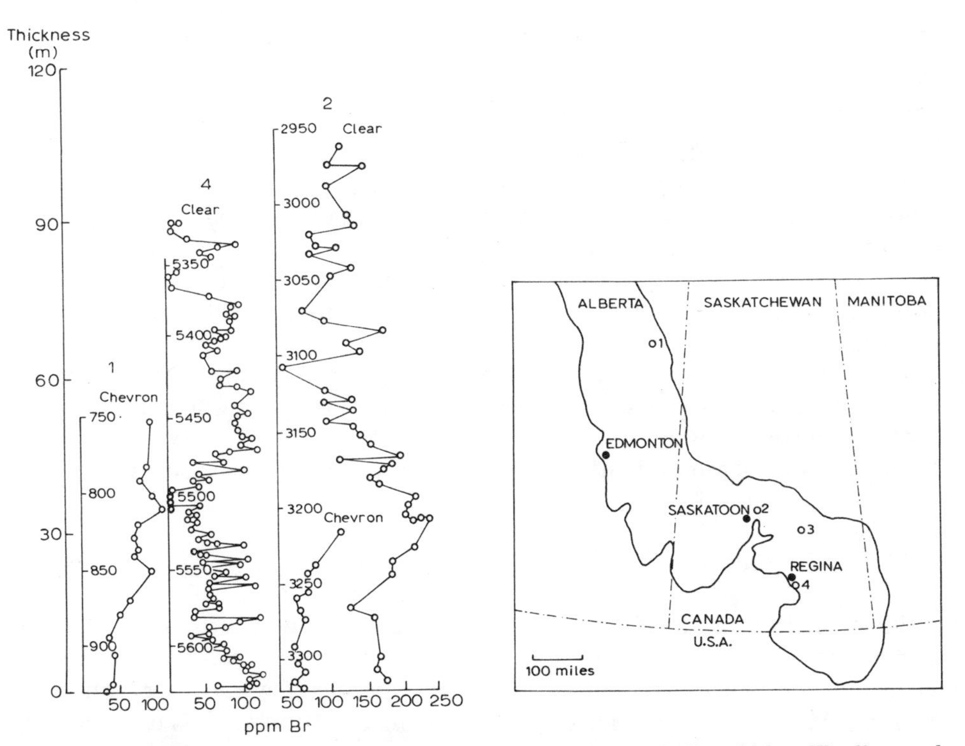

Fig. 2-13. Different bromine contents of chevron and clear halite. (After Wardlaw and Schwerdtner, 1966.)

Mineral identification as in Fig. 2-12. Profiles from three cores through the lower part of the Prairie Evaporites at the following locations (see Fig. 2-14): *1* = Fort McMurray, northern Alberta; *2* = Osler, central Saskatchewan; *4* = Regina, southern Saskatchewan.

Fig. 2-14. Location map, showing the present edge of the Prairie Evaporite salt. (After Wardlaw and Schwerdtner, 1966.)

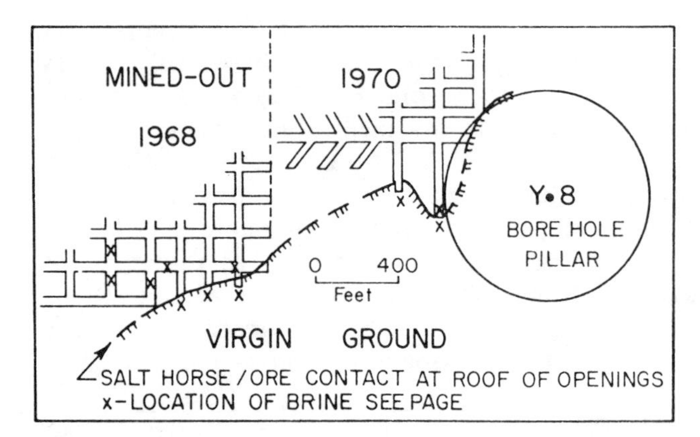

Fig. 2-15. Local brine seepage, IMC-K 1, Saskatchewan (Baar, 1974a).
$MgCl_2$ brine was trapped in pockets between halite crystals such as shown in Figs. 2-12 and 2-13. The term "salt horse" indicates a barren area in the potash bed, consisting of rock salt.

different degrees in various parts of the basin of deposition. It is emphasized that highly concentrated $MgCl_2$ brines are occasionally encountered in the potash mines which operate in this deposit; Fig. 2-15 demonstrates such a situation, the brines seeping out of cavities between halite crystals. For details, the reader is referred to Baar (1974a); for brine analyses, see Baar et al. (1971). The brines were classified as Devonian brines as their $MgCl_2$ content cannot be derived from minerals which are still present in the deposit.

Conservation of such and similar petrographic features must not be expected in salt deposits which suffered extensive deformation and recrystallization; particularly Braitsch (1971) cautioned repeatedly against attempts to explain petrographic features observed in salt domes as caused by the original precipitation.

2.2.3 Fluid inclusions

Fluid inclusions are frequently encountered in evaporite minerals and rocks. For practical considerations, it is extremely important whether such inclusions are isolated from each other, or whether they are interconnected; in the latter case, such inclusions would migrate down pressure gradients when such gradients develop.

Fluid inclusions in primary salt crystals such as hoppers or chevrons apparently could not escape from the host crystals. Similar inclusions, including gas inclusions, are frequently present in secondary salt minerals which formed deep below the surface under certain geological conditions, and in recrystallized salt rocks; such inclusions may also be trapped in isolated intercrystalline cavities.

Many salt and potash deposits contain gas inclusions which are isolated from each other; the mining of such deposits is plagued by sudden outbursts of possibly large quantities of such gases which expand their volume, filling adjacent mine openings. In some cases, gases released by such outbursts blew out of the mine shafts and caused casualties at the surface. The rock mechanics of such outbursts are dealt with in detail in sub-section 4.5.1.

Fluid inclusions in interconnected cavities pose different problems in mining operations; in salt rocks, the total volume of interconnected cavities is usually limited, as shown in the example of Fig. 2-15. Much larger volumes frequently resulted from tectonic events which caused fracturing, fracture and fissure zones possibly extending into waterbearing formations. Rock-mechanical processes caused by mining may tap such fissure zones; for this reason, such inclusions are dealt with in some detail in section 4.6.

Isolated brine inclusions in salt rocks cause no major problems in mining operations if such inclusions are free from gases. The geological conditions

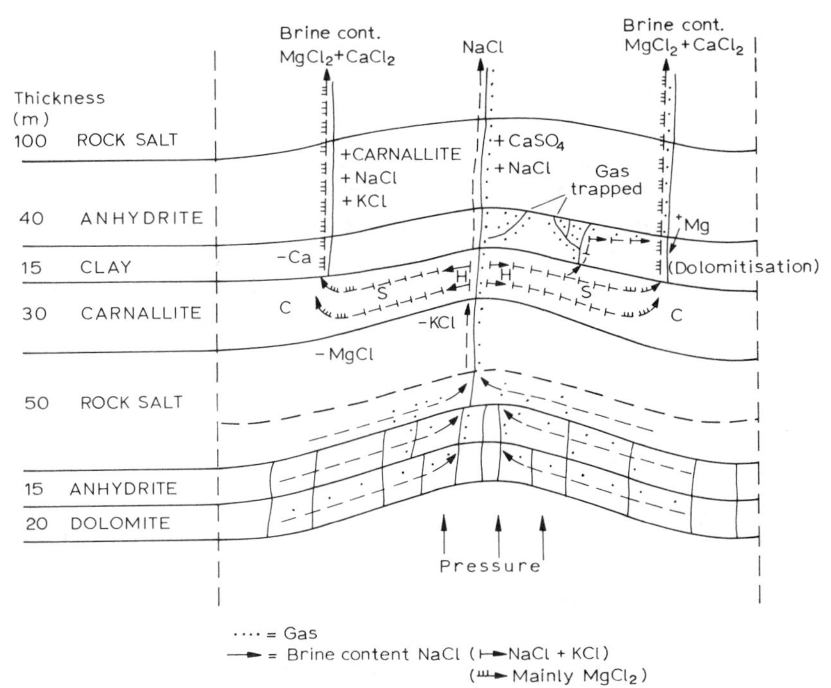

Fig. 2-16. Alterations of the Stassfurt potash bed after consolidation (Baar, 1958, 1966c). Schematic section showing dissolution (—) of potash salts and crystallization (+) of corresponding "reaction" salts at the indicated locations.

Brines and gases originated below the rock salt and were trapped in suitable structures until tectonic events caused vertical fractures. Carnallite (*C*) was replaced by sylvite (*S*), and finally by halite (*H*). Gases were occluded in "reaction" salts and trapped in dead-end fractures in the shale bed above the potash bed.

for the inclusion of gases in salt deposits are outlined shortly in the following.

Fig. 2-16 shows schematically the geological conditions in marginal areas of the Zechstein basin which provided the prerequisites for the inclusion of gases: tectonic events created passage ways (fissures) through the salt sequence, allowing gases and brines to escape from their host formations below the consolidated — and therefore impermeable — salt sequences. When contacting potash beds, the brines caused alterations of the potash beds, dissolving the most soluble potash minerals and precipitating less soluble salts; as the result, carnallite was replaced by sylvite and halite. In cases where such processes extended laterally into potash beds, sylvite was the first potash mineral to be contacted and dissolved, and to be reprecipitated when the brines contacted carnallite.

Such processes of "retrograde metamorphism" (Borchert and Muir, 1964) or "solution metamorphism" (Braitsch, 1971) are subject to the known physico-chemical principles outlined in sub-section 2.1.2.1. So they can be calculated quantitatively in cases where the original composition of the respective potash beds is known.

For the conditions of Fig. 2-16, such calculations were presented by Baar (1952a); details of such calculations were discussed by Kühn (1955b) and by Braitsch (1971).

As gases, if present, are occluded only in what Borchert and Muir (1964) termed "reaction" minerals — minerals which formed during intracrustal processes of solution metamorphism — the relative amounts of reaction minerals in the newly formed salt rocks, compared to the amounts of minerals incorporated from the original gas-free rocks, are extremely important in assessing the potential danger of gas outbursts. It has been shown by Baar (1958) that reaction minerals can make up 100% of secondary salt rocks, depending on rock-mechanical deformations caused by the loss of volume of carnallitic potash beds. Such losses may exceed 50% of the original volume of carnallitic potash beds; the resulting deformations are similar to those caused by extraction mining of potash beds.

Although gas inclusions in salt rocks are extremely important in some salt and potash mining districts for their effects on mining operations, they received little attention in textbooks which are now available in English translations, e.g., Borchert and Muir (1964) and Braitsch (1971). This may be due to the fact that, at the time when the German originals were written, these matters were still under controversial discussion; see Gimm and Pforr (1964) with contributions by Obert, Ignatieff, Panek, Baar. More recent publications (e.g., Gimm, 1968) indicate that the views expressed by Baar (1954d, 1958, 1962) have been generally accepted.

As the formation of gas inclusions in salt rocks requires special conditions, as outlined, such inclusions are rare in salt sequences without potash salts, as no reaction salts could form. This is the reason why the salt domes of the

Gulf Coast are free from gas inclusions, with the exception of some cases where gases were occluded in recrystallizing halite, or secondary halite which may have crystallized from migrating solutions due to cooling. In contrast, the salt domes of northern Europe are loaded with pockets of salt rocks with abundant gas inclusions; as a matter of fact, several potash mines were abandoned at the beginning of this century because of the problems caused by gas outbursts (Gimm, 1968, p. 553).

To date, no gas inclusions have been encountered in the Prairie Evaporites, in spite of extensive formation of reaction salts; this is attributed to the absence of gases during the alteration processes which occurred shortly after the deposition of the potash beds (Baar, 1974a). Most other potash deposits which suffered relatively little tectonic disturbance contain gas inclusions at certain locations where the prerequisites had been met; this applies to the Russian deposits of Solikamsk and Soligorsk, to the Upper Rhine deposits, and particularly to the Zechstein deposits of all series which contain potash beds (Gimm, 1968).

The composition of the gases occluded in salt rocks varies considerably, depending on the local conditions. In some cases, gases of different origin may have mixed prior to being occluded. Most analyses reported in publications indicate mixing with air after the gases had been released into mine openings.

For these reasons, no attempt at presenting results of gas analyses will be made here; frequently reported gases are CH_4, CO_2 and N_2 as main components of gas mixtures occluded in salt rocks. The composition of gases of the second type defined above is usually similar at any particular location; in fact, outbursts of occluded gases frequently release gases of the second type, and it is difficult, if not impossible, to determine the relative amounts attributable to either one of the two types.

It is emphasized that the presence of fluid inclusions in salt deposits, disregarding the question of how they were trapped, clearly indicates the impermeability of salt deposits under sufficient cover of overburden, provided that the impermeability had not been disrupted since the time of inclusion by events similar to those which enabled the fluids to enter the salt.

The rock-mechanical aspects of gas outbursts are discussed in Chapter 4 (sub-section "Outbursts of gas and salt").

2.3 TEXTURE AND STRATIFICATION OF UNDISTURBED DEPOSITS

2.3.1 Primary and diagenetic features in shallow-water deposits

It is a matter of definition and opinion to draw the exact boundary between strictly primary crystallization and diagenetic crystallization or recrystallization. The problems become apparent from Figs. 2-12 and 2-13:

it could be argued that only chevron halite is truly primary, while the clear halite and other components are not; on the other hand, in cases of predominant crystallization at or near the bottom of a deep basin, the minerals would be termed primary, although associated with hopper crystals of halite which had formed earlier at the surface of the brine body. In this text, primary and diagenetic crystallization, and resulting features, are dealt with together.

Consolidation of salt deposits by diagenetic crystallization and recrystallization takes place rapidly compared to other sediments, particularly in cases where temporary exposures due to lowering of the brine level occur. This has been emphasized by Valyashko (1956) with reference to Recent salt deposits forming in intracontinental depressions.

Garrett (1970) gave the following description of the process of consolidation of sea water-type deposits in basins less than 6 m deep: over the normal day—night temperature cycle, much of the salt deposited in the evening redissolves during the day, and subsequently recrystallyzes (this cycle would be longer for deeper basins). This tends to consolidate the deposit, and with the normal packing effect caused by the consolidation and filling-in of later deposited salts, the void volume progressively decreases. A typical example of such porosity values might be: 40 vol.% initially, 30% when under 15 cm of salt, 20% under 30 cm of salt, 15—20% under 60 cm of salt, 5—10% under 6—12 m of salt, and then gradually to nearly zero with deeper covering.

Several lines of evidence suggest similar figures for salt deposits which formed under deeper brine cover (Baar, 1960); particularly the irregular bromine profiles at the basis of thick salt sequences (see sub-section 2.1.2.2) indicate that vertical brine movements through consolidating deposits ceased after deposition of 20—30 m of salt. In this way, fluids present or originating underneath the salt were forced to migrate laterally out of their host formations, leaving the salt and potash deposits unaffected until, eventually, abrupt earth movements created vertical passage ways (Baar, 1958).

It should be obvious from the foregoing that the textural features of evaporite deposits may vary considerably, depending on the depths of the brines from which the evaporites formed: with brine depths over 6 m, the effects of diurnal temperature changes lost their importance gradually, with regard to vertical brine movements in the consolidating deposit. Seasonal temperature changes remained effective regardless of the brine depth, as shown by Dellwig (1955) for the Salina salt of Michigan; the salt was deposited in layers or bands of three distinct types: (1) cloudy layers of inclusion-rich pyramidal shaped hopper crystals of halite; (2) clear layers of inclusion-free halite; and (3) laminae of anhydrite and dolomite. In cases where large amounts of non-marine material were carried into an evaporating basin, such material sedimented with the laminae of anhydrite and dolomite, see p. 26.

However, in shallow water the conditions were different, not only during deposition and diagenesis of salt rocks as documented in Figs. 2-12 and 2-13, or during sabkha conditions, if sabkha conditions were at all as important in the early phases of evaporite deposition as emphasized in recent publications. Richter-Bernburg (1955) in particular has published convincing evidence of anhydrite and dolomite textures which show most of the petrographic characteristics ascribed to sabkha deposits.

Richter-Bernburg's (1955) photographs show textures of anhydrite cores from "rapidly sedimented anhydrite banks" which formed according to Fig. 2-17; in shallow water, the relative concentration of $CaSO_4$ becomes higher than in deep water because of the greater relative volume loss by evaporation; assuming water depths of 20 and 200 m, respectively (h_1 and h_2 in Fig. 2-17), and yearly evaporation of 2 m of water, the loss in volume is 10% and 1% at the respective locations. As a result, the more concentrated water will precipitate anhydrite in shallow areas and flow into the deeper part of the basin, where eventually halite crystallizes due to cooling (see Fig. 2-6 for confirmation by bromine profiles).

Wardlaw and Reinson (1971) published numerous photos of anhydrites with similar textures; the anhydrites formed on the top and at the slopes of carbonate banks and reefs in the Prairie Evaporite basin of western Canada: "transitions from agitated open-marine waters to quiet restricted conditions apparently occurred abruptly, as did changes from oxidizing to reducing conditions; in an offbank direction, halite appears to be a facies equivalent of anhydrite". Confirming in this way earlier conclusions based on bromine investigations (Wardlaw, 1963), these authors reject the sabkha hypothesis proposed by other writers, emphasizing the similarity between their findings and those of Richter-Bernburg (1955). The evidence of repeated exposures of salt deposits in marginal areas (Fig. 2-10), and related features (Baar et al., 1971; Baar, 1974a) confirm rather abrupt changes from deep-water to shallow-water conditions.

The resulting textural differences between shallow-water and deep-water deposits apparently are primary and diagenetic, as no indication of any alteration of the overlying potash beds by ascending solutions has been detected to date (Baar, 1974a). To the contrary, Fig. 2-12 and 2-15 suggest extensive seepage of concentrated solutions into the deposit; these solutions may

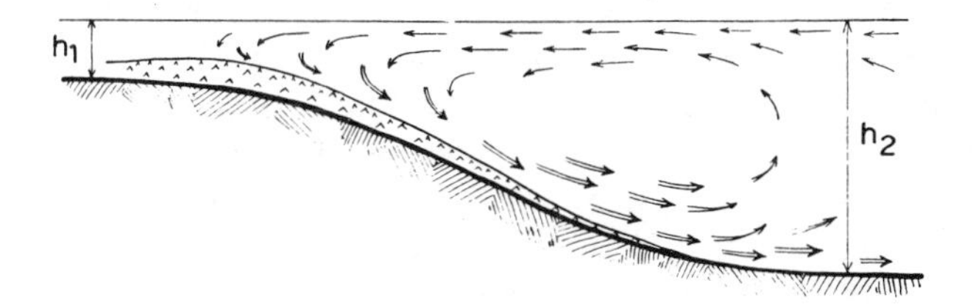

Fig. 2-17. $CaSO_4$ precipitation in shallow water. (After Richter-Bernburg, 1955.) h_1, h_2 indicate the water depths; the arrows indicate currents.

have contributed to dolomitization processes in the underlying carbonates as suggested for shallow areas of the Zechstein basin (Baar, 1958).

2.3.2 Stratification of deep-water evaporites

Kupfer (1974) pointed out that using the words "shallow" and "deep" without qualification can be misleading; with reference to recent publications and controversies regarding depositional environments of evaporation of sea water, he emphasized that these environments are transitional, and what is shallow to one person may be deep to another. Kupfer provided undisputable proof of his point.

In this text, the criteria discussed in the foregoing sub-section are applied: several lines of evidence indicate that vertical brine movements in consolidating salt deposits ceased after thicknesses of approximately 30 m of sedimented and cemented salt had been reached, brine depths to 50 m are termed shallow; any greater depths are deep. This qualification disregards Kupfer's proposed classifications of depths of salt *basins* and of environments; it rather acknowledges Kupfer's other statement that depths and environments are transitional: any deep basin — over 50 m deep — has marginal shallow parts along its shoreline, and any deep basin may intermittently or finally become a shallow basin. This has been shown in subsection 2.1.2. Hence, the terms "shallow" and "deep" as used in this text refer to *brine depths* rather than to *basin depths.*

In the foregoing, some characteristics of deep-water evaporites in the Michigan basin were quoted after Dellwig (1955). Referring to different characteristics of shallow-water deposits in marginal areas of this basin, and of adjacent basins, Dellwig and Evans (1969, p. 956) agree to a model developed by Rickard: the shallow-water salt was deposited in water less than 30 m deep, whereas the deep-water salt developed in water at least twice as deep, probably more.

Fig. 2-18 shows typical laminations observed in deep parts of the Zechstein basin (Richter-Bernburg, 1955). Thicker laminae with increased clay content are believed to reflect the 11-year sun-spot cycle. The relative thicknesses of the respective sediments between the laminae are given as follows: carbonates 0.06 mm, anhydrite 0.8 mm, halite 80 mm. Wardlaw and Schwerdtner (1966) refer to similar annual precipitates in the lower half of the Prairie Evaporites as seasonal layers: thin laminae of anhydrite, less than 1 cm thick, alternate with halite units of about 2—10 cm thick.

Apparently, the annual evaporation yielded approximately equal thicknesses of halite during the phase of halite deposition in both the Zechstein and the Prairie Evaporite basins. At the beginning of potash deposition in these two basins, the brine depths differed greatly: shrinkage of the brine volume and related regressions of the shore line of the Prairie Evaporite basin are reported in sub-section 2.1.2.3. As a result, potash beds were deposited

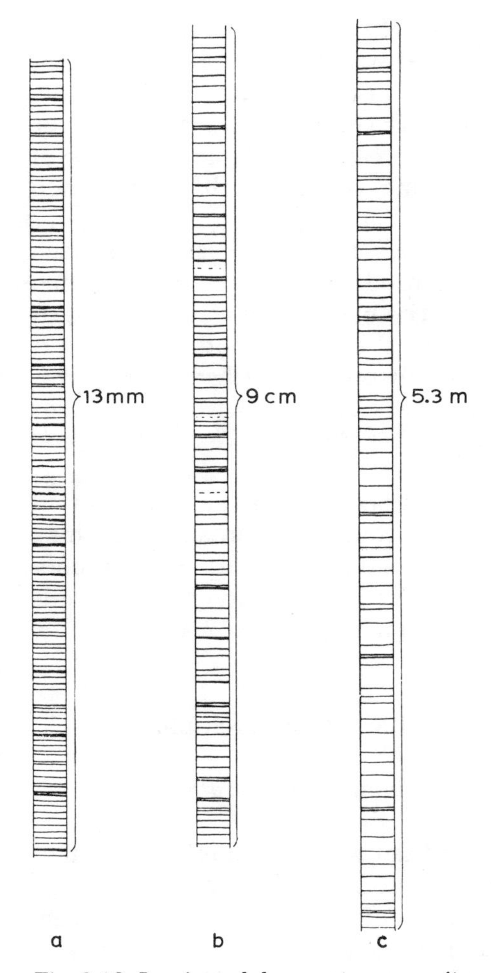

Fig. 2-18. Laminated deep-water evaporites. (After Richter-Bernburg, 1955.)
Corresponding thicknesses of: a. carbonates of phase I, Fig. 2-1; b. sulphates of phase II; c. rock salt of phase III.

under shallow brine cover: even in deeper depressions, the brine depths over potash beds during their deposition may not have exceeded 50 m. The resulting differences in the stratification of the potash beds are significant: seasonal layers of clay and anhydrite are present in deep parts of the basin (Fig. 2-11), some of the clay beds exceeding 10 cm in thickness; in shallow parts of the basin the stratification was destroyed shortly after deposition.

In the Zechstein basin, comparable regressions of the shore line did not occur during the deposition of the Stassfurt potash bed above the laminated rock salt of Fig. 2-18. The thickness of the potash bed, if preserved as primary carnallitic potash, is 30—40 m throughout the basin. It has been shown by calculations (Boeke, 1908; Braitsch, 1971) that the average depth of brines after the deposition of the potash bed was over 100 m; such depths result from the requirement of keeping the remaining $MgCl_2$ in solution. Therefore, the Stassfurt potash bed qualifies as a deep-water deposit, and the seasonal lamination is present in marginal areas. Occasional additions of diluted surface layers of less-concentrated brines resulted in the precipitation of halite beds as mentioned in sub-section 2.1.2.3.

In marginal areas, some of the clay laminae of central basin areas reach thicknesses up to 10 cm, suggesting conditions similar to those reported above for deeper parts of the Prairie Evaporite basin. Such clay layers are of particular importance for their rock-mechanical effects in both the Stassfurt potash bed and the Prairie Evaporite potash beds (see Chapter 00). This statement also holds true for the potash beds mined in two other major potash mining districts, the Soligorsk district of Belo-Russia and the Upper Rhine district of Alsace, France.

In cases where lateral alterations of potash beds occurred as schematized in Fig. 2-16, bed separations due to loss of volume of carnallitic salt may have provided voids for crystallization of secondary minerals. Such secondary stratification with salt and/or potash layers in the "wrong" sequence is quite common in some deposits. This is shown in sub-section 2.4.3.

2.4 CHANGES TO TEXTURE AND STRATIFICATION AFTER CONSOLIDATION

2.4.1 Relative deformability of salt rocks and interbedded strata

The different behavior of various types of salt rock and interbedded strata of other rocks has been observed by numerous investigators in salt and potash mines. In general, salt rocks tend to deform plastically by flowage, while intercalated carbonates, anhydrite and "clay" exhibit elastic behavior until they rupture. "Clay" beds encountered in deeply buried evaporite sequences have been completely dehydrated during compaction; therefore, they are, in fact, dry shales or mudstones, and they behave accordingly when deformed after consolidation.

The latter point is emphasized because, from time to time, writers are misled by the term "clay" and postulate plastic behavior, as would be expected if the shales and mudstones encountered in mines were wet, as found in near-surface deposits. In addition, when exposed to moist air in the mines or at surface (drill cores), these rocks tend to adsorb water and to change their behavior towards increased plasticity. Some published controversies (e.g. Baar, 1953a,c, with discussions) testify to such misinterpretations of the behavior of "dry" saliferous clay.

According to Borchert and Muir (1964, p. 248), ease of deformation in response to tectonic stress increases — among dry salts and associated sediments — in the following order: limestone/dolomite—marl—"dry" saliferous clay—anhydritic sediments—kieseritic rocks—halite—sylvite—carnallite—bischofite. When two or more members of this series are interbedded, boudinage structures, joints, and faults tend to develop in the more competent of the two.

This statement confirms the order ot plasticity assigned to the most common sediments encountered in salt mines by other authors, e.g., Lotze (1957, table 31, p. 326) and Baar (1953c). Some figures to support this general statement are shown in the following subsections.

2.4.2 Submarine slumping questioned

Richter-Bernburg (1955) proposed submarine slumping of consolidated evaporites as the mechanism to produce peculiar features such as folding of individual layers interbedded between more competent strata, or brecciation of carnallitic salt. Others disagree, particularly Borchert and Muir (1964), offering different explanations; the discussion of some cases is still going on (Dellwig and Kühn, 1974).

Carnallite breccias are frequently encountered in carnallitic potash beds, even in cases where the general layering is essentially horizontal; controversies in attempts to explain their origin have continued over more than half a century; apparently, there is still some disagreement among geologists; for details, see Lotze (1957, pp. 319 ff.) and Borchert and Muir (1964, pp. 247 ff.).

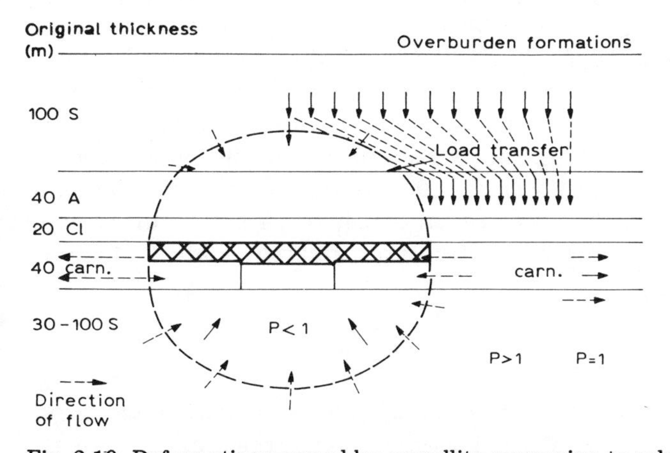

Fig. 2-19. Deformations caused by carnallite conversion to sylvite (Baar, 1958).

The original rock volume is reduced by approximately 50% as shown in Fig. 2-16. The dashed arrows indicate the directions of plastic deformations of salt rocks.

S **= rock salt;** ***A*** **= anhydrite;** ***Cl*** **= clay (shale);** ***P*** **= overburden load, overburden load according to the depth assumed as 1 unit of pressure.**

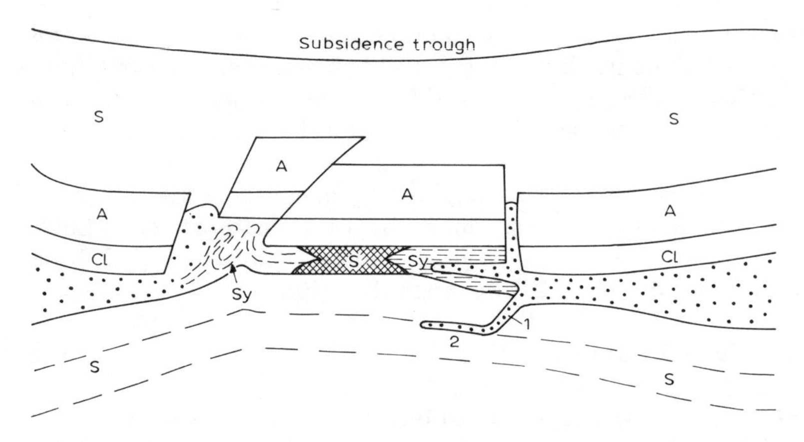

Fig. 2-20. Deformations caused by carnallite conversion to sylvite (Baar, 1958).

Final state after re-consolidation, shown in a schematic section of geological conditions encountered after events schematized in Fig. 2-19.

Carnallite stippled; *Sy* = sylvinite; *1* = former shear fracture; *2* = former bed separation void.

On the basis of evidence from underground observations in numerous potash mines, Baar (1958) suggested the mechanism schematized in Figs. 2-19 and 2-20: due to the loss of volume of carnallitic salt in the course of alterations as shown in Fig. 2-16, lateral deformations caused its brecciation, and the folding of sylvinite; brecciated carnallite may also have been squeezed into bed separations in the overlying and underlying rock-salt strata, into shear fractures, and into voids between blocks of anhydrite. The mobility of brecciated carnallite was increased by brines which also enabled extensive recrystallization, and crystallization of carnallite in open fractures.

To avoid misinterpretations: this is one mechanism of brecciation of carnallite; it has been emphasized because of certain analogies to the redistribution of overburden load after partial extraction of salt and potash beds in mining operations; these analogies are discussed in more detail in Chapter 5. Similar brecciation of carnallite and of other salt rocks has occurred more frequently in salt domes and similar structures, due to extensive tectonic deformations.

2.4.3 Salt crystallization in voids due to bed separation

Another phenomenon related to the loss of volume by carnallite conversion to sylvite also has remarkable analogies in mining: the separation of salt beds along planes of weakness provided by interbedded layers of clay, opened voids which may have been filled with secondary salts crystallizing from migrating solutions. Fig. 2-21 is a schematic representation based on actual geological conditions frequently encountered in the Werra potash

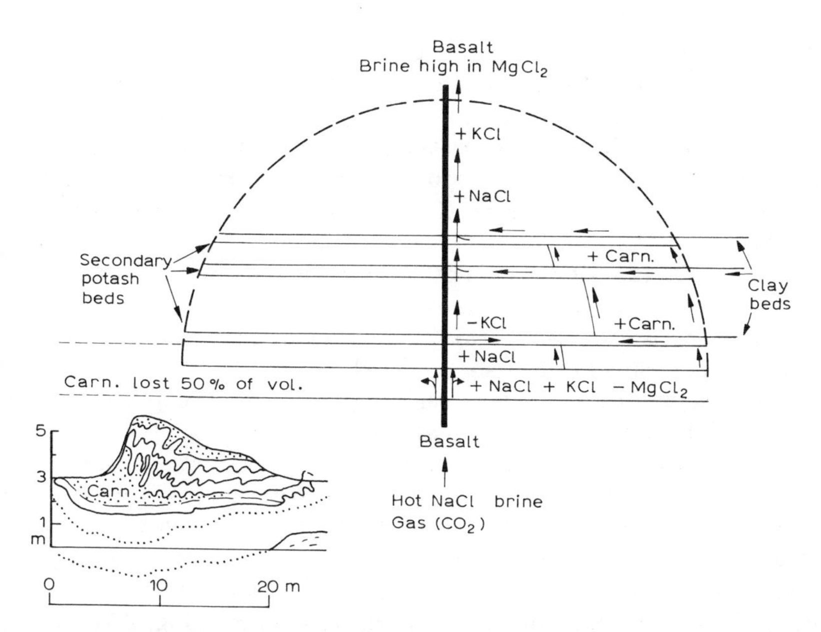

Fig. 2-21. Crystallization of secondary salts in bed separation voids.

Right side: migration of brines and gases as indicated by arrows; (—) carnallite conversion resulting in approximately 50% loss in rock volume; (+) crystallization of indicated salts, occlusion of gases.

Left side: present situation; carnallite was squeezed out of abutment zones as shown in Fig. 2-20; schematic sections (Baar, 1958). Inset shows folding as observed by Hoppe (1958).

mining district of Germany. Vertical fractures were caused by tectonic events associated with volcanism; hot brines penetrated laterally into existing potash beds, causing alterations similar to those sketched in Fig. 2-16. As the ascending brines varied in composition, depending on the amount of KCl and $MgCl_2$ dissolved in the potash beds, the composition of the layers of secondary salts which crystallized in voids is also very much variable, ranging from halite to sylvite and/or carnallite. For further details, the reader is referred to Baar (1958).

In the Werra district, the ascending brines were accompanied by gases (mainly CO_2); large amounts of these gases were occluded in secondary salts as outlined in sub-section 2.2.3. In cases where the layers of secondary salts, above the mined potash beds, contain gases, the effects on the rock mechanics of gas outbursts are remarkable, as shown in sub-section 4.5.1. It should be pointed out here that secondary crystallization of layers of salt rocks, be it halite, sylvite or carnallite, is an important prerequisite for occlusion of gases; another prerequisite is the temporary existence of bed separations. These prerequisites are not met by primary thin potash layers as postulated for the Werra district by geologists, e.g., Hoppe (1958). Consequently, such relatively thin layers with occluded gases are indicative of the mechanism shown in Fig. 2-19; this has been confirmed by bromine investigations which show high values in layers of secondary salts, while normal levels of bromine contents above the potash horizon indicate dilution of the brine body to such an extent that primary potash deposition became impossible (Baar, 1963). In this regard, the descriptions by Gimm (1968) are still inconclusive, although remarkable adaptations of his earlier published opinions to those published by Baar (1954d, 1958) cannot be overlooked. Secondary salts which crystallized in voids in less mobile evaporites and shales may also be loaded with gas inclusions; see Baar (1958) for evidence from salt domes of northern Germany where small outbursts occurred.

Gas inclusions in layers of secondary salts certainly are exceptional for the prerequisites which must be met. In many cases, the secondary origin of relative thin layers of halite, sylvite or carnallite is not as evident. This holds, for example, for layers of pure sylvite in sylvinitic potash beds, for layers of pure carnallite in, above or below potash beds, and for layers of rare minerals such as bischofite and tachhydrite which are considered secondary by most investigators. Also, the conversion of layers of kieserite into polyhalite and finally into anhydrite, under conditions as shown in Fig. 2-16, has been observed and documented in the literature, e.g., Baar (1960). $CaCl_2$ contents of the brines which caused such alterations were particularly effective in converting kieserite ($MgSO_4 \cdot H_2O$) into anhydrite ($CaSO_4$) within short distances presently observed in the mines. For more details regarding such reactions, the reader is referred to d'Ans (1947) or Braitsch (1971). Borchert and Muir (1964), deriving the solutions involved in such processes mainly from anticipated dehydration of gypsum, tend to underestimate the importance of $CaCl_2$ solutions.

$CaCl_2$ solutions result from reactions between $MgCl_2$ solutions and rocks which provide the calcium. Such solutions have been encountered in virtually all major evaporite basins, particularly in many mines in which the Zechstein potash beds are being, or have been, extracted. The Salina basin (Sorensen and Segall, 1974), and the Prairie Evaporite basin with limited seepage of such solutions into potash mines (Baar, 1974a), represent other examples.

In most cases of seepage of $CaCl_2$ brines into mines, the brines discharge from isolated reservoirs. The inflow rates decrease with time. Eventually, such inflows will terminate completely. There is no danger of mine flooding, provided that appropriate action is taken immediately. Detailed data regarding the actual development of such inflows is presented in Chapter 4.

2.4.4 Effects of slight tectonic deformations of salt deposits

Bed separations in near-horizontal salt deposits — e.g., those which result from the conversion of carnallite — need to be triggered by tectonic events

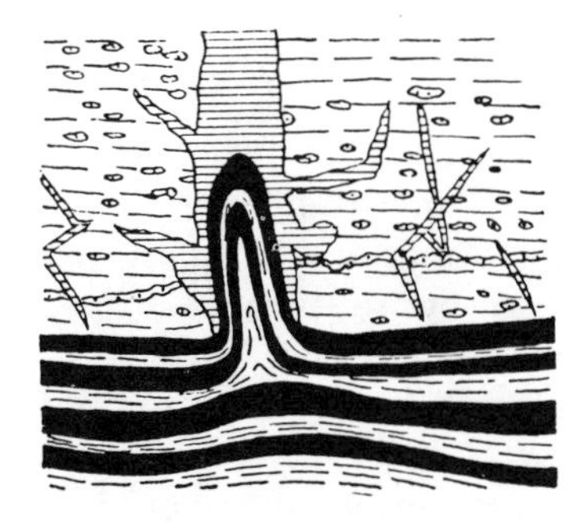

Fig. 2-22. Intrusion of salt into fractured marl. (After Lotze, 1957.)
Rock salt layers (black) with marl-anhydrite laminae (dashed) flowed into fracture as it opened in dolomitic marl with knots and bands of anhydrite. Other fractures filled with fibrous salt.

Fig. 2-23. Intrusion of salt into marl bed. (After Lotze, 1957.)
Salt (black) and sylvite (stippled) flowed into gap similar to Fig. 2-22. *DA* = dolomitic anhydrite; *DM* = dolomitic marl.

to cause vertical fractures through which fluids can travel. However, the mechanisms which actually cause the bed separations, are clearly atectonic and comparable to similar mechanisms caused by extraction mining. Frequently, tectonic deformation of salt deposits did not cause or involve brine migrations and related features.

Fig. 2-22 and 2-23 (Lotze, 1957, figs. 196 and 198) are of particular significance. They show dolomitic marl beds, and anhydrite beds, ripped apart by tectonic forces related to the deepening of the Upper Rhine graben after deposition and consolidation of the evaporites. In sharp contrast to such elastic reaction, the salt layers deformed plastically, closing the gaps as they developed and maintaining in this way the impermeability of the evaporite sequence.

Many more figures in Wagner's (1955) and Lotze's (1957) studies as well as in many other publications demonstrate the plastic reactions of salt rocks in cases where beds of dolomite or anhydrite responded elastically, by fracture, to tensional or shear stresses.

The extraction of salt and potash beds causes tension zones above the boundaries of mined-out areas. Such tension results in widening of cracks and fissures in competent rocks such as dolomite and anhydrite beds. It is extremely important to ensure, by appropriate development of extraction mining areas, that only plastic reactions occur in the salt rocks between extracted areas and competent formations. This is of vital importance in conventional mines where an impermeable seal must be maintained against waterbearing formations above or below the mine workings; for details, see Volume 2.

2.4.5 Effects of severe deformations on texture and stratification

The order of plasticity given in sub-section 2.4.1 is based on observations in severely deformed evaporite sequences. One important difference emerged: the "competent" members of evaporite formations — limestone/dolomite, anhydrite, marl and shale — responded by rupture to excess deformation. No matter how intensive the fragmentation may have been, e.g., during diapirism, the individual fragments suffered hardly any textural changes.

Lotze (1957, fig. 171) made an attempt at reconstruction of the original thickness of an anhydrite bed in a salt dome; the individual fragments are scattered throughout adjacent salt masses.

Thin seasonal laminae of the type shown in Fig. 2-18 usually were not destroyed during extensive flowage in salt domes. This holds for the North-European salt domes as well as for the Gulf Coast domes. However, in shear zones between different spines, as shown by Kupfer (1974), the salt may have incorporated up to 50% of "foreign" material, mainly clay; as a result, the "dirty" salt in such shear zones, which may be 3—300 m wide, has "physical properties distinct from those of salt and may make for hazardous mining conditions" (Kupfer, 1974, p. 215). Locally, shear zones "may contain large brine cavities. Petroleum leaks, gas pockets, and other mine problems may be associated with them" (Kupfer, p. 224).

These observations resemble those reported in sub-section 2.2.3 from salt domes in northern Europe; for the reasons given, the effects on the texture of salt rocks were much more severe in the Zechstein domes which contain potash beds.

Marl and shale beds are particularly prone to fragmentation, as shown in Figs. 2-22 and 2.23. In some salt deposits, such layers may account for up to 50%, or more, of the rock mass which consisted originally of horizontal layers. In cases where such deposits suffered extensive deformations, e.g., in diapirs or Alpine-type wide overthrusts, the marl and shale layers are brecciated to degrees which may not allow the reconstruction of the original succession. The salt, after frequent recrystallization, may exhibit fabrics with preferential grain orientation.

Studies of preferred orientations in salt deposits which had undergone extensive deformation, "have prooved exceedingly fruitful" (Borchert and Muir, 1964, p. 247). However, "often it is only the last formed texture that can be recognized"; therefore, some caution appears to be indicated regarding conclusions drawn by some investigators from petrofabric analyses, e.g., by Clark and Schwerdtner (1966) for potash rocks from the mines near Esterhazy, Saskatchewan: preferred orientation of halite and sylvite grains is postulated "as a result of syntectonic crystallization" related to anticipated fold axes. However, since folding is absent and many primary features have been preserved — see Figs. 2-10 to 2-14 — preferred orientation, if really present, must have originated in a different way.

Schwerdtner and Morrison (1974) found natural fabric patterns for halite and sylvite observed in an anticline "incompatible with the model patterns" expected from syntectonic recrystallization. Apparently, these matters are still very much under discussion among theoreticians. More reinterpretations (Schwerdtner, 1974) may be expected; interested readers are referred to the above-mentioned publications, and to Borchert and Muir (1964) for further references. For practical purposes, grain-size investigations appear more relevant than investigations of possible preferred orientations; see Kühn (1968) for numerous photos of recrystallization patterns observed in salt domes.

2.5 STRUCTURAL GEOLOGY OF EVAPORITE DEPOSITS

2.5.1 Faulting and continental rifting

Kinsman (1974) describes the major evaporite deposits of the world in plate-tectonic terms as follows:

(A) Those formed on a single lithospheric plate atop continental lithosphere: the evaporites of intracontinental, intracratonic or interior basins. (B) Those formed between two lithospheric plates, either converging or diverging. The latter are considered the most important continental-margin evaporite deposits.

Regarding evaporite deposits of the interior basin environment (category A), Kinsman states that "it are these evaporite deposits which are best known and which are generally thought of as the major evaporite deposits of the world. This conclusion is in need of modification in the light of more recent findings of extensive, thick continental margin evaporite deposits." And: "Rifted continental margins of Triassic to Quaternary age border nearly the entire North and South Atlantic Ocean, much of the Indian Ocean, the Red Sea and possibly much of the Arctic Ocean and intracontinental rift valleys are well developed in East Africa. Examples of thick evaporite deposits in intracontinental rift valleys related to the rupture of continental plates are the Sergipe—Alagoas Basin evaporites of Brazil, and possibly analogous deposits of Gabon; the Dead Sea evaporites; and the Danakil Basin evaporites of Ethiopia."

These evaporites, of course, have been known for decades; it appears that Kinsman's hypothesis was primarily triggered by the recent discovery of thick evaporites underneath the Red Sea. It remains to be seen whether Fig. 2-24 is indeed in need of future modifications; presently available information clearly points to the overwhelming importance of evaporites of Kinsman's category A. From a practical point of view, category-B deposits are negligible.

Disregarding the source of the forces which caused rifting, or formation

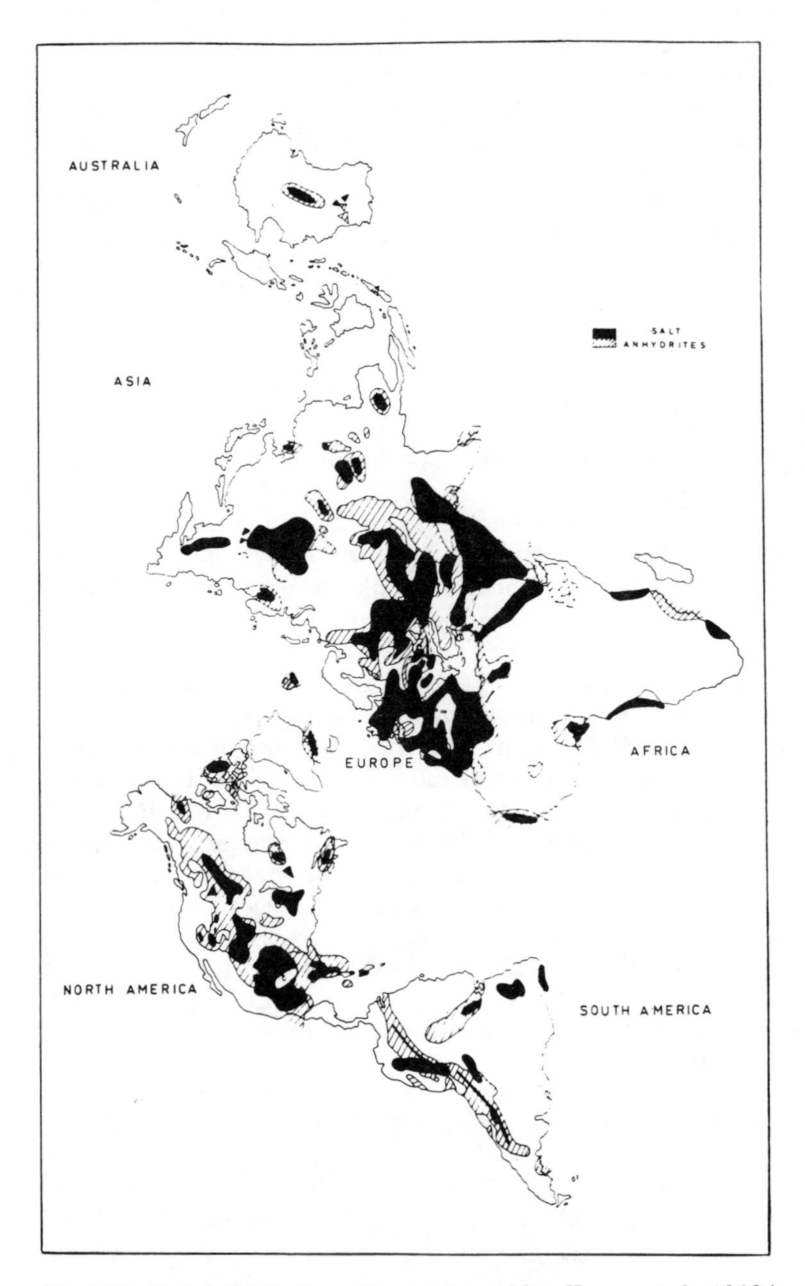

Fig. 2-24. Global distribution of evaporites. (After Kozary et al., 1968.) Composite distribution through geologic time.

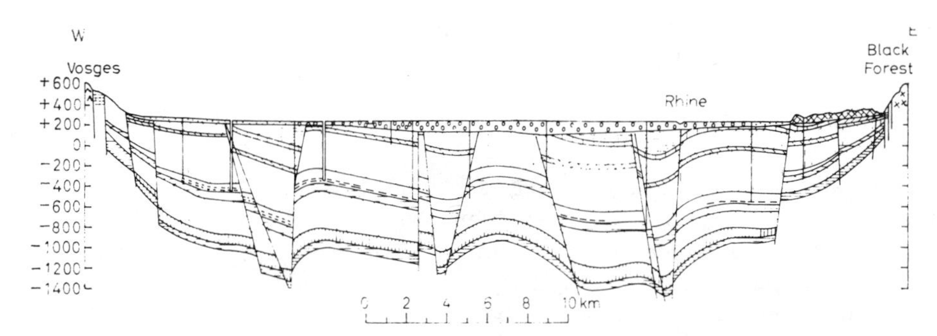

Fig. 2-25. Effects of block faulting on evaporites. (After Borchert and Muir, 1964.)
Section of the Upper Rhine graben; potash beds indicated by dashed lines.

of grabens, or block faulting in general, the effects on consolidated evaporite deposits appear similar: unless subsequently compressed by lateral forces, individual blocks including evaporite successions exhibit little, if any, internal disturbance; some blocks may have been tilted, others simply subsided as shown in cross-sections of the Rhine graben (Figs. 2-25 and 2-26).

Similar expansion tectonics (Kupfer 1974, p. 209) probably contributed to evaporite formation and deformation in the Gulf Coast area, individual basins being formed prior to the rifting apart of Africa and South America. The Maritime basins of Eastern Canada show certain similarities to the Rhine graben, as emphasized by Baar (1966c) with reference to other publications.

Kupfer summarized his observations as follows (p. 211): "All this is com-

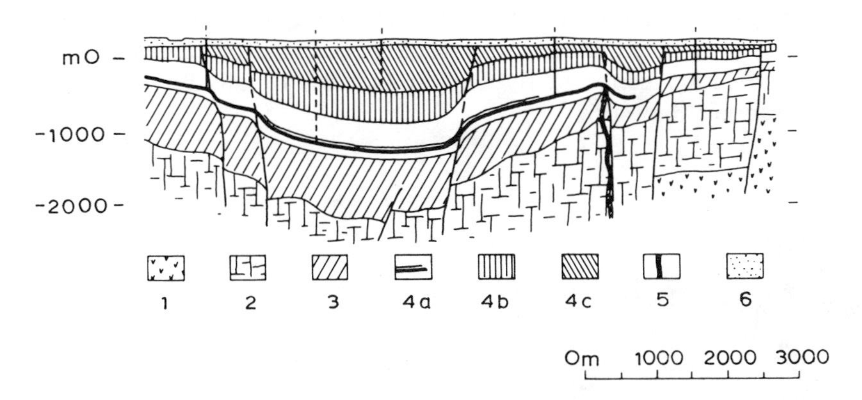

Fig. 2-26. Effects of block faulting on evaporites. (After Kühn and Dellwig, 1970.)
Another interpretation of the section shown in Fig. 2-25, eastern part.
1 = basement; *2* = Triassic and Jurassic; *3* = Eocene; *4* = Oligocene; *a.* salt sequence with potash beds, *b.* Middle, *c.* Upper; *5* = basalt dyke; *6* = Quarternary.

patible with the doctrines of sea-floor spreading and plate tectonics, but not proof; several other hypotheses are equally viable".

2.5.2 Folding

Large-scale folding of evaporite sequences has had little effect on the salt members, as would be expected from their plasticity. The contrasting behavior of competent members such as anhydrite or dolomite beds is demonstrated in Fig. 2-27; it shows anhydrite blocks which suffered little internal deformation, the lateral shortening being accomplished by slipping along fractures.

Fig. 2-27 has been published many times as an example of different behaviors of salt beds and anhydrite beds in the Stassfurt anticline (Lotze, 1957; Borchert and Muir, 1964, with reference to previous publications).

Fig. 2-28 is a section through the Stassfurt anticline, showing another feature often related to anticlinal deformation: near the crest of the anticline, groundwater action has removed all of the salt and much of the anhydrite, leaving cap-rock equivalent to the leached residues atop the salt domes.

Fig. 2-29 shows alterations of the potash bed caused by groundwater action: kieseritic carnallite has been transformed into kainite. Mining of kainite during the first decades of potash mining has resulted in many mine floodings; the rock-mechanical aspects of such floodings are discussed in more detail in Volume 2, as they are also of great significance in other mining districts.

In cases where folding of evaporite sequences occurred at depths which exclude groundwater action, faulting may have affected the competent members similar to Fig. 2-27. Some steep "second-order" folds (Evans and Linn, 1970) in the Cane Creek anticline (Fig. 2-30) of the Paradox Basin seem to be indicative of faulting in the overlying competent member; such faulting, in turn, may be responsible for rock-mechanical problems such as rockbursts and collapses reported by Dreyer (1969). Details of these relationships are given in Chapter 5.

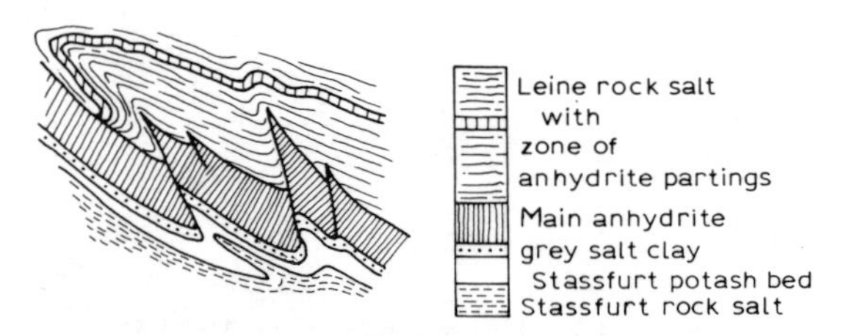

Fig. 2-27. Different reactions of salt rocks and anhydrite. (After Borchert and Muir, 1964.)

Folding of salt, brittle fracture of anhydrite bed.

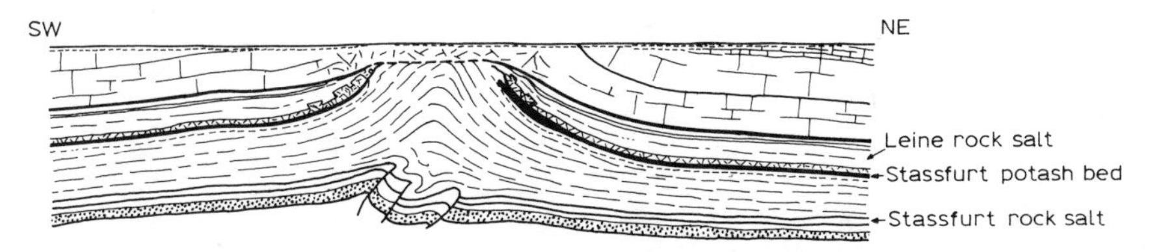

Fig. 2-28. Section of the Stassfurt anticline at Stassfurt, Germany. (After Lotze, 1957.)

The term "diapiric folding" has been used to characterize anticlinal structures in which mobility of salt rocks has resulted in relatively simple deformations.

With increased deformation within the evaporite cores of anticlines, the structures become more complex, exhibiting transitional stages leading to the extremely complex structures observed in salt domes.

Fig. 2-31 shows numerous elongated salt stocks in northwestern Germany (Sannemann, 1968; see also Trusheim, 1960); many of them may have started out as anticlines. "The subsequent wave-front-like growth of the salt-stock families took place in a purely halokinetic way, that is, by the movement of salt under the influence of gravity. The absolute rate of the horizontal wave-front-like flow of the salt over great distances averages about 0.3 mm a year" (Sannemann, 1968, p. 261).

2.5.3 Diapirism

Modern investigators appear in agreement regarding the need for a triggering mechanism to initiate salt diapirism. Tectonics movements, resulting in faulting of competent formations are usually invoked. It may be emphasized with Kupfer (1974) — see preceding sub-section — that expansion tectonics

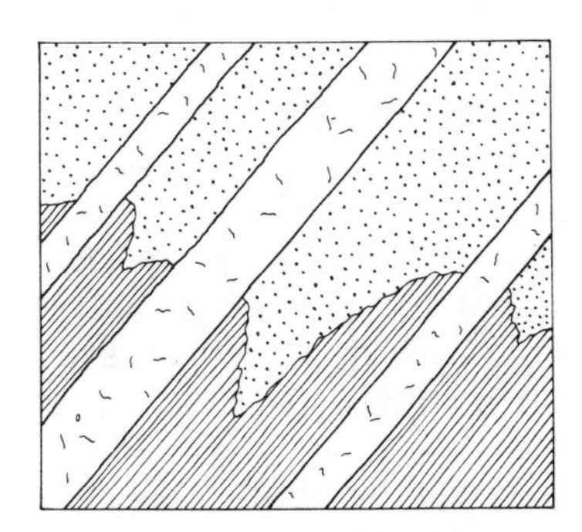

Fig. 2-29. Effects of groundwater, Stassfurt potash bed. (After Lotze, 1957.)
Rock salt beds are not affected; potash beds are converted into kainite (stippled).

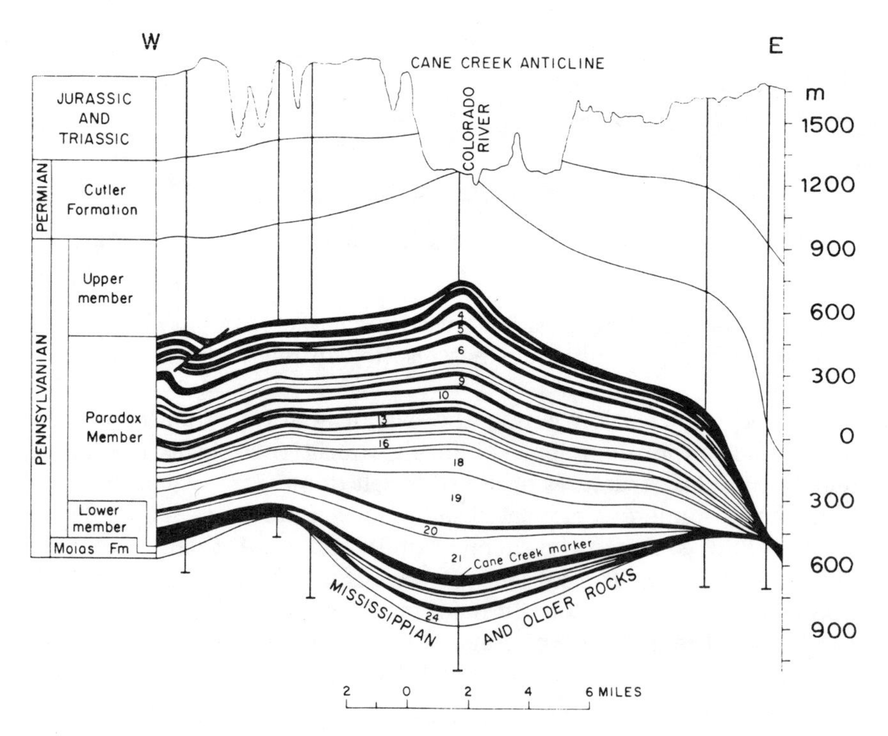

Fig. 2-30. Section through the Cane Creek anticline, Utah, U.S.A. (After Hite, 1970.)
The dark zones in the Paradox Member represent clastic-penesaline marker beds; the intervening units, several of which are numbered, are salt beds.

provide the triggering of halokinetic movements of salt as shown in Fig. 2-31. Upwelling of salt masses results in further expansion of overlying formations.

Introductory statements by Borchert and Muir (1964, p. 237) refer to obsolete hypotheses which call for "tectonic compression" as the cause of complex folding and brecciation of salt beds. However, these authors contradict such hypotheses clearly, and their statements may be quoted for their clear ideas (l.c., p. 243): "The upwelling of salt domes may be caused almost entirely by the load of the overburden. At depths of between 2500 and 3000 m the temperature is about 100°C and the pressure over 600 kg per cm^2. Under these conditions salt is very plastic; it is in fact about as soft as butter on a hot summer's day. And with increasing depth of burial, it becomes even more mobile." Their explanation of salt dome formation appears to be in agreement with Trusheim's (1960) halokinesis.

Another quotation from Borchert and Muir (1964, p. 239) is equally instructive: "Evaporites may exhibit intrusive characteristics where they are cut by faults. They are squeezed up the fault plane to form a fissure filling whose thickness decreases gradually away from the main salt body. The pressure required to produce this flowage is exerted by the overburden. Fault intrusions are particularly common in horst and graben terrain. At the time of intrusion, the fault is an open one. If the area is later subject to a compressional phase, the fault filling is squeezed as in a vice. Very complex structures may be produced, for the salt will flow away along the fault plane — more or less in the direction of minimum stress." And: "Large-scale injection

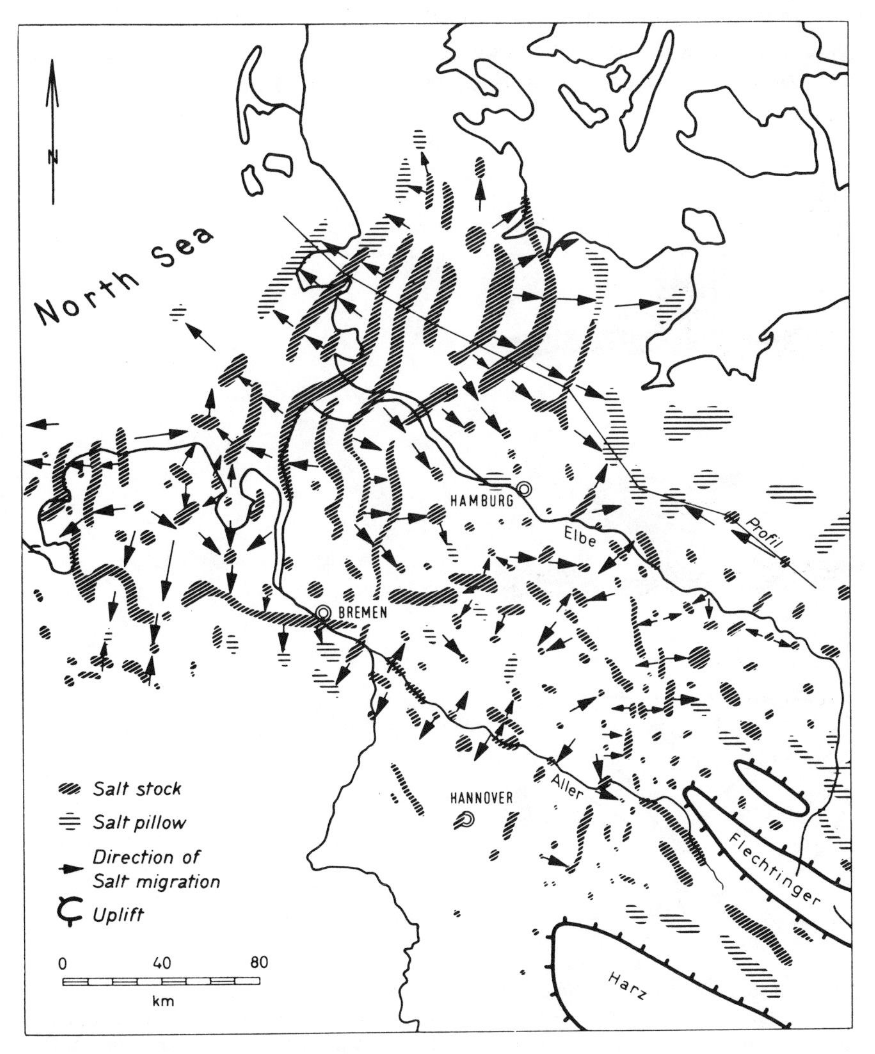

Fig. 2-31. Directions of salt migration, northwestern Germany. (After Sannemann, 1968.)

of salt up fault planes is often accompanied or followed by the development of grabens as blocks of the overburden subside."

Some figures may serve to illustrate the variability of structures involving flowage of salt up fault planes.

Fig. 2-32 shows Lotze's (1957) observations on salt diapirs in Spain, and his explanation of the mechanism that squeezes salt into fault planes: subsidence of overburden formations results in widening of the loosened ground, as the salt is flowing due to differential lithostatic stresses. This is the mechanism shown on a much larger scale in Fig. 2-31.

Fig. 2-33 (Lotze, 1957, fig. 124) represents a section of the Leine graben of West Germany; as several potash mines have been operating in this area, the geological conditions as shown have been known for many decades. Salt intrusion into the faults contributed considerably to the formation of this graben zone.

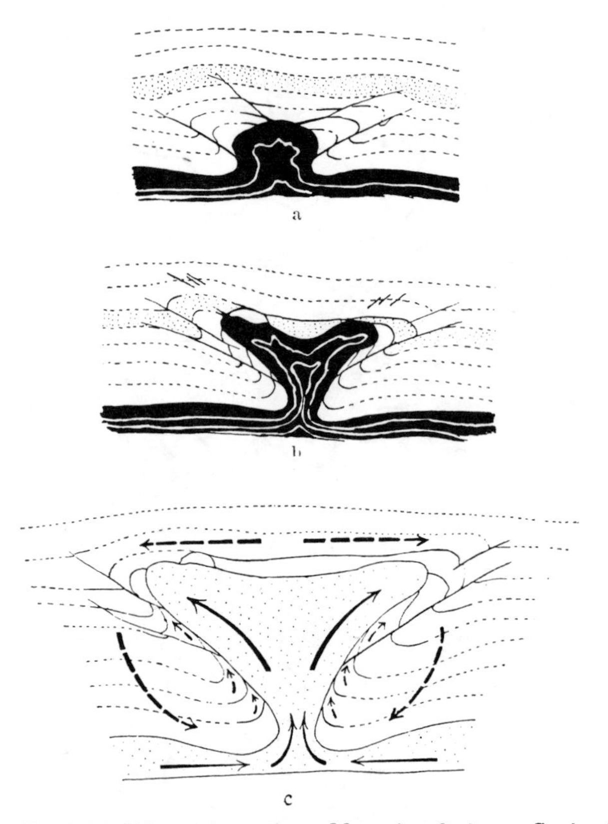

Fig. 2-32. Schematic sections, Murguia salt dome, Spain. (After Lotze, 1957.)

Salt (black) with clay laminae. Interpretation of deformations: a. intermediate; b. final state of deformations; c. directions of flow of salt (stippled) indicated by arrows, the dashed arrows showing the movements in overlying formations.

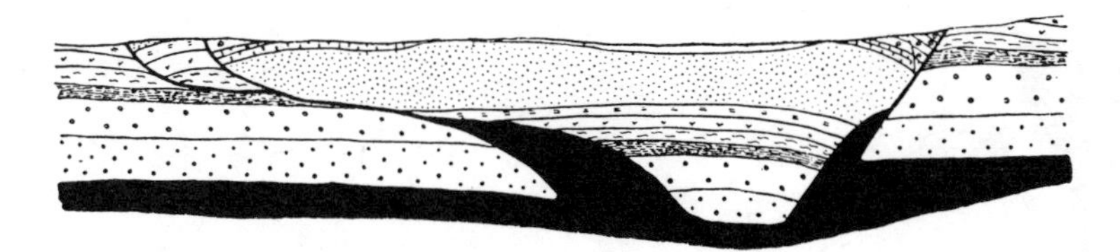

Fig. 2-33. Section through the Leine Graben, western Germany. (After Lotze, 1957.) Zechstein salt (black) intruding into thrust faults.

Borchert and Muir (1964) and Lotze (1957) show many more sections from other areas of faulting, the salt members exhibiting similar flowage. More excellent work has been done in many other areas of the world; the results were presented in *Geol. Soc. Am. Spec. Pap.*, 88 (1968); the reader is particularly referred to the papers by Liechti, Tortochaux, Bentor, Stöcklin, Dunnington. The mechanism of intrusion of Gulf Coast salt has been dealt with in detail by Kupfer (1970, 1974) with reference to numerous other publications.

Fig. 2-34 (Lotze, 1957, fig. 164) illustrates schematically the type of flowage structures encountered in North-German salt domes; evaporites which had been pushed above the present level of the top of the dome, were dissolved by groundwater, leaving residues consisting mainly of gypsum fragments and insoluble material.

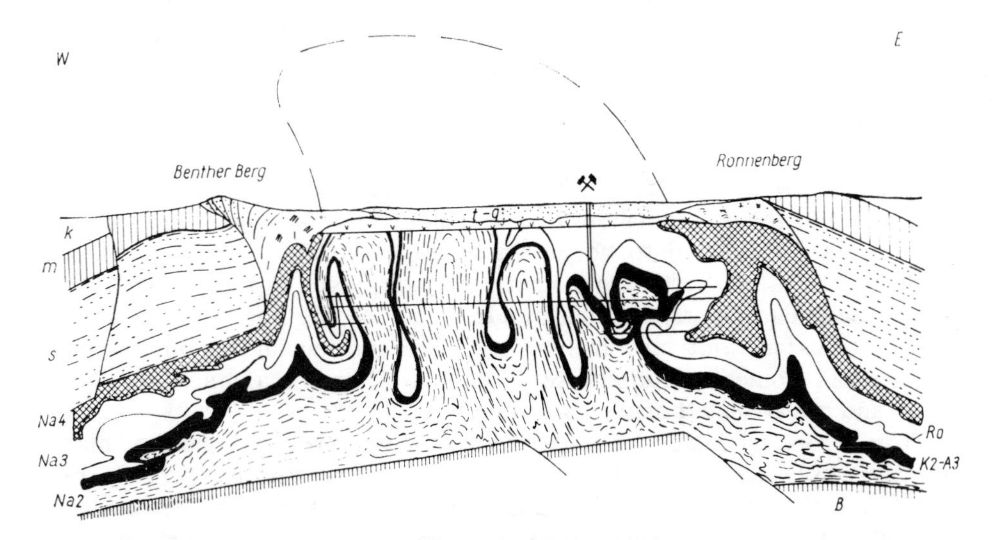

Fig. 2-34. Internal structures in a typical Zechstein salt dome. (After Lotze, 1957.)

Section through the Ronnenberg potash mine with levels from 560 to 850 m. *s*, *m*, *k* = Triassic formations; *t*, *q* = Tertiary and Quaternary formations; *Na2* to *Na4* = Zechstein rock salt; *A3* = Main anhydrite; *K2* and *Ro* = potash beds; *B* = basement formations.

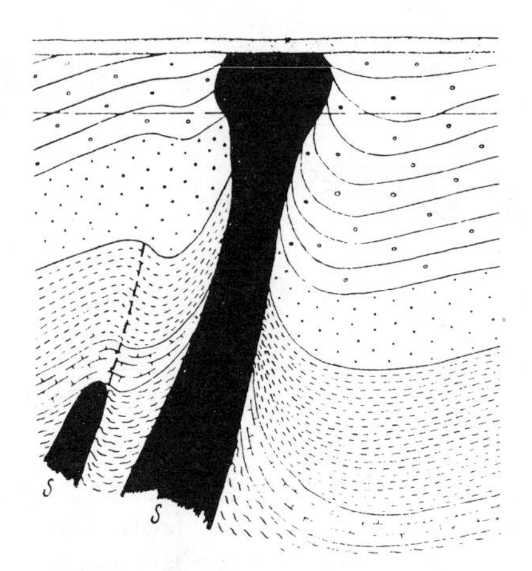

Fig. 2-35. Salt diapir of Floresti, Romania. (After Lotze, 1957.)

Fig. 2-35 (Lotze, 1957, fig. 121) shows another phenomenon frequently observed on salt domes: near surface, the diameter is considerably larger than at depth. Some investigators suggested that such "mushrooming" may have been caused by glacier-like flowage of salt at surface (Lotze 1957, p. 282), similar to what is observed on Recent salt domes. These observations are of particular interest regarding the flowage of salt into underground openings in salt deposits excavated in mining operations; for this reason, such observations are dealt with in some more detail in the following subsection.

2.5.4 Salt glaciers

Stöcklin (1968, pp. 158—159) may be quoted as follows: "The Middle East is one of the classic regions of salt geology. Salt features, which in other parts of the world had to be pieced together during decades of toilsome work in dark mine shafts, are magnificently displayed in full daylight in the Middle East. Particularly in Iran, hundreds of outcropping salt domes reveal salt tectonics and salt movements." . . ."The spectacular agglomeration of exposed salt domes in Iran and the Persian Gulf area is a unique case in the world and has attracted many observers."

From a rock-mechanical point of view, the numerous salt glaciers, magnificently displayed in full daylight as shown by air photos (e.g., O'Brien, 1955), are the most spectacular *and important* features; the lessons to be learned from these salt glaciers cannot be pieced together by short-time measurements and observations in mines, not to mention laboratory-time experiments under inadequate conditions of loading and confinement.

The salt glaciers in Iran are flowing down from salt domes which rose to

Fig. 2-36. Salt glacier flowing out of crater, Iran. (After Lotze, 1957.)

about 1200 m above the surrounding plains; although considerable winter rains carry away salt in solution, and the glaciers remove solid salt, the domes are maintaining their height. As the salt dome mountains represent the highest elevations in some areas, it may be concluded that such domes are still rising, the losses by dissolution and glacier flowage being at least compensated by salt flowage from depth.

In some cases, of which Fig. 2-36 is representative, salt flowage from sources at unknown depths apparently is no longer sufficient to make up for losses at surface. As a result, a large crater is left, with a salt glacier flowing out of the crater at one side.

O'Brien (1955, p. 810) emphasized the significance of these observations: the plasticity of salt should be related to its viscosity since the time-dependent deformations of salt masses apparently are a function of the viscosity. Given stress differences exceeding those required to initiate plastic reactions of solid salt bodies, the mobility of salt masses resembles the mobility of fluids subjected to short-time application of differential stress. The salt glaciers of Iran, flowing downhill from the summits of most salt dome mountains, are considered convincing proof of the fact that salt is flowing even under its own weight only.

No information is available regarding the depths to the bottom of the valleys through which the salt glaciers flow out of the pipes formed by the rising salt. Such information would allow determination of the stress differences required to cause lateral flow of salt bodies in the direction of the minimum principal stress; in the case of the Iranian salt glaciers, the minimum lateral stress at the outlets of the dome pipes apparently equals the atmospheric pressure. Equivalent conditions are created in mining operations. In section 4.2, it is shown that the reaction of salt masses to excavation resembles, in principle, the flow of salt glaciers out of their confinement; the conditions of restraint, of course, are different, and this causes different rates of flow or creep; the latter term is commonly used in mining.

The significance of the existence of numerous salt glaciers and of their flow under apparently minimal loads, as emphasized by O'Brien (1955), went largely unnoticed in most publications and textbooks which pretend to deal with the rock-mechanical behavior of salt rocks in situ, particularly in mining. Borchert and Muir (1964), and W. Dreyer (1974) confine themselves to short repetitions of what had been stated by Lotze (1957), overemphasizing results of laboratory investigations and disregarding the mechanical behavior demonstrated by Recent salt glaciers.

Trusheim (1960, p. 1527), in particular, stressed observations on apparent fossil salt glaciers which may be quoted: "Since salt was able to move upward against the confining force of overburden, the head of the diapir could also break through strata whose specific gravity was less than that of salt. The diapir could rise for a period until the gravitative equilibrium of the rock system surrounding the salt was reached. If the cover of lighter sediments was very thick, and the salt replenishment was limited, the diapir could remain stationary at a certain level for any length of time. On the other hand, in the presence of a thin cover of lighter material the salt could reach the surface and flow out."

"Such *'extrusions'* are known in both fossil and recent form. In northern Germany these outflows seem to have taken place predominantly under water. Presumably, the greater part of the flap or mushroom-shape overhangs must be regarded as *remains of 'salt glaciers'*, which extruded below a veneer of unconsolidated sediments (Gripp 1958, p. 255), rather than as protrusions of the diapir stage."

"In connection with these outflows breccias have also been observed. These coarse masses of detritus, with blocks more than half a meter in diameter, partly lie on top of the pierced salt and have partly been intercalated as rock flows in the flank sediments. It seems that the debris of the broken roof is entrained by the salt paste in process of extrusion, and remains as *'saline moraines'*, after the salt has been leached out. These extrusive breccias represent a special type of solution breccias."

Borchert and Muir (1964, p. 252), in their shortened version of Lotze's (1957, p. 282) report on salt glaciers, are in apparent agreement with such conclusions drawn from Recent salt glaciers; their statements may also be quoted:

"Occasionally, as in Iran or northern Spain, salt domes or diapirs manage to extend right to the surface. As they approach it, they dome up the overburden and so may finally emerge on the top of a mound standing a hundred feet or so above the surrounding area. If the climate is humid, the exposed salt is usually dissolved away, and a crater is produced in the mound. But if the climate is arid, the upwelling salt is preserved and flows slowly down one side of the mound and onto the surrounding plain. This surface-flowing salt behaves rather like a glacier in that it submerges minor topographic features lying in his path. Those which are too high to be completely covered are left standing rather like 'nunataks' *amid a sea of salt.*"

"The salt does not, however, possess quite the erosive power of ice. Lees (1927) and Harrison (1930) give full accounts of the Persian salt glaciers."

It appears rather difficult to reconcile such observations — see Fig. 2-36 for "nunataks" amid a salt glacier which flows around them — with creep or flow limits for salts as postulated from laboratory experiments—see Chapter 3.

2.6 REVIEW OF SALT MINING METHODS

2.6.1 Introduction

Evaporites other than salt and potash are rarely extracted in underground operations; if evaporites such as anhydrite or gypsum are mined underground, the mines are operating at shallow depths. As the rock-mechanical properties of such deposits are similar to the ones of other sedimentary deposits, and greatly different from those of salt deposits, mining of evaporites other than salt is not included in the following.

Regarding the variability in geological structures as outlined in the foregoing section, it is apparent that no single mining method is applicable to all salt and potash deposits. To the contrary, the most suitable methods of mining under given sets of geological conditions are quite different; some resemble methods applied in the extraction of other tabular deposits such as coal seams, some resemble methods applied in hardrock mining, and some are applicable exclusively in salt deposits because of their solubility.

Solution mining methods take advantage of the solubility of salt rocks which, on the other hand, requires extreme caution in conventional "dry" mining operations: any leakage of water into salt and potash mines must be prevented, as the inflowing waters tend to dissolve salt along their passage ways, enlargening them to extents which eventually would allow unlimited discharge of waters from overlying aquifers, particularly if such aquifers are connected to groundwater aquifers. For example, the history of potash mining in Germany is the history of many futile battles against water inflow; some lasted only a few days, some continued over decades — and only very few were successful. The history of salt mining in countries where salt has been mined extensively for centuries, even millenniums, is very much the same — and the newcomers are catching up, repeating all kinds of mistakes made elsewhere. Learning how to mine salt deposits certainly is very costly if newscomers start out with laboratory concepts, disregarding facts which are of vital importance.

2.6.2 History of salt and potash mining

Nobody knows for sure where and when underground salt mining started. Fatal accidents in underground salt mines have been dated back many thousand years: in the Hallstatt salt mine in Austria, miners who apparently had been killed by accidents, were found completely encased in recrystallized salt; the time of their accidental death could be determined from the tools they had used: most of the tools were made of bronze, hardly any of iron. There are speculations that mining salt in that area may have begun during the New Stone Age. This also holds true for the beginning of salt mining in Romania. Written documentation of underground mining activities in Austria and Romania is available from the time the Romans conquered these areas. The famous Wieliczka salt mine in Poland has been in operation for about 1000 years; now, it is a tourist attraction, as is also the case with some old mines in the Alps (Kramm, 1973).

Salt mining began to boom when rock salt became an important raw material in the chemical industry; this applies to conventional dry mining as well as to solution mining. Solution mining through boreholes drilled from surface is a relatively new technique compared to solution mining in conventional mines which has been practiced for centuries, particularly in Austria where such techniques have been used for at least 1200 years. The simple reason for the development of solution-mining techniques in Austria was the lack of pure rock salt; salt with impurity contents up to, or more than, 50% is dissolved in the mines in large solution cavities, the insolubles settling to the bottom of the cavities while concentrated brines are evaporated outside the mines.

The same principle is being applied in modern solution-mining operations which use boreholes from surface. Some disadvantages result from the need for evaporation of the water in cases where solid salt is produced, and from the need for brine disposal facilities in cases where storage caverns are constructed.

Potash mining is relatively new compared to salt mining; it began some 100 years ago at Stassfurt, Germany, and about 50 years ago in other countries. In the two countries with the largest potash production at the present time, the U.S.S.R. and Canada, extensive potash mining began only in recent years: in 1963, potash production commenced in both the Soligorsk district of Belo-Russia and the Province of Saskatchewan, Canada. The production capacity of these new potash mines exceeds considerably previous standards and records; for example, the capacity of the potash operations of the International Minerals and Chemical Corporation (Canada) Limited (IMC), with two shafts, equals approximately the capacity of each of the formerly leading countries (East Germany, West Germany, France).

2.6.3 Room-and-pillar mining

The mining method in rock salt mines and in most potash mines is the room-and-pillar method. The original idea behind room-and-pillar mining was to extract certain percentages of the ore beds by excavation of openings, and to leave pillars as deemed necessary for permanent support of the overburden. At shallow depths, this can be done, as demonstrated by the 1000 years old Wieliczka mine in Poland, provided the support pillars are made strong enough.

With increasing depths and related overburden loads, stronger pillars are required. To achieve the increasing pillar strength required, restrictions on extraction ratios and pillar sizes — widths and heights — become increasingly stringent with increasing depths.

In cases where the ore reserves are limited — this is the normal case in most countries because of the limitations to owned mining properties or leases — restrictions on extraction ratios mean cut-backs in economy and life-time of mining operations.

The design of support pillars left after partial extraction of salt and potash beds is frequently based on civil engineering principles which call for stable structural elements, i.e., for mine pillars with sufficient strength to support the deadweight of overburden formations. Apparently, civil engineers engaging in mine design believe in the applicability of salt-rock parameters determined in standard laboratory tests. Such belief, however, is not justified as shown in detail in following Chapters.

It must be emphasized that salt pillar design on the basis of laboratory strength parameters involves some basic errors which resulted in series of catastrophies in salt and potash mining: In Germany, where potash mining began over 100 years ago, one potash mine shaft per year was lost by flooding (Baumert, 1955); the largest rock burst known in world mining until 1958, was caused by inadequate design of a potash mine several square kilometres of which collapsed (Spackeler et al., 1960; Gimm and Pforr, 1961, 1964). In 1975, another earthquake caused by inadequate design of a potash mine shook Europe again.

Some writers (e.g., Baumert, 1955) attempted to blame these series of catastrophic events in salt mining on cosmic events such as sun-spot activity. However, all of these and similar disasters were caused by inadequate mine pillar design and related dangerous hypotheses applied to mine development. Such consequences of the application of erroneous hypotheses clearly demonstrate the magnitude of basic errors in these hypotheses, no matter whether they call for square pillars or for pillars which should be as long as possible.

The reported observations on salt glaciers — sub-section 2.5.4 — demonstrate that salt pillars must not be expected to behave like elastic material until their nominal laboratory strength is exceeded. The plasticity of salt rocks makes entirely different design principles mandatory in room-and-pillar mining of salt deposits, contrasting to room-and-pillar mining in elastic rock.

The recent development of special mining equipment, particularly for potash mining, calls for long pillars rather than square pillars as customarily designed in room-and-pillar layouts. In most Canadian potash mines, the disadvantages of the standard square or rectangular design encouraged the introduction of long rooms and pillars. The scanty information available on the new Russian potash mines near Soligorsk indicates similar trends.

In all but one of the Canadian conventional potash mines, the ore is cut by mining machines with rotating cutting heads (rotors); the resulting cross-section is given its final shape by trimming devices attached behind the rotors.

In modernized mines, the ore is transferred directly to extensible conveyor belts which unload onto stationary conveyor systems leading to the mine shafts. This material-handling system requires straight rooms and pillars, the only limitation to their length being set by the maximum economic length of the extensible conveyors.

Similar mining methods with long rooms and pillars were developed in those Canadian mines which had been using the standard loader-shuttlecar haulage equipment between mining machine and stationary conveyors. As a matter of fact, rock-mechanical problems in standard room-and-pillar panels forced the change.

It is puzzling to see recent developments in East and West Germany following exactly the opposite trend: there, the previous slusher haulage from the face to the main haulage system called for long rooms and pillars, the lengths being dictated by the capacity of the slusher. The recent adoption of loader-shuttlecar haulage systems resulted in standard room-and-pillar design with square pillars, as they customarily go with shuttlecar haulage.

2.6.4 Other mining methods

Originally, leaving pillars in salt mines was thought necessary to keep the overburden undisturbed in its original position; with increasing depths of mining, additional extra large pillars were frequently left. The general belief was that such pillars would serve as strong abutments for stable stress arches which would safely support the overburden. Some still believe in such design principles, see Chapter 5.

However, other methods of mining with effective control of overburden subsidence proved, many decades ago, that the above principles are obsolete. Provided the thickness of extracted tabular potash beds does not exceed certain limits set by geological conditions, the total extraction of the ore has been shown feasible in various potash mining districts. The volume of the resulting surface subsidence trough comes close to the volume of the extracted ore, quite similar to the conditions observed in coal mining districts. Backfill can reduce the surface subsidence so that controlled subsidence becomes feasible in mining potash beds of considerable thickness.

One method of controlling the overburden subsidence is the second pillar mining; it has been practiced for decades in the potash mines of New Mexico, U.S.A., without causing any known problems. After first mining in standard room-and-pillar patterns, the pillars are mined, leaving only small remnants deemed sufficient to provide for safe conditions during the second mining. The subsiding overburden slowly crushes the pillar remnants. Somewhat larger pillar remnants are left when approaching the boundary of a second mining panel, to achieve smooth transitions between non-extracted areas and almost completely extracted areas. The thickness of the extracted potash beds rarely exceeds 2 m.

Similar high extraction can be achieved in one single step if shale beds make up the immediate roof, provided the thickness of the shale which is made to cave is sufficient in relation to the thickness of the extracted potash bed. The basic design is a standard room-and-pillar design. The pillars are standard size at the advancing face; they are reduced to small remnants along predetermined lines of caving. Caving is introduced by blasting the remnants away. Such methods have proved feasible in the French potash mines and in certain German potash mines where the prerequisites are met.

Another alternative method of second mining for increased extraction ratios has been successfully applied in German potash mines: after the first mining of long rooms, the rooms are backfilled with refinery waste. The waste is transported hydraulically, using saturated brines. After the brines have drained off, the backfill becomes rather solid due to crystallization processes. Second mining allows the extraction of some or all of the pillars left after the first mining; the new rooms are also backfilled for limitation of the final amounts of subsidence.

Longwall extraction, as practiced in coal mining, is also applied in potash mines, where the prerequisites exist, e.g., in France. Subsidence control is achieved by intentional caving; roof control at the face is provided by hydraulic support systems.

The mining methods employed in steeply dipping potash deposits in German salt domes are of local importance; in many aspects, they resemble methods developed in steeply dipping ore bodies in hardrock mines. Considering the plasticity of salt rocks and the volumes of locally extracted potash ore bodies, the rock-mechanical effects may be assumed extensive; however, measurements have not been published and observations are kept secret, following old tradition in German potash mining.

Chapter 3

PHYSICAL PROPERTIES AND MECHANICAL BEHAVIOR OF EVAPORITES

3.1 GENERAL REMARKS

Salt rocks possess unique physical properties, and exhibit unique mechanical behaviors as shown by geological features dealt with in the preceding chapter. In contrast, the mechanical behavior of evaporites other than salt rocks resembles that of "normal" sedimentary rocks.

This is the reason why it is imperative to distinguish salt rocks from other evaporites, such as anhydrite or dolomite, when dealing with physical properties and mechanical behavior. Such distinction is particularly necessary in analyses of deformations which are caused by the removal of salt from deeply buried evaporite deposits, no matter whether the salt is removed by conventional mining, by solution mining, by natural salt solutioning, or by flowage.

Apparently, salt rocks in situ are highly ductile and rather easily deformed by what is usually termed creep. Such plastic behavior is demonstrated by salt glaciers and by flowage patterns observed in salt domes.

The contrasting behavior of non-salt evaporites, particularly of anhydrite beds, resembles the elastic, or near-elastic, behavior of most materials dealt with in civil engineering, and in mining engineering concerned with what is usually called hardrock mining: such materials and rocks exhibit elastic behavior until their strength is exceeded, and failure occurs more or less suddenly and violently without any noticeable preceding plastic deformation.

After failure and intensive fracturing, formations which consist of elastic rocks may exhibit what is often called pseudo-plastic behavior. Displacements between individual fragments and blocks may occur rather easily, particularly in response to tensional and/or shear stresses. Such displacements cannot be predicted from laboratory investigations on drill cores as the fracture patterns rather than the strength of individual cores determine the mechanical behavior of the formation.

At or near the surface, rational design can be based with some confidence on information provided by modern methods in geological engineering; the reader is referred to pertinent textbooks of civil engineering, and to related publications on slope stability.

In mining operations deep below the surface, information required for the rational prediction of the reaction of overburden formations to extraction mining is frequently not available. This holds particularly true for the first mining operations in newly developing mining districts. On the other hand, in mining districts where mining has been going on for long periods of time, careful observations and measurements have provided the information needed for reliable prediction of overburden reactions. It goes without saying that the actual geological situations, which may differ between individual mines, must be taken into consideration. This is shown in more detail in Chapters 4—6 for salt and potash mining.

The reason why the in-situ behavior of salt rocks cannot be predicted from most laboratory tests on cores or other samples is different: obviously, salt rocks in situ exhibit plastic behavior; however, in the laboratory, the very same salt rocks exhibit elastic behavior under most testing conditions used to determine the mechanical behavior of samples or models made to represent underground structures; when loaded too fast in excess of their strength, salt rocks are destroyed like any other rock without having shown measurable plastic deformations. This is the case particularly in uniaxial compression tests to determine the compressive strength.

Strength parameters obtained in such ways must not be used in the design of underground openings and support pillars in salt deposits; the reasons are outlined as follows:

(1) The natural creep limits — limits of elastic behavior — of salt rocks are extraordinarily small compared to most other rocks encountered in mining operations. For various reasons which are to be outlined shortly, it is difficult to determine the true creep limits of salt rocks precisely in the laboratory. First of all, however, it appears indicated to clarify some terms used by various writers.

The general behavior of salt rocks in situ may be termed elasto-plastic; this means that the behavior is elastic unless differences in the principal stresses exceed the limit of elasticity. It should be noticed, and it must be borne in mind that the difference in principal stresses initiates creep deformation, no matter how the difference is created.

Other terms used for the limit of elastic behavior are “creep limit”, “flow limit”, “yield limit”, or “octahedral shearing strength”. If and when the stress difference exceeds this limit, salt rocks in situ begin to deform plastically by creep, and continue to creep until the limit of elastic behavior is re-established at any given point.

In room-and-pillar mining with pillars left for support of the overburden, normally the limit of elastic behavior is exceeded in salt-rock pillars. In such cases, constant loading conditions result in constant creep rates. This important statement is based on in-situ measurement data presented in Chapter 4.

(2) As salt rocks in situ behave plastically if and when stress differences in excess of the limit of elastic behavior occur, the terms “yield limit” and “octahedral shearing strength” must not be applied to salt rocks with the

meaning of these terms in elasticity theories. In elasticity theories, yielding by shearing implies structural damage to the rock mass; heavily fractured, yielded rock exhibits little, if any, supporting capacity until it is re-compressed.

It must be borne in mind that the plastic deformation of salt rocks in situ does not cause structural damage or fractures. As a matter of fact, one of the fundamental principles in the design of salt-mining operations is to prevent such structural damage, and to ensure that inevitable creep deformations do not result in fracturing.

(3) The main reason why strength parameters, which are determined on salt rocks by standard testing procedures, must not be applied in mine design is the well-established fact that such testing causes strain hardening in salt rocks. Strain hardening is also termed "work hardening" (e.g., Shlichta, 1968, p. 608), "strenghtening" (Borchert and Muir, 1964, pp. 259 ff.), or just "hardening" (Dreyer, 1972, pp. 2ff). In loading tests in the laboratory, creep rates of salt rocks decrease in time as the result of strain hardening. In situ, the long-term creep deformation of salt rocks is not affected by strain hardening, as demonstrated by constant creep rates under constant conditions.

Only very few laboratory researchers found constant creep rates of salt-rock samples and pillar models under constant loading conditions. Most theoreticians postulate that the decreasing creep rates caused by strain-hardening in laboratory tests also occur in situ, and they are applying such laboratory parameters to the design of openings and pillars in salt deposits. The expected decreasing creep rates, of course, do not materialize.

However, it is an exceptional case when a theoretician admits discrepancies between expected and actually measured creep rates in salt deposits; McClain (1964, p. 494) emphasized the discrepancy between laboratory and in-situ data which "was so large (two orders of magnitude) that no correlation was possible. In fact, it is felt that these results constitute a serious indictment of the validity of applying laboratory test data to underground behavior in plastic materials".

Some theoreticians are using rather questionable methods in attempts to adapt their in-situ measurement data to hypotheses which are based on laboratory testing, or merely on theoretical assumptions. Erroneous hypotheses applied in the design of salt and potash mining operations repeatedly resulted in major disasters. Nevertheless, such hypotheses are promoted again and again in textbooks and in numerous publications. This makes it necessary to show in some detail the differences between laboratory and in-situ behavior of salt rocks.

3.2 THE MECHANISM OF CREEP OF SALT ROCKS

3.2.1 Single rock-salt crystals

Common rock salt or halite is the most abundant salt mineral; it received almost exclusive attention in research dealing with the mechanism of creep in salt deposits. It must be emphasized that only a few evaporite deposits consist of rock salt plus small amounts of what is usually called insoluble material, such as clay, anhydrite, etc.; as most of the salt deposits which are exploited contain large amounts of other salt minerals, and of non-salt material, the creep mechanism of rock salt is frequently not the only factor to be considered in applied rock mechanics of evaporite deposits. Sylvite, the main component of most exploited potash beds, has a crystal lattice which resembles that of halite. Most investigators assume the same creep mechanism for both halite and sylvite crystals.

The following description of the mechanism of deformation of rock-salt crystals is quoted from Shlichta (1968, pp. 606 ff.); references are not included as, "in addition to a vast literature, numerous treatises and reviews are extant on dislocations and plastic deformation in general and on the deformation mechanism operative in crystals having the sodium chloride structure".

"Plastic deformation occurs by the movement of edge and/or screw dislocations in the plane of shear. These commonly occur as mixed edge-and-screw dislocation loops, so that shear occurs by the expansion of these loops. Since each dislocation can account for only one atomic layer of shear, macroscopic deformation requires that large numbers of dislocations be generated. These dislocation sources are thought to be pinned dislocation segments (Frank-Read sources) or cross-slipped dislocation jogs. Since all the dislocations from a single source move in more or less the same plane, slip is observed to occur in discrete slip bands."

"If the dislocation loops could expand unimpeded until they were annihilated at the surface, easy glide would continue indefinitely. There exist, however, numerous barriers (precipitates, low-angle boundaries, the crystal surface, forests of grown-in dislocations, and intersecting dislocations from other slip planes) which cause the dislocations to pile up, just as in a traffic jam, until the first dislocation can 'climb' around the obstacle (by vacancy diffusion or cross-slip) or until sufficient stress is built up to overcome it. These phenomena cause work-hardening."

"Fracture results when sufficient local stress is built up to nucleate and propagate cleavage cracks. The ultimate extent of plastic deformation depends on the amount of work-hardening that occurs before the fracture-stress threshold is reached."

"Thus, we see that the stress-strain curve or any one crystal will depend quite drastically on the density, distribution, and impurity atmosphere of the dislocations and sources initially present (and, hence, on the mechanical and thermal history of the crystal), on the presence of impurities, color centers, precipitates, low-angle boundaries, and surface irregularities, and also on the rate and manner in which the deforming stress is applied."

Fig. 3-1 (Shlichta, 1968, fig. 4) shows the six slip systems observed in halite crystals. Borchert and Muir (1964, p. 256) emphasize that these six slip systems "act as shear planes, along which translation of the lattice structure occurs very easily. In individual crystals, this translation may vary from a few Angströms up to several millimetres."

The latter writers point out that "the same is true of the individual grains of a polycrystalline halite deposit. Variation in the direction of applied stress therefore makes little difference to the flow behavior of halite beds." Fig. 3-2 shows the effect of temperature on the creep limit of rock-salt crystals. As sylvite crystals exhibit the same behavior, the above statements equally apply to sylvinite. It should be pointed out that bedding planes and similar discontinuities make considerable differences to the creep of salt and potash beds; this is shown in detail in Chapters 4—6.

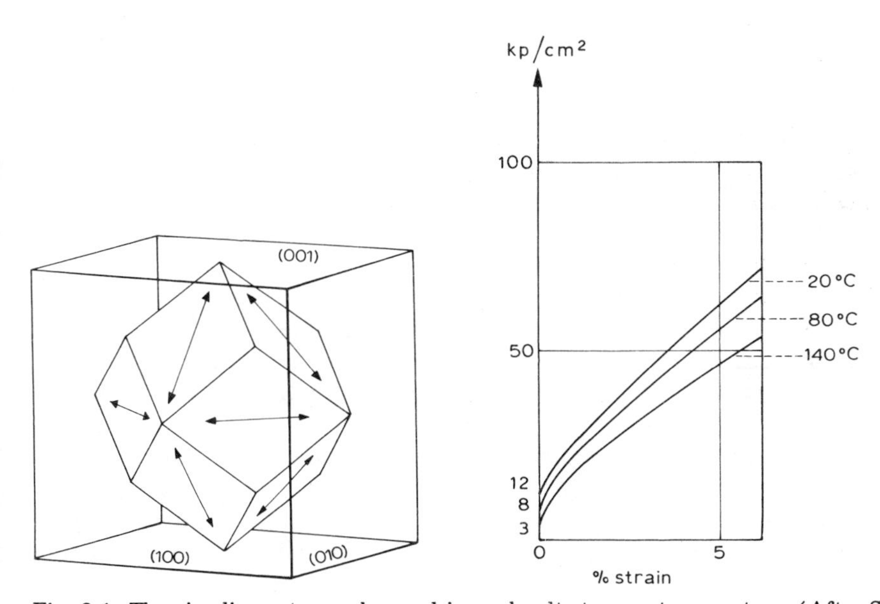

Fig. 3-1. The six slip systems observed in rock salt at room temperature. (After Shlichta, 1968.)

Fig. 3-2. Stress—strain curves of rock salt crystals at various temperatures. (After Dreyer, 1972, pp. 9, 10, fig. 2.)

3.2.2 Single carnallite crystals

Hardly any research work has been done to elucidate the deformational mechanism of carnallite crystals. Translation planes are known to develop very easily in carnallite crystals, as shown in publications (e.g., Baar, 1953a). Dreyer (1972, p. 57) emphasizes that carnallite crystals "are usually strongly twinned". Easy glide along such translation planes makes carnallite crystals readily deformable, provided that sufficient time is available. Sudden changes in stress conditions cause elastic reactions of carnallite, as outlined in detail in sub-section 3.3.2.

3.2.3 Deformation of polycrystalline salt rocks

"If a polycrystalline material is to be ductile, each grain must be capable of deforming into an arbitrary shape, defined by the shapes of the adjoining grains; this requires that each crystal have five independent (i.e., orthogonal) slip systems (Von Mieses, 1928). But the six slip systems of salt are equivalent to only two independent slip systems; hence, polycrystalline salt should be brittle at room temperature. Experiments do indeed show only a small amount of initial plasticity (apparently due to localized cross-slip within slip bands), and an extremely high rate of work-hardening and brittle fracture. Much work remains to be done.

Since room-temperature deformation occurs by cross-slip, there is reason to believe that slow creep might occur far more extensively and with much less work-hardening than is observed in rapid laboratory-time deformation. Extensive steady-state creep might occur by simultaneous deformation and recrystallization. There is considerable evidence that badly deformed salt crystals recrystallize at room temperature in the course of a few weeks; this might account for the large, clear crystals frequently found in salt domes." (Shlichta, 1968, pp. 609—610).

Borchert and Muir (1964, p. 262) also emphasize that "recovery and recrystallization play an important part in producing natural flowage of large halite masses, since they are continually making deformation easier".

With reference to careful measurements in mines, Borchert and Muir (1964, p. 276) point out "that mining produced immediate flowage. All these measurements confirm that the superior stability of large mine workings is due not to any unusually great strength of the evaporites themselves, but rather to the fact that the amount of plastic deformation accompanying any stress increase is then at a maximum. The ability of halite deposits to reduce the stress increase by deforming plastically is of vital importance."

In the foregoing quotation, the word "evaporites" is to be replaced by "salt rocks" for the obvious reason that evaporites such as anhydrite do not exhibit plastic deformation; the term "stress increase" is to be changed to "stress difference" as the difference between maximum and minimum stress at any point causes plastic deformation, no matter how such a stress difference is created. In the laboratory, stress differences are usually created by loading test specimens or models; in situ, however, the first stress difference to cause plastic deformation is created by the excavation of an opening in salt rocks which are subjected to high original stresses; these depend on the depth below surface, and on other conditions to be dealt with in following chapters.

This is an extremely important difference between laboratory conditions, and in-situ conditions which initiate plastic deformation in mines and in other man-made openings in salt deposits. Most laboratory researchers do not realize how important this difference is. In fact, it is this difference that must be recognized and utilized in mining salt rocks at great depth. Neglect of the fact that the excavation of an opening in salt rocks at great depth inevitably causes what has been termed "stress relief creep", frequently resulted in dangerous mistakes in the design of support pillars in mines: the "immediate flowage" produced by excavation is not indicative of stress increase by which deformation is produced in the laboratory; to the contrary, such immediate flowage is indicative of stress decrease within the zone of plastic deformation around any new opening. Stress-relief creep causes considerable reductions of the load supported by pillars; neglect of this fact has repeatedly caused major disasters particularly in potash mining. This is shown in Volume 2; it is emphasized here because it originates from erroneous assumptions of creep limits of salt rocks.

3.3 CREEP LIMITS OF SALT ROCKS

3.3.1 Rock salt and sylvinite

Borchert and Muir (1964, p. 255) pointed at the various definitions of creep limits: "in the testing laboratory, the yield point is taken where the permanent deformation is just 0.2%; this, however, is a completely arbitrary, though technically meaningful definition".

"Basically, however, even the smallest non-elastic deformation should be described as 'flow'. So, in physics, 0.001% of plastic deformation is usually taken as the yield point. In the course of days or weeks, such a minimal deformation may produce measurable deformation of a mine pillar, and given geologic time would even cause kilometers of flowage in evaporites."

"Since the establishment of a minimal permanent deformation depends on the sensitivity of the measuring equipment employed, true yield points cannot be given precisely, but 10 kp/cm^2 may be taken as the approximate elastic limit of single crystals or of a completely recrystallized mass of halite. Any greater stress (should read 'stress difference'!) will produce flowage."

Dreyer (1972, pp. 78—79) is repeating these statements, but reports "the technical yield limit of 74 kp/cm^2" for rock salt under what is called quasi-static loading conditions. Quasi-static stress-strain curves for rock salt and sylvinite "run nearly horizontally at pressures above 200 kp/cm^2" (Borchert and Muir, 1964, p. 258, after Dreyer, 1955). These curves indicate that higher stresses are required to produce creep deformations.

The same authors (l.c., p. 266, table 12) are listing "measured biaxial stresses" at working faces and pillar surfaces of various mines; the salt rocks are halite or sylvinite; the limits of elastic behavior were determined by "the point of inflexion method". The postulated increases in vertical stresses at the surface of the exposed salt rocks are listed as ranging from 15 to 86%, with absolute stress values from 150 to 390 kp/cm^2. The same values are listed by Dreyer (1972, p. 196, table 35), who claims that "the measurements described provide, for the first time, a means by which stresses in-situ can be quantitatively assessed".

It is apparent that such high values cannot be reconciled with the value of approximately 10 kp/cm^2 that "will produce measurable flowage", according to the same writers. Observations on salt glaciers, and actual results of underground measurements dealt with later, definitely show the high creep limits mentioned above inapplicable to the in-situ behavior of salt.

Odé (1968a, p. 574) emphasizes that "in general the stress—strain curves obtained in these tests (usual compressive strength tests), even when extrapolated to a so-called quasi-static case (Dreyer, 1958), are of little use in a study of the steady-state creep properties of salt". With reference to triaxial tests performed by various investigators, Odé (l.c., p. 575) states "that a differential pressure between 50 and 100 bars was required to induce permanent deformation in salt. This value can be lowered considerably by using more sensitive methods of measurement, such as observation of the sudden appearance of slip bands in crystals (Obreimov and Shubnikow, 1927) which occurs at much lower differential stress. This lower limit has been called the optical elastic limit."

Summarizing his review of creep properties according to the result of numerous quoted publications, Odé (1968a, p. 560) concludes "concerning the plasticity of rock salt, that is, whether rock salt has a true yield limit: if rock salt has a yield limit, this limit must be low". This conclusion is in exellent agreement with earlier literature surveys, e.g., by Spackeler and Sieben (1944) with reference to publications until 1936; as the creep limits of rock salt can be increased by repeated loading and unloading to almost any desired value up to the uniaxial compressive strength, only the lowermost values reported from laboratory testing can be regarded as representative of true creep limits. These values are of the order given above: 10 kp/cm^2, or even less, depending on the composition of any specific rock salt. The significance of such very low natural creep limits had to be emphasized repeatedly in recent years (Baar, 1972a, 1973) in view of the much higher values applied by various authors in calculations of the plastified zones around underground openings in salt rocks, and in mine pillars.

Some of the creep limits of various salt rocks (except carnallite) used by various authors in calculations of creep processes are listed as follows (in kp/cm^2; approximate values in cases where calculated from other original units):

Gimm (1968) and Höfer (1964): 150;

Serata (1966): 105 (by 1972, gradually reduced to about 40 for the same rock salt);

Coolbaugh (1967): 105;

Albrecht and Langer (1974), and Langer and Hofrichter (1969): 75;

Dreyer (1972), p. 466 and table 84): "the values of the circumferential stresses determined in-situ by using the compensating method" are listed and compared with "calculated elastic and plastic circumferential stresses" around a shaft in rock salt. At the depth of 825 m, the respective values in kp/cm^2 are: measured 231, calculated elastic 119, calculated plastic 380. These values indicate creep limits of several hundred kp/cm^2.

The contrast to the natural creep limits emphasized elsewhere in numerous publications by Dreyer and Borchert is evident.

Other authors, e.g., Măndzić (1974), completely ignore creep of salt rocks, taking the uniaxial compressive strength as effective strength of salt rock pillars; in one particular case, the value of 500 kp/cm^2 was found, and assumed to allow safe and economic mine design.

It is apparent that such contrasting values of vital design parameters require a more detailed discussion.

3.3.2 Carnallite and carnallitic salt rocks

Some writers (Gimm, 1968; Dreyer, 1974a) consider the mechanical behavior of mine pillars consisting of carnallite and/or carnallitic salt rock responsible for heavy rockbursts experienced in German potash mines. No efforts are known to determine more precisely the creep limits of such rocks, the time required for creep, and to find out whether or not strain-hardening of such rocks occurs in the laboratory.

Extensive laboratory testing has been devoted to determinations of the compressive strengths of pure carnallite and carnallitic salt rocks under standardized conditions of loading, etc.; the strength parameters obtained in this way are meaningful only in cases of elastic behavior, and "under the specified conditions, but they do not allow any inferences for creep under other conditions" (Odé, 1968a, p. 560).

It has long been known that carnallite and carnallitic salt rocks respond elastically in an explosion-like manner to sudden loading and unloading; this was demonstrated by Schmidt (1943) in the laboratory, see Baar (1970a, p. 280): "One of his most important findings was the answer to the question: 'Why carnallite can behave in two very different ways; either deforming plastically very easily without failure; or failing abruptly by brittle fracture without measurable plastic deformation?' To constant limited loads, carnallite responds by plastic creep. However, when the same carnallite is loaded or unloaded too fast, it will fail suddenly 'by brittle rupture'."

"His conclusion was that plastic pillar deformation in a mine is by no means hazardous, even when it results in some slabbing off the surface of a pillar. This slabbing is due to plastic creep of the interior pillar which results in shortening in height, while the pillar surface tends to remain full height."

"Therefore, formation of slabs at the surface of a pillar should be regarded as an indication of safe conditions with respect to a rockburst hazard in that the pillar is deforming plastically. This excludes the hazard of sudden failure."

The same type of explosion-like failure of laboratory specimens was reproduced by Dreyer (1974a, pp. 120—122); disregarding, not even mentioning the identical observations reported above, Dreyer arrived at the contrary conclusion: carnallite is storing considerable energy in cases where lateral constraint on mine pillars disallows sufficient plastic deformation; this condition is assumed to exist in mine pillars which are only, say, 3—5 m high, but 10 or more m wide; in such cases, sufficient plastic deformation should be made possible by reducing the size of the pillars, but increasing the number of small pillars as required by the theoretical ultimate pillar strengths to be calculated on the basis of compressive strength values. Dreyer (l.c.) refers to the confirmation of his conclusions by Gimm (1968), and claims priority.

The most recent earthquake (June 23, 1975) caused by potash mining in the Werra district of central Germany leaves little doubt of the practical application of these hypotheses which are based on the following laboratory values of ultimate compressive strengths:

pure carnallite:

(Dreyer, 1972, p. 61)	66 kp/cm^2	
(Dreyer, 1974, pp. 120, 121)	65 kp/cm^2	("quasi-static" loading)
	41 kp/cm^2	(standard loading)
(Gimm, 1968, p. 180)	62 kp/cm^2	("quasi-static" loading)

carnallitic rock salt (50% carnallite, 42% rock salt):

(Dreyer, 1972)	130 kp/cm^2	("quasi-static" loading)

Gimm (1968, p. 183) refers to the same value obtained by Dreyer (1955). Both authors compare these values with values of about 400 kp/cm^2 for rock salt.

Such application of laboratory parameters to the design of mine pillars is typical of extremely dangerous practical applications promoted by some theoreticians.

If such comparison of compressive strength values has any meaning with respect to the creep limits, it tells that the creep limits of pure carnallite and of carnallitic salt rocks are lower than those of rock salt. Since the creep limit of rock salt is not known precisely — all appear to agree on natural creep limits less than 10 kp/cm^2 — there is little justification for further theoretical arguments in view of statements such as "carnallite will flow by crystal twinning at very low stresses" (Dreyer, 1972, p. 136).

Instead, the significance of some observations in carnallite mines which have been in operation for many decades, is emphasized with Baar (1959b): with reference to numerous previous publications, Höfer (1958a) reported on extensive near-horizontal voids surveyed for decades above mined-out areas with carnallite pillars regarded adequate for overburden support. These voids across mined-out areas of limited extent were up to 50 cm wide; this definitely excludes any contribution to pillar loads from formations above the voids.

The pillar loads can be calculated from the thickness of the anhydrite below the voids (about 50 m), and the extraction ratio; with extraction ratios of 50—75%, the average pillar loads were in the range of 25—50 kp/cm^2. Evidently, the loss in pillar height by plastic deformation was exceeding the convergence of the mined-out area for many decades, leading to the opening of the voids.

As these observations prove beyond doubt that relatively small average pillar loads cause long-term plastic deformations of carnallitic salt rock pillars, the quoted values of ultimate compressive strengths (Dreyer, 1972, 1974; Gimm, 1968) cannot be correct; the application of such values in mine design has repeatedly proved extremely dangerous as shown in more detail in Volume 2.

3.4 STRAIN HARDENING IN THE LABORATORY BUT NOT IN SITU!

3.4.1 The significance of strain hardening

Strain hardening of salt rock samples and pillar models has been observed in numerous laboratory investigations, including "quasi-static" loading, loading and unloading cycles, etc.; the question is: does strain hardening occur under normal mining conditions? The answer is a definite "no", contrary to what most hypotheses based on laboratory research predict, and contrary to what some authors claim as achieved in mining operations by application of certain "methods of stabilizing underground openings" for introduced strain hardening in situ.

The significance of the question of whether or not strain hardening in situ occurs, and whether it can be enhanced by certain methods, seems to be obvious; as Abel (1970, p. 207) put it: "rock mechanics is not a magic wand which can be waved over a 50-ft. pillar to turn it into a 150-ft. pillar".

In other words: if strain hardening does not strengthen mine pillars as expected in hypotheses applied to mine design, there is frequently no remedy to inevitable consequences; if pillars are too small, and fail to provide the assigned support by deforming continuously at unexpected, high rates, the potential hazards in conventional mining as well as in solution mining must not be underestimated. This has been emphasized repeatedly by various authors referred to by Baar (1959a, 1966a, 1970a, 1972b); published controversies (e.g., Piper, 1974) show the need for detailed discussion of the strain-hardening problem.

Many theoreticians postulate mine pillar behavior in accordance with a so-called ideal creep curve. When samples are loaded, three stages of creep occur after the initial instantaneous elastic deformation:

(1) primary or transient creep at decreasing rates; (2) secondary or steady-state creep (also called pseudo-viscous or quasi-viscous creep) characterized by constant creep rates; (3) tertiary creep at increasing rates, leading to failure by rupture.

Due to the effects of strain hardening, most of the numerous published strain-time curves, obtained in the laboratory by loading salt-rock samples and pillar models, resemble ideal creep curves; only the initial parts are obtained in cases where failure is not achieved by applying loads less than the "ultimate strength" of the respective samples. In such cases, continued strain hardening causes continuously decreasing creep rates until, at least in laboratory time, no further creep is measurable.

Among many laboratory investigators, only Obert (1964) found that "creep rate in salt and potash model pillars becomes constant after a relatively short transient period". However, Obert's data were re-interpreted by others to fit a power law as postulated by most investigators (King and Acar, 1970, p. 227). Plotted on log-log diagrams versus time, such continuously decreasing creep rates result in a straight line with a certain angle to the time axis. Fig. 3-3 is an example of misleading premature publications to "prove" power laws valid for in-situ creep. Some writers who postulate similar relationships continue to refer to the original figure, although it has been shown erroneous in various publications (Baar, 1972a, p. 35, fig. 3; 1974b, p. 291, fig. 1). It is of particular concern that some writers, e.g. Dreyer (1974a, p. 162), continue to refer to the original of Fig. 3-3 as "proof" of their hypotheses. Such attempts to prove erroneous hypotheses resulted in the published controversies at the 4th International Symposium on Salt (Piper, 1974).

In the strongest way possible must be stressed again that Abel's (1970) above quoted statement must be taken more seriously by some theoreticians.

Since the postulated effects of strain hardening in situ are the subject of ongoing controversies, it appears indicated to resort to quotations from publications of advocates of strain-hardening.

Borchert and Muir (1964, pp. 277—279) are expecting "end values" of plastic deformation, listing laboratory values which "plot as a beautifully asymptotic curve whose maximum deformation value of 0.366% would be

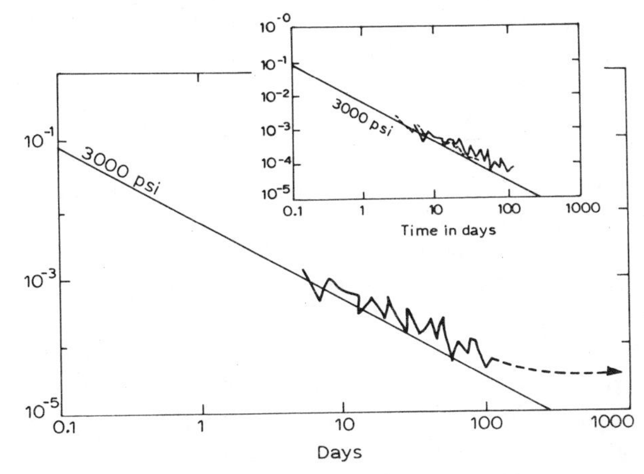

Fig. 3-3. Predicted and measured closure rates of single openings, IMC-K 1.

The 3000-psi line "represents the closure rate for an isolated opening at 3000-ft. depth. The exponent, minus 0.85, of the last term (in the formulas developed by Dr. S. Serata) is the slope of the line" (Coolbaugh, 1967, p. 71 and fig. 7).

The "somewhat erratic" line shows the closure measured in one of the test rooms in the mine (erroneously identified as shaft closure rate in the original figure—Coolbaugh, 1967, p. 72). "As the rooms become older, the amounts of movement become progressively smaller and more difficult to measure. Slight errors in measurement produce noticeable short-term variations in the slope of the creep-rate line" (Coolbaugh, p. 72).

The dashed line, added to the original figure, shows the factual closure rate according to curve 1 of Fig. 5-1. The measured closure rate has been constant since 1964. In log-log diagrams, constant convergence rates are represented by parallels to the time axis.

Fig. 3-3 was published repeatedly (Baar, 1971b, fig. 3; 1972a, fig. 3; 1974b, fig. 1) to show that some writers are using rather questionable methods in attempting to "prove" premature hypotheses.

achieved only after the stress had been applied continuously for about two years. This is typical of very coarsely crystalline and relatively pure halite." They emphasize that "deformation may continue for several months when the relatively small pressure of 100 kp/cm^2 is applied. This again brings about the near-impossibility of determining quasistatic breaking strengths."

These authors refer to their research (l.c., p. 276); however, the same data is referred to by Dreyer (1967, pp. 101—115) in many more details, and again by Dreyer (1972, pp. 126—141); this appears to indicate that it is Dreyer's work that led to the above quoted postulated "end values" of plastic deformation due to strain hardening. The "asymptotic end values" are called "asymptotic limiting creep strains" (Dreyer, 1972, p. 129). The "maximum deformation value of 0.366%" quoted above is the "asymptotic strain at constant stress", or the "total strain" (Dreyer, 1972, p. 132). Dreyer's "quantitative formulation of the hardening of rock salt" (l.c., p. 105), using a "strain-hardening coefficient (which) must be determined by experiment" (l.c., p. 106), is based on the laboratory work referred to.

Dreyer (l.c., p. 111) emphasizes that "no quantitative measurements have yet been made to determine the strain-hardening coefficient for hardrock salt and carnallite. Nevertheless, the general shape of the stress-strain curves of these rocks indicates that the strain-hardening coefficient is greater for hardrock salt and *always smaller* for carnallite than it is for halitic rock salt." (Hardrock salt appears to be rock salt with certain contents of anhydrite and/or other impurities; halitic rock salt appears to be pure salt).

Heavy criticism has been levelled by Borchert and Muir (1964, pp. 271—275) and by Dreyer (1972, pp. 274—279) on researchers who arrived at different conclusions with respect to the effects of strain hardening; particularly, Kegel (1957) was attacked for his apparent disagreement with the newly developed "ultimate strength" concept based on Dreyer's laboratory investigations (1955 is the year of the first publication of this concept).

Borchert's (1959) and Dreyer's (1967) original German texts are literally identical; the respective English translations differ only slightly, focussing on what is translated as "fatigue deformation" which, according to Kegel (1957), takes place in time to the effect that the "ultimate strength" of mine pillars is continuously reduced when compared with compressive strength values obtained in the laboratory.

This criticism, dragged through textbooks and other publications for almost two decades, is definitely unjustified as it merely results from misinterpretation of Kegel's (1957) publication: Kegel refers to the lowering of the strength of strain-hardened laboratory samples with time due to recrystallization and recovery; this means that the theoretical supporting capacity of mine pillars is decreasing with time in cases where laboratory values of about 200 kp/cm^2 are used to calculate the ultimate pillar strength. To explain this mechanism to laymen, Kegel compared human fatigue, the process of getting tired with time when carrying heavy loads, with the actual supporting capacity of mine pillars. There is definitely no point in perpetuing scholarly arguments based on Kegel's use of a word which has a defined meaning in specialized papers. It should be noticed that Dreyer (1974a, pp. 140—161) was finally forced to acknowledge "proof of constant creep rates" in mines.

It is apparent that creep at constant rates cannot be reconciled with increasing strengths due to strain hardening, and with strain hardening coefficients to arrive at "asymptotic end values", "asymptotic limiting creep strains", "total strains", whatever term may be used. In-situ creep rates do not decrease with time according to exponential creep laws with creep rates plotting as straight lines on log-log diagrams (e.g., Dreyer, 1972, fig. 102).

In contrast to Dreyer, Serata (1974, 1976) is continuing to advocate his so-called stabilizing methods by which strain hardening in situ is allegedly achieved by forcing salt rocks to undergo certain predetermined amounts of creep. The concept reads as follows (Serata, 1968, p. 305): "Any opening made in a salt media at a depth below 3000 ft. (900 m) will suffer a large amount of plastic closure. However, regardless of the initial rate of closure, the creep rate also decreases exponentially to virtually zero as the opening reaches a stable condition. The total cumulative closure of the opening approaches the *ultimate value* beyond which no further closure is possible." (p. 306): "The plastic closures in formations other than halite follow the pattern of the halite closure. The basic characteristics of closures in potash, carnallite and other materials will be identical except for the values of the *ultimate closure.*"

The practical application of these hypothetical concepts is described as follows (Serata, 1972, p. 102; 1974, p. 53): "Two pre-stress openings are excavated, each of which forms an envelope of increased stress. These two openings are placed sufficiently close together so that the ground between these two openings is also subjected to additional compression for strain-hardening. Then, through the strain-hardened ground, a third opening is excavated." The author continues to claim successful application of these principles in Canadian potash mines, and elsewhere. It is shown in the following chapters that the contrary took place: considerable creep at constant rates developed in entry systems designed according to these hypotheses; the expected strain hardening did not materialize.

Serata's hypothetical concepts resemble those emphasized by Höfer (1958a, pp. 103—105); these concepts were completely dropped in most East German publications since, in 1958, a severe rockburst had shown them untenable.

All such concepts which are based on ultimate strengths to be achieved by strain-hardening, cannot be reconciled with surface subsidences due to continuing pillar deformations by creep; surface subsidences above some potash mines in Germany have been measured over more than 50 years—see Volume 2 for details.

In spite of such undisputable proof of pillar deformation over long periods of time, pillar design is still based on so-called maximum allowable plastic deformation, with values of about 2% given by Höfer (1958a) for sylvinite pillars. According to Serata's quoted principles, the predetermined amount of plastic deformation would ensure sufficient strain hardening; at the same time, not too much of strain hardening is to be achieved, as the right fracture strength to make the pillars yield as required to eliminate so-called primary stress envelopes, must also be achieved.

In these hypotheses, superficial spalling off the pillar walls is regarded indicative of pillar failure caused by excessive loading; the "loading capacity" which must not be exceeded in competent support pillars, is calculated on the basis of "fracture strengths" or "breaking strengths" determined in the laboratory (Dreyer, 1972, pp. 312—320).

Referring back to subsection 3.3.2 it is re-emphasized that mine pillars which deform continuously at constant rates provide full support for the overburden, controlling the subsidence rates; visable damage (spalling) may look alarming to theoreticians, but is usually an indication of safe conditions.

Another so-called creep law derived from ideal creep curves by some writers (Höfer, 1964; Uhlenbecker, 1971; many others), is also shown invalid by creep at constant rates: in situ, creep at constant rates is not indicative of impending failure; this is shown in the following sub-section.

3.4.2 Constant creep rates under constant loading conditions

Obert's (1964) laboratory investigations of the deformational behavior of model pillars made from salt, trona and potash ore, and attempts to re-interpret his data to fit theoretical creep laws (see p. 76), deserve particular attention; some of his findings and conclusions may be quoted as follows.

"In underground excavations in *competent* rock the pillars may remain intact and support the structure indefinitely; in some *incompetent* rock, especially in *salt and potash mines*, the pillars usually undergo continuous plastic and/or viscoelastic deformation (creep). The creep rate depends on the average pillar stress (which in turn depends on the extraction ratio) and is usually limited to a value such that the pillar deformation and attendant effects will not interfere with mining operations."

"A number of researchers have investigated the strength of model pillars where the limiting strength was determined from the load at which the model pillar ruptured or fractured with a sudden release of applied load, that is, as a brittle material."

"Although the load on the mine pillars is not constant at the time it is being formed, it is virtually constant over the larger part of its effective lifetime. As the load on the pillar prior to reaching this constant value is less than that after the pillar has assumed its maximum load, any design based on constant load conditions would be on the safe side."

"Höfer (1958a) measured the in-situ lateral deformation in a number of potash pillars over periods up to 500 days. In general, his results show that the in-situ strain rate is constant for any specific average pillar stress."

"Axial pillar deformation measurements taken in the Hattorf mine, Philippsthal, Germany, over an 8-year period also show that in areas undisturbed by mining the creep rate is virtually constant."

"None of the salt and potash models failed like a brittle material, that is, by sudden fracture with a decrease in load bearing capacity. Also, these models did not exhibit a tertiary phase, that is a terminal phase in which the strain rate accelerated. The models loaded at low stress levels showed no discernable indication of failure (spall) on the pillar surface. At intermediate stress levels, the surface bulged and there was some spall; and at high stress levels there was heavy spall in all instances. However, when the latter group of specimens were wedged apart at the mid-plane, it was found that the diameter of the solid core was at least 94% of the original diameter. (Some tests) were terminated after the strain exceeded 25%."

"Although field data are meager, there is a general agreement between model pillar and in-situ pillar measurements. From all indications the creep rate in salt and potash model pillars and mine pillars becomes constant after a relatively short transient period."

"Because of the tendency for evaporite minerals to creep rather than fracture if the lateral constraint is high enough, if the width-to-height pillar ratio in mine pillars is sufficiently large (4 or greater for salt and potash), the possibility of a chain-reaction type of failure is excluded."

It may be emphasized that the word "strain hardening" does not occur in Obert's paper; the measured constant creep rates to strains exceeding 25% leave no room for strain hardening which occurs under conditions of standard testing with rapidly increasing loads, and with loading and unloading cycles.

Since Obert's publication, the results of numerous in-situ measurements of creep in salt and potash mine pillars became available; many of them were published. None of these in-situ results contradict Obert's statement that the creep rate becomes constant after a relatively short transient period.

It is interesting to note that the word "strain hardening" neither occurs in Dreyer's (1969, 1974a) evaluation of constant creep rates which were measured by Wieselmann (1968) in the Cane Creek potash mine near Moab, Utah. Comparison of Fig. 3-4 and 3-5 shows one important difference, emphasized by Baar (1973): in Fig. 3-5, the initial creep at decreasing rates is stress-relief creep, while Fig. 3-4 shows the transient creep caused by loading of stress-relieved samples. The exponential decay of creep rates until constant creep rates develop, is similar for stress-relief creep, and transient creep under constant loads. This has been confirmed by numerous in-situ measurements to which Dreyer (1974a, pp. 140—161) refers in his "proof of constant creep rates" in mines.

However, it appears that Dreyer (1974a) has not yet realized that creep at constant rates demands the elimination of any strain-hardening coefficients from creep formulas presented in the rest of his publications. As a matter of fact, the acknowledgement of constant in-situ creep rates is taking all his previous work, which is based on strain hardening, "ad absurdum", to use his own words in criticizing Obert's (1964) work (Dreyer, 1967, p. 87; 1972, p. 289).

The most recent rockburst (23 June, 1975) in a potash mine, which apparently was designed on the theoretical expectation of decreasing creep rates from strain hardening which did not materialize, might prevent the application of such erroneous hypotheses in cases where serious consequences are inevitable. Typically, the proof of the absence of strain hardening in situ, as established by in-situ tests upon request of theoreticians in 1957, has been simply ignored by the same theoreticians, although this neglect had caused the flooding of another potash mine (Baar, 1959a, 1966a). These in-situ

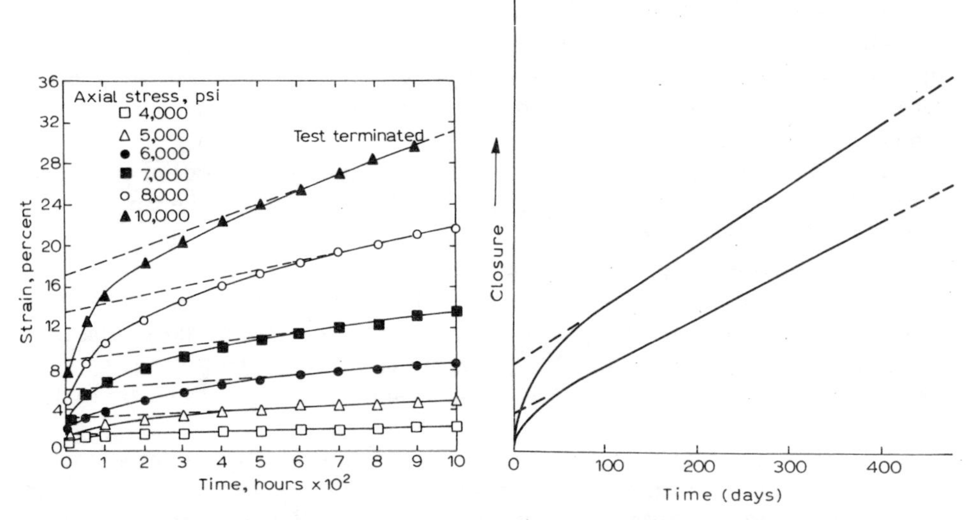

Fig. 3-4. Strain-versus-time diagram, Kansas rock salt. (After Obert, 1964, fig. 11.) Deformation of pillar models under constant loads as indicated.

Fig. 3-5. Typical vertical and horizontal closure curves. (After Wieselmann, 1968, and Dreyer, 1969, p. 443.)

Rectangular entry, 3 m high and 5.4 m wide, conventionally mined (drill and blast), Cane Creek potash mine near Moab, Utah; mining depths exceeding 900 m (Wieselmann, 1968).

The same figure is shown by Dreyer (1969) with slightly different dimensions: height 3.3 m, width 6 m, mining depth 1000 m. The vertical closure rate became constant at 1.25 mm/day after 60 days since mining; the horizontal constant closure rate of 0.94 mm/day was measured after 35 days since mining.

tests, and subsequent tests to disprove the strain-hardening-in-situ hypothesis, are described in the following sub-section.

3.4.3 Proof by in-situ investigations: no strain hardening of salt rocks

3.4.3.1 Tests in a German potash mine at the depth of 820 m. Tests in a German potash mine were conducted in 1957 as no agreement could be reached between the hypothetical stresses up to 390 kp/cm^2 reported by Borchert and Muir (1964) and Dreyer (1972), see sub-section 3.3.1, and conflicting results of measurements with hydraulic cells (Baar, 1959a, 1966a), and of deformation measurements (Wilkening, 1958, 1959).

The hydraulic stress measuring method developed during the preceding years is described in more detail in sub-section 3.8.2. Here it appears sufficient to know that artificial inclusions of hydraulic oil were established in slightly inclined bore holes into the salt rock at various distances from under-

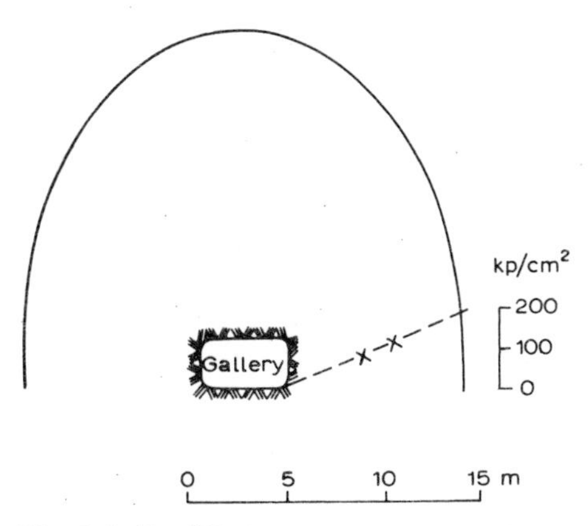

Fig. 3-6. Equilibrium pressures in sealed sections of boreholes in rock salt, depth 820 m, entry in virgin ground two years old (Baar, 1966a).

ground openings, simply by sealing the deapest 30 cm of the respective boreholes; such cells, filled with oil, were connected to the mine openings through the seals by high-pressure tubes. This arrangement made it possible to close the measuring systems, and to change the cell pressures arbitrarily.

It had been established by long-term observations that the cell pressures reached an equilibrium pressure which was equal at equal distances from mine openings, provided equality of other conditions. Fig. 3-6 shows such equilibrium pressures at the indicated distances from a single entry in rock salt in virgin ground; apparently, there was a regular increase in quasi-hydrostatic stresses from the opening to the boundary of the creep zone which had been established by deformation measurements. These results, of course, could not be reconciled with stress accumulations at or near the opening (circumferential stresses or stress envelopes) as postulated in hypotheses referred to above, particularly those advocated by Dreyer (1972) on the basis of laboratory investigations.

The possibility of stress-relief creep into new openings was strictly rejected; it was claimed that very different equilibrium pressures would develop if further strain hardening would be introduced around the cells by forcing the surrounding salt to creep continuously, e.g., by lowering the cell pressures to zero, or by increasing them to much higher values. This objection was checked out over periods of several months by alternately, every few days, lowering the cell pressures to zero, and increasing them to nearly 200 kp/cm^2.

The reactions were recorded with recording pressure gages connected to the measuring systems; Fig. 3-7, upper part, shows the remarkably fast readjustment of varied pressures to equilibrium pressures which had been established over several months previously. The lower part of Fig. 3-7 shows the reaction record of another measuring system in which the pressure was

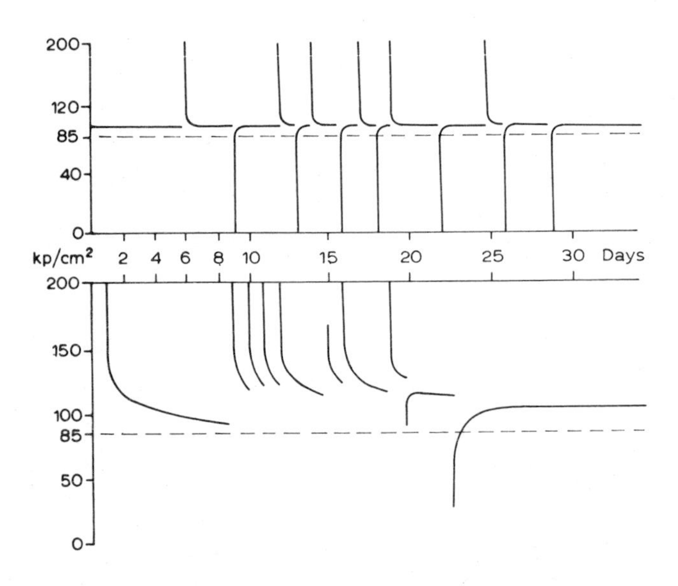

Fig. 3-7. Recorded reactions to intentional alterations of equilibrium pressures in sealed boreholes, conditions as in Fig. 3-6 (Baar, 1966a).

repatedly increased, but not released. In this way, a somewhat higher equilibrium pressure was established temporarily; the record shows that the salt rocks around the cell began to re-establish the original equilibrium pressure when manipulations ceased. Apparently, no strain hardening had taken place around the measuring cells.

The conclusion drawn from these experiments and related deformation measurements may be quoted (Baar, 1966a, p. 29): "Strengthening by plastic deformation does not play an important part under mining conditions, contrary to sample behavior under laboratory conditions. Even repeated stress alterations as shown in Fig. 3-7 do not affect the plastic behavior of the involved rock. Under such conditions, comparable to conditions after excavating an opening, obviously no raising of yields points and no strain-hardening of involved rock takes place."

3.4.3.2 Tests in a Canadian potash mine at the depth of 950 m. Similar tests were carried out in the IMC-K 1 potash mine in Canada in 1969. Serata's (1968) hypotheses, on which the pillar design was based, called for increased pillar strengths by strain-hardening; at that time the assumed creep limits were still of the order of 900—1200 psi (approximately 63—84 kp/cm^2), and these limits were expected to be raised considerably by strain-hardening near the openings where the creep maximum had been measured as reported by Zahary (1965) and Coolbaugh (1967). Serata (1968, correspondence with

IMC) raised objections to pillar stress measurements with U.S.B.M. copper cells; his objections resembled those reported above regarding in-situ measurements with pressure cells.

Copper cells were installed at depths of 1.5 and 3 m into pillars in the shaft safety pillar. The test results may be quoted from Baar (1972a, p. 61):

"Over periods of weeks, equilibrium pressures built up which were in good agreement with those found in another potash mine (Baar, 1966a). Reactions to intentional changes in pressures also were similar: equilibrium pressures once established were re-established by creep of the surrounding potash, no matter whether the cell pressures were increased or decreased from outside. These tests also confirmed earlier conclusions (Baar, 1966a): in the creep zone around openings in salt rocks, local near-hydrostatic stresses develop which increase from the opening to the boundary of the creep zone."

3.4.3.3 Roof loading tests in carnallitic salt rocks. At the mine referred to in the foregoing, "roof failure became a very serious problem when an area of high carnallite concentration was encountered because the carnallite reduces the strength of the material substantially" (Serata and Schultz, 1972, p. 39). "Mass failure is caused by a high concentration of the ground stress in the immediate roof media. The stressed mass yields and then expands to failure."

The carnallite content of the salt rocks referred to may be as high as 10%; carnallite occurs between halite and sylvite crystals, as shown in Fig. 2-12. The roof loading tests had to be made as Serata (1968, IMC correspondence) suggested "that intergranular loosening and/or bed separation along anticipated stress arches might contribute to roof failure, in particular in wide openings where the carnallite bond between halite and sylvite crystals might deteriorate" (Baar 1972a, p. 58).

Details of the test conditions can be found in the publication referred to (Baar, 1972a, pp. 58—60). Here, the conclusions may be quoted:

(1) Creep does not cause any structural damage in the deforming salt rocks. No intergranular loosening occurs in roof strata even if complete stress relief to the limits of elastic behavior takes place.

(2) The wider a room, the safer it is with regard to shear slabbing.

(3) Repeated loading and unloading in situ does not change the creep behavior of salt rocks.

It is emphasized that creep deformation could be achieved only to depths of about 5 m into the roof; the recorded creep curves were reproducible, which shows that neither loosening nor strain-hardening occurred. These results were acknowledged by Serata (1969, IMC correspondence); as a result, wider rooms instead of narrower ones (Serata and Schultz, 1972, p. 39) were introduced at this mine (Mraz, 1973) to eliminate the problems of roof failure by shearing. It should also be emphasized that wider rooms did not "stabilize" the salt rocks around openings; to the contrary, the initial stress relief creep rates, and the final constant creep rates under constant loading conditions, increased considerably; this is shown in Chapters 4 and 5.

3.4.4 Elasticity parameters insignificant in salt deposits

Elasticity parameters, such as Poisson's ratio, derived from uniaxial compression tests, are insignificant for apparent reasons: such parameters apply to elastic behavior, which is observed in the laboratory after strain-hardening; in situ, the salt rocks exhibit plastic behavior as soon as the low creep limits are exceeded. Therefore, the determination of elastic constants (e.g., Dreyer, 1972, pp. 10 ff.) may be regarded as an interesting academic exercise; such constants, however, must not be used in calculations of pillar strengths and stress accumulations near underground openings. As Dreyer (1972, p. 124) put it: "elastic analyses are bound to yield erroneous results", see the complete quotation of this important statement in the following section.

3.5 LOW TENSILE AND SHEARING STRENGTHS OF SALT ROCKS

Some facts may appear puzzling: the compressive strength of salt rocks received much attention in recent publications although it has no practical value whatsoever in the design of salt mining operations. To the contrary, mine pillar design based on laboratory values of compressive strengths of salt rocks frequently proved diastrous; this is shown in detail in Volume 2.

In contrast, the uniaxial tensile and shearing strengths, which dominate the visible and measurable deformations near underground openings, are given little, if any, attention in publications which deal with the rock-mechanical aspects of salt extraction from deeply buried deposits.

Most of the roof failure problems in salt and potash mining, be it conventional or solution mining, are related to the extremely small tensile and shearing strengths of salt rocks; this is shown in detail in subsequent chapters.

Here, the scarce literature data on tensile and shearing strengths is reviewed. It goes without saying that the shearing strength under triaxial stress conditions equals the creep strength, provided that creep can eliminate excess stress differences as they develop under normal mining conditions. However, when excess forces are applied suddenly, they may cause elastic reactions. The low creep limits of salt rocks, and the relative speed of creep disallow tensile stresses of any significance under triaxial stress conditions in situ.

As strain hardening does not affect the creep of salt rocks in situ, Dreyer's (1972, p. 124) statement must be borne in mind: "it is emphasized once more that elastic analyses are bound to yield erroneous results... The increase of the octahedral shear stress with the vertical stress was examined both in uniaxial and triaxial strain experiments... the strength increase with confining pressure is insignificant. At a (vertical) stress of 260 kp/cm^2, the octahedral shear stress reaches 5.92 kp/cm^2."

Referring to his own investigations and to other publications, Dreyer (1967, pp. 37—39) is listing the value of 6 kp/cm^2 as "critical" shear stress; the corresponding load is "only approximately 12 kp/cm^2", see Fig. 3-2.

"Any greater stress (difference!) will produce measurable flowage" as quoted in sub-section 3.2.3 from Borchert and Muir (1964) and Dreyer (1967). These authors do not mention any values of tensional strength of rock salt.

The uniaxial tensile strength of pure carnallite is given as 10 kp/cm^2 by Dreyer (1974a, p. 121); it is not explained how this relatively high value is to be reconciled with different values presented by Gimm (1968, p. 180).

It has been known for decades (Spackeler and Sieben, 1944) that the uniaxial tensile strength of salt rocks is only about 4—8% of the compressive strength; consequently, the tensile strength of pure carnallite with the compressive strength of 41 kp/cm^2 (Dreyer, l.c.) should be about 1/3 of the quoted value. Such differences are significant for various reasons.

The tensile strengths are particularly important in cases where gases are occluded in salt rocks; in undisturbed salt, the gas pressure equals the lithostatic pressure as shown in the preceding sub-section; it causes excess tension at newly created surfaces in mine openings, resulting in brittle fracture of the occluding salt, the gases expanding in an explosion-like manner. This mechanism, emphasized by Schmidt (1943) and Baar (1953a, 1954a), is dealt with in more detail in sub-section 4.5.1.

The tensile strengths are also significant in the propagation of fractures near underground openings where stress relief creep eliminated stress differences in excess of the creep limit. Increased pressures in fractures easily widen the fractures and make them advance according to the principle of least resistance.

Such effects are to be expected particularly in cases where the tensile strength is next to zero, e.g., along bedding planes or fractures which were caused by tectonic events, but were not eliminated by recrystallization. Such fractures are known in certain salt deposits, as outlined in Chapter 2, section 2.4.

It makes no difference whether excess tensile stresses at the tip of advancing fractures are caused by fluids which migrated into the fractures in the process of rock-mechanical deformations caused by mining, or whether they are caused by fluids forced into the fractures in application of hydrofracing methods, or pressure grouting methods. Particularly the latter possibility is often overlooked, and such mistakes have repeatedly resulted in water inflows in salt and potash mines which forced the abandonment of the mines.

Some case histories are dealt with in Volume 2.

Tensile strengths of anhydrite and limestone are much higher than those of salt rocks, corresponding to the much higher values of the compressive strengths which may exceed 1000 kp/cm^2. For example, Gimm (1968, p. 180) compared the uniaxial compressive strength of 550 kp/cm^2 for anhydrite with a tensile strength of 42.6 kp/cm^2, equal to 8% of the compressive strength; applying the same percentage to the compressive strength of 755 kp/cm^2 of another anhydrite found by Dreyer (1972, p. 292), this anhydrite would possess a tensile strength of 60 kp/cm^2.

It is apparent from these values of tensile strength that an evaporite sequence made up of anhydrite and rock-salt beds is considerably re-inforced against tensional forces which act in the direction of the bedding planes; eventually, anhydrite beds rupture as seen in Figs. 2-22 and 2-23. After the rupture of interlayers of rocks with relatively high tensile strength, and when tensile forces are acting perpendicular to bedding planes, the tensile strength of the rock mass equals the strength of the weakest member; this may be a layer of salt with extremely low tensile strength, or a bedding plane along which the tensile strength is next to zero. These relationships are extremely important in applied rock mechanics in salt deposits.

3.6 THE PERMEABILITY OF EVAPORITE DEPOSITS

3.6.1 General remarks

Standard permeability tests on cores from evaporite sequences are useful only in cases where the cores had not been damaged during and after drilling. This prerequisite may be met by cores of anhydrite, dolomite or limestone, i.e., by rocks which exhibit elastic or near-elastic behavior during the process of drilling. It is emphasized that the values obtained in such permeability tests may not be applicable to in-situ conditions for reasons which are apparent from the foregoing: in cases of fractured near-elastic rocks, and particularly in cases where karstification of such rocks along fractures has taken place, the permeability of single cores is insignificant with respect to the permeability of the formation.

Standard permeability tests on salt-rock cores are usually of no use for the cores are damaged when taken out of their triaxial in-situ stress field; such damage may be caused by stress relief deformation that results in intergranular loosening. Additional damage may be caused when the samples are prepared for testing.

For the reasons given, some geological facts, which testify to the absolute impermeability of salt deposits under certain conditions, should be given preference over questionable laboratory results. Some of the geological facts which are considered proof of the absolute impermeability of salt rocks below certain depths are outlined in sub-section 2.2.2: fluid inclusions, such as gas inclusions, have been preserved in salt deposits at relatively shallow depths for many millions of years; apparently, these fluids are trapped practically forever — unless escape routes are provided by tectonic events, or by mining operations. It is not necessary that cavities filled with fluids are hit directly by drilling or by advancing underground openings, to the contrary: in many cases, rock-mechanical processes caused by excavations in salt deposits provided for escape routes for large amounts of trapped fluids. Some case histories are dealt with in Volume 2.

3.6.2 Salt rocks in situ are impermeable at depths exceeding about 300 m

The figures given in section 2.3 for depths at which the vertical permeability of newly deposited salt sequences becomes nil for all practical considerations, may appear contradictory to the above depth limit of approximately 300 m. It should be borne in mind, however, that lateral migration of fluids may have been possible for much longer periods of sedimentation, until sufficient compaction of non-evaporitic lateral facies equivalents of salt rocks had taken place.

The 300-m figure given in the heading is based on the following observations during drilling and shaft sinking into salt deposits:

(1) Numerous drill holes from surface and from temporary bottoms of sinking shafts hit pockets of gases occluded in Zechstein salt of northern Europe. Such incidents occurred in salt domes as well as in anticlines shortly after the drill holes and/or shafts had passed through the caprock made up of solution residues which are usually highly permeable. Some of the spectacular incidents reported in the literature are shortly described in the following. It is emphasized that the gas and brine pockets referred to did not have any connections to fracture zones in anhydrite beds which, in turn, may have been connected to aquifers in the caprock; such fracture zones are known to occur at depths greater than 300 m; when hit in mining operations, the resulting inflows usually cannot be stopped as the inflow rates tend to increase with time.

The isolated pockets referred to in the following apparently did not receive any recharge; the outflows decreased to zero within time periods of various lengths; all data may be supplemented with more details to be found in the reference publication (Löffler, 1962).

1886: the shaft Aschersleben II was flooded when it had reached the depth of 300 m; a pilot hole, drilled from the temporary bottom of the shaft into the underlying Stassfurt rock salt, hit a gas pocket with H_2S—CH_4—N_2 gases under apparently high pressure: for two hours, NaCl brine was ejected to the height of houses by the pressure of the gas mixture. The shaft was abandoned.

1887: the shaft Leopoldshall III at Stassfurt, having been sunk through the caprock, hit H_2S-containing gases while sinking to the total depth of 412 m through the Stassfurt rock salt; in 1887, four miners were killed by gases, and in 1889, seven more were killed. According to Gimm (1968, p. 547), gases were encountered during sinking various other shafts in the Stassfurt area since 1856. In all instances, the gases were encountered at short distances from the top of the salt deposit.

In some cases, gases were encountered in fracture systems of anhydrite close to the top of the salt anticline; such fracture systems apparently had connections to the groundwater as the gases were followed by water. In other cases, fracture systems to connect the mines to overlying aquifers were caused by inadequate mining methods. As the result of the combined effects of mining and salt dissolution by inflowing waters, surface subsidences of 7—8 m occurred within the city limits of Stassfurt, averaging 24 cm per year during the decade from 1900—1909, and 13 cm per year during the following decade. Large areas of the city were destroyed.

(2) In the Werra potash mining district of central Germany, large amounts of CO_2 gases were encountered in 1895 at the depth of 206 m in rock salt when the shaft Salzungen reached that depth (Gimm 1968, p. 547). Numerous other outbursts of gases occurred in the same district during mining at depths greater than 300 m (see Chapter 4, pp. 00—00), for more details.

(3) The published records of gas occurrences in German potash mines for the time period 1907—1917 (Gropp, 1918) provide data and comments on 106 gas occurrences at depths of about 300 m, and at greater depths. Many of these occurrences caused casualties, particularly in salt dome mines in the Hannover area. Several potash mines were abandoned for this reason. The list of gas occurrences encountered near the top of salt deposits in Germany, and elsewhere, could be supplemented with many more cases reported in the literature. Salt rocks of relatively small thicknesses kept the gases and other fluids trapped for many millions of years.

The conclusion that salt rocks in situ are absolutely impervious at depths below about 300 m appears inescapable.

3.7 PLASTICITY OF SALT ROCKS—TEMPERATURE EFFECTS

3.7.1 Equivalent viscosity

Plastic behavior as exhibited by salt rocks differs from viscous behavior as shown by fluids and amorphous material, such as certain clays and sands which consist of grains without crystalline bonding. Many researchers have been trying to apply laws of viscosity to the plastic behavior of salt rocks which is usually termed creep. In the following, some statements by Odé (1968a, pp. 552 ff.), who devoted an extensive literature search to this particular problem, may be quoted.

"Unfortunately, the slow flow or creep of solid materials, including rocks and salt, is exceedingly complex. In particular the behavior of crystalline salt masses probably closely resembles that of metals, which have been much studied. Consequently, values of the equivalent viscosity (certain parameters which indicate the viscosity of a Newtonian fluid) cannot be simply assigned. They are functions of temperature, confining pressure, differential stress, time, etc.; in the case of salt, other variables — the amount of water, impurities such as gypsum, etc. — must be considered."

"The extremely slow flow inferred from studies of salt flowage in salt mines and domes indicates that the equivalent viscosity of salt is many orders of magnitude larger than the viscosity of well-known materials such as honey and tar. However, we must realize that the concept of equivalent viscosity is nothing more than a crude measuring stick, because the physical process called viscosity in a liquid such as water has little in common with the slow creep in a polycrystalline salt sample. Almost all modern work deals with the creep of metals, but so far few systematic investigations have been made for rock materials. It might appear that a discussion of creep of rock salt on the basis of results obtained for metals would not be significant. Apart from a difference in lattice structure between certain metal crystals and sodium chloride, there seem to be essential differences which are best expressed by the terms ductile and brittle. Brittle and ductile behavior depends on the circumstances of the test such as confining pressure, strain rate, temperature, and interstitial fluid pressure (Heard, 1960)."

"However, . . .rock salt and some other ionic crystals owe their apparent brittleness to secondary causes and thus are closer to metals than has formerly been thought. Rinne (1926) pointed out some similarities between salt and metals."

Odé (l.c., pp. 553 ff.) proceeds in his review with discussions on the creep laws proposed by numerous authors, beginning with Andrade (1914), and particularly referring to Fig. 3-8 which shows typical strain/time curves under different constant uniaxial stresses:

"The early parts of these curves can be fitted with reasonable accuracy to an equation of the form:

$$\text{(deformation) } e = A + B \log t + Ct, \tag{1}$$

an equation applied to the creep of steel by Weaver (1936). The term A represents the instantaneous deformation (elastic), $B \log t$ represents the transient flow, and Ct represents the steady-state flow."

"The law:

$$e = B \log t \tag{2}$$

was first proposed by Phillips (1905), who observed the deformation of rubber, glass, and metals. This law, the so-called logarithmic creep law, can be written in the form:

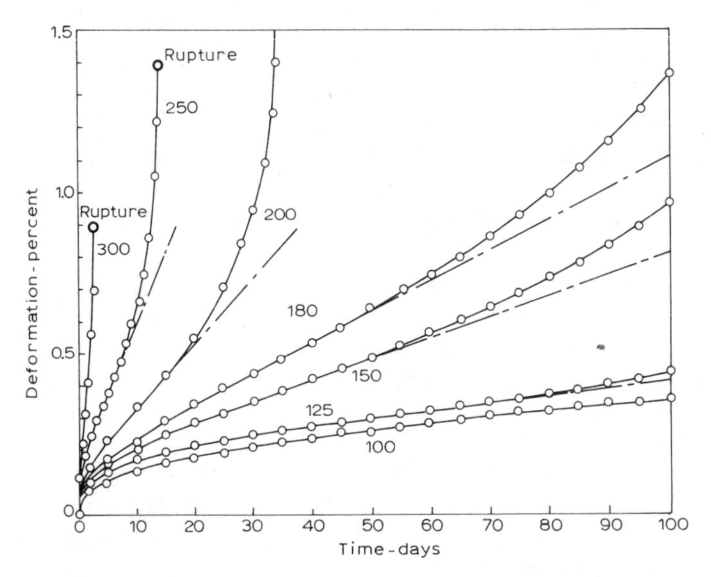

Fig. 3-8. Creep (strain/time) curves for water-saturated gypsum under different constant uniaxial stresses (shown in bars for each curve) at atmospheric conditions. (After Odé, 1968, from Griggs, 1940.)

(deformation rate) $\dot{e} = B\,t^{-1}$ (3)

and leads to the conclusion that immediately upon application of the load the creep rate is indefinitely large, even at low temperature (Cottrell, 1952).

In 1910, Andrade proposed to write for the transient creep $e = B\,t^{1/3}$, or:

$$\dot{e} = \frac{B}{3}\,t^{-2/3} \tag{4}$$

Thus, eqs. 2 and 3 both appear to be special cases of the general formulation:

$$\dot{e} = B\,t^{-n} \tag{5}$$

According to Cottrell (1952), the value of n is usually less than unity at faster creep rates."

It is emphasized in section 3.4 that these creep laws may apply to laboratory conditions which cause strain-hardening; thus, it is by no means surprising when modern investigators find the same creep laws. However, it must be borne in mind that, apparently, the initial stress relief creep also follows eq. 5, and the subsequent creep at constant rates represents the term Ct of eq. 1.

Odé (1958, pp. 576 ff.), after discussion of numerous published results of laboratory investigations since 1894, arrives at the conclusion that "the tests discussed so far are not suitable for a determination of the equivalent viscosity of salt. Only a few measurements are available for this determination. The first controlled creep test on halite was made by Griggs (1939). A prism (1 by 2 cm) cleaved from a single crystal was loaded in a small creep tester to 61 kg/cm^2, and its creep was observed for 42 days at room temperature. This load corresponds to approximately 250 m of overburden. The value of this equivalent viscosity would be $2.6 \cdot 10^{17}$ poises."

Summarizing the literature data suitable for calculation of equivalent viscosities under different testing conditions, Odé (1968, table 6, p. 581) notes values between 10^{18} (Weinberg, 1927) for rock salt and 10^{11} (Höfer, 1958) for sylvite.

Referring to measurements of convergence rates in salt mines, Odé (1968, p. 583) states: "Unfortunately, most reports give no data from which the equivalent viscosity of the salt can be estimated. Remarkably enough, one of the very earliest reports (Busch, 1907) admits another approximate estimate of the salt viscosity... The corresponding viscosities are ... $2 \cdot 10^{16}$ poises at 500 m and $2.3 \cdot 10^{17}$ poises at 300 m depth."

Regarding the effects of temperature on the equivalent viscosity of salt, Odé (1968, pp. 576 ff.) refers to series of tests with artificial aggregates of salt made by compacting pure NaCl powder (LeComte, 1960): "Confining pressures were 0, 200, and 1000 bars; the temperature varied from 25° to 198°C; the axial stress was 35, 60, 69, and 138 bars; and the duration of tests was usually between 80 and 200 hours. Therefore, his creep curves seldom extended into steady-state creep. For this reason, LeComte's values of the equivalent viscosity are probably low." Temperature increase of various degrees increased the creep rate by factors of 4—5 to 22. "Confining pressure appeared to lessen the effect of temperature increase."

"Results similar to those by LeComte were obtained by Serata and Gloyna (1959); the equivalent viscosities (10^{15}—10^{16} poises) . . .are much smaller than those of Griggs (10^{17} poises)."

Considering the variability of reported and estimated equivalent viscosities —10^{11}—10^{18} poises—it is apparent that the conditions of creep are extremely important.

Odé's (1968, p. 556) estimates are based on the assumption "that the creep of salt takes place by steady-state creep". In this case, "it is possible, once the differential stress Δs and the creep rate $\dot{e}$ are known, to estimate the equivalent viscosities of the salt by:

$$\eta_{eq} = \Delta s / 3\dot{e} \tag{6}$$

In general, the equivalent viscosity will be a function of temperature, confining pressure, and differential pressure. It has been found that the steady-state creep is strongly dependent on temperature. This suggests that thermal activation plays a major role in creep phenomena."

With respect to the latter statement, the effects of recrystallization and recovery dealt with in sections 3.2 and 3.3 are extremely important. In cases where strain hardening due to rapid loading in the laboratory had increased the equivalent viscosities, the thermal re-activation of much lower equivalent viscosities results in much higher creep rates under constant loads and equal differential stresses.

Thermal activation indeed plays a major role in creep phenomena in situ. As a matter of fact, thermal activation has long been known as the cause of mine disasters in certain German potash mines where thermal activation of plastic deformability of salt rocks takes place in the following way (Baar, 1959a, with reference to papers the publication of which was suppressed; see Fulda (1963) and Michalzik (1966) for references to controversial reports):

In 1907, the first in-situ tests were made in potash mines to hydraulically backfill mined-out panels with refinery waste; this method is used in a great many mines; it was, and still is expected (Gimm, 1968, pp. 370 ff.) that the supporting capacity of mine pillars is greatly improved in this way. However, the contrary is the case as the refinery waste is hot, about 100°C when leaving the plant; the temperatures around the backfilled rooms may

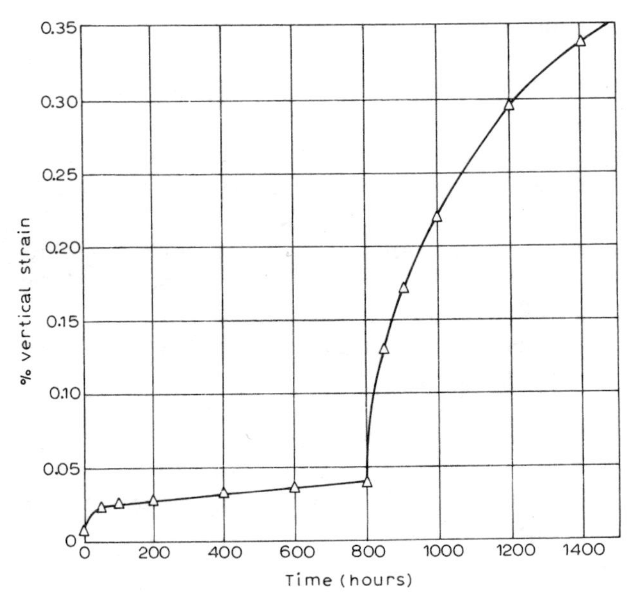

Fig. 3-9. Deformation of pillar model at 22.5°C and 100°C. (After Lomenick and Bradshaw, 1969.)

The rock-salt model was loaded to 4000 psi (280 kp/cm^2) at room temperature for about 800 hours; then, the temperature was increased to 100°C.

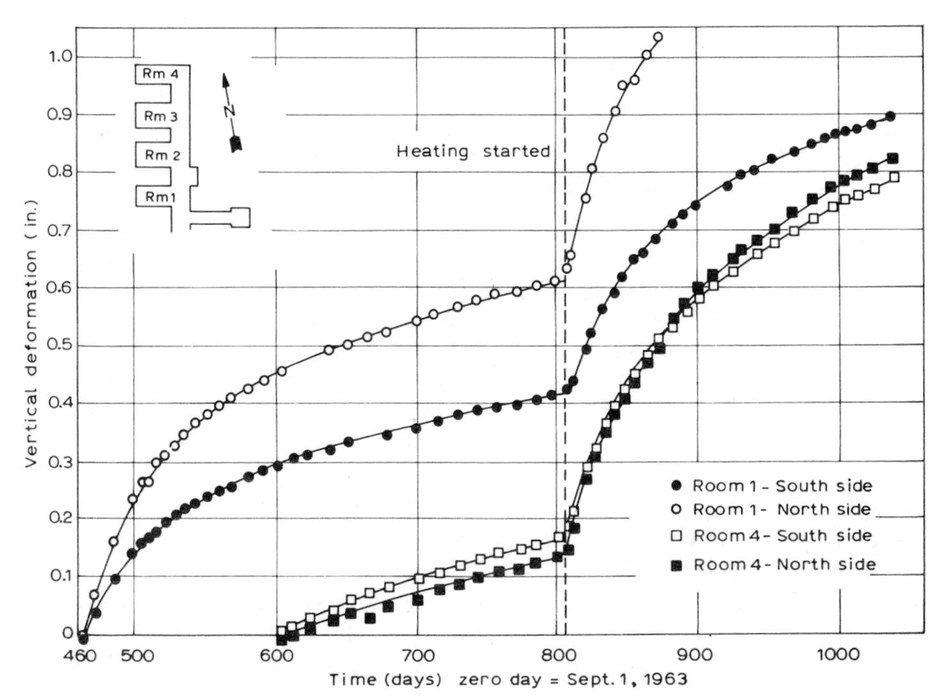

Fig. 3-10. Accelerated creep caused by heating of a salt-mine pillar; peak temperatures were about 150°C. (After Empson et al., 1970.)

be increased to about 60°C. As a consequence, the supporting capacity of the pillars, already greatly reduced by stress-relief creep, is further reduced by the thermal activation of the creep deformation of the pillars. The resulting accelerated creep caused numerous gas and water inflows, and rockbursts, at the time when the backfilling was in progress, or immediately after it had been completed (Baar, 1953a, 1954a). Such activation of creep by increase in temperature caused the loss of the potash mine referred to in Figs. 3-6, 3-7, and 3-12. More details are presented in Volume 2.

The effects of heat on the deformation of salt mine pillars was demonstrated by Empson et al. (1970) referring to pillar heating experiments in the Lyons mine, Kansas, during Project Salt Vault (see also Lomenick and Bradshaw, 1969, and Bradshaw and Lomenick, 1970). Fig. 3-9 shows the thermal activation of creep in a model pillar. Fig. 3-10 shows effects on floor-to-ceiling convergences around a heated pillar at the Lyons mine.

The accelerated creep seen in both Figures is not caused by thermal stress as suggested in the publications referred to, but rather by considerably increased plasticities resulting from the elevated temperatures; other measurements results gathered at this demonstration site also confirm that the pillars had been stress-relieved by stress-relief creep prior to the heating experiments. This is shown in section 5.5.

3.7.2 Migration of brine inclusions

Another observation during Project Salt Vault is described as follows (Empson et al., 1970, pp. 460, 461): "It was found that small quantities of water, trapped in so-called "negative crystals" or brine-filled cavities, tend to migrate toward a heat source and into the disposal hole (for heat-generating radioactive waste). Migration of brine-filled cavities occurs because of the slight increase in solubility of sodium chloride in water with increase in temperature. Salt dissolved on the warmer side of the cavity diffuses to the cooler side where it is redeposited. This phenomenon did not constitute a problem in the demonstration, since the water was removed by an off-gas system."

Some details of tests to determine the effects of trapped moisture on the deformational behavior of salt samples are reported by Bradshaw et al. (1968, pp. 647—651): "Most bedded salt contains trapped moisture which is released with violence at temperatures above about 250° C"... "The salt 'explodes' at nearly the same temperature regardless of the rate of heating." These statements refer to salt samples from the Hutchinson and Lyons mines, Kansas; the water contents are listed as ranging from 0.08—0.29 wt.%, with one exceptionally high value of 1.1 wt.%.

No "explosion" of salt-dome salt occurred at temperatures up to 400° C. High-temperature in-situ tests resulted in salt shattering, "when the salt at the periphery of the heated hole reached a temperature of about 280° C. As the 280° C temperature front moved away from the hole, salt continued to fracture, releasing trapped water. About 190 ml of water was actually collected during a test in a pillar (no shale layers) from a fractured salt volume estimated to be 1.8 cubic feet. More than 190 ml was actually released by the shattering, but it was not possible to get an accurate estimate of the amount not collected."

Apparently, the shattering resembles the brittle fracturing of salt rocks with occluded gases dealt with in more detail in section 4.5. As a matter of fact, salt samples with occluded gases are known to "explode" when placed into sunshine, in which case the gas expansion with temperature increase causes the tensile stresses to rupture the salt.

The thermomigration of brine-filled cavities in individual salt crystals toward the heat source could also release some moisture; these phenomena were investigated by Anthony and Cline (1974, pp. 313—331) on a single crystal of KCl 3 × 3 × 10 mm with less than 30 ppm of impurities: a cylinder of brine was produced in the salt crystal by drilling a 0.3-mm diameter hole, filling it with deionized water, and sealing it with wax to prevent evaporation. "A thermal gradient of 20° C/cm was applied along the length of the salt crystal for one week. During the first few days, the brine cylinder simultaneously migrated and disintegrated into a myriad of small brine droplets with a size range of 2—150 microns."

"The isolated brine droplets continued to migrate up the temperature gradient until they came to the hot end of the crystal. There at the crystal surface, the brine in the liquid droplets partially evaporated until salt precipitation generated by the evaporation resealed the droplets. The resulting biphase vapor-liquid droplets then *turned around and migrated down the temperature gradient back into the crystal.*"

In the above quotation, the quoted thermal gradient of 20° C/cm appears to be incorrect as the authors refer to a thermal gradient of 3° C/cm in their following text, concentrating on the thermomigration of vapor-liquid droplets back into the crystal, and emphasizing that "very small liquid droplets do not move" (l.c., p. 314).

This observation might be indicative of brine movements along intracrystalline slip systems (see Fig. 3-1) during the recrystallization processes which were initiated by dissolution of salt along, and around, the 0.3-mm hole drilled into the crystal, and filled with water. Also, the previous stress history of the small crystal, including possible effects of the drilling, may have affected the crystal structure. For these reasons, the practical significance of the theoretical deductions appears questionable.

With reference to previous papers, Anthony and Cline (1974, pp. 319, 320) are applying their findings "to the storage of nuclear wastes in salt formations" as follows:

"The thermomigration of liquid-brine droplets into nuclear waste burial crypts. Natural salt regularly contains about 0.4 atomic percent water in the form of small brine inclusions several hundred microns in diameter" (with reference to Bradshaw et al., 1968, as quoted above; as these writers refer exclusively to salt from the Hutchinson and Lyons mines, Kansas, such local conditions *must not* be generalized!).

"We have shown that such liquid brine droplets will migrate up a temperature gradient towards higher temperatures in salt... We estimate that the flow rate of water into the nuclear waste burial crypt (Lyons mine, Kansas) will be $2.4 \cdot 10^{-7}$ moles H_2O/sec. This is equivalent to an inflow of 140 cm^3 per year into the crypt. Recent results (reference: McClain et al., 1967) from Project Salt Vault show a water inflow into the waste crypts of 365 cm^3 per year, in reasonable agreement with our calculated value." These statements contradict the above quotation from Bradshaw et al. (1968), who derive the water inflow from shattering due to heating.

"The transport of radioactivity away from the burial crypt by the thermomigration of biphase vapor-liquid droplets. A further possible result of the self-heating of nuclear wastes is that contaminated vapor-liquid biphase droplets generated on the walls of the burial crypt may transport radioactive wastes away from the crypt into the salt formation."

"Assuming that the selfheating of the nuclear wastes maintain the value of

300° C for the equivalent of two years, we find that vapor-liquid droplets containing a foreign gas could only transport radioactive wastes a distance of 10 cm into the salt formation. On the other hand, vapor-liquid droplets containing only water vapor in the vapor phase could carry nuclear wastes a distance of 600 cm into the salt formation in two years."

"This figure, however, ignores the pronounced ability of grain boundaries to trap migrating droplets."

Anthony and Cline (1974, p. 320) arrive at the following conclusion regarding the possible escape of radioactivity to the external environment from the Lyons, Kansas, nuclear waste burial facility with an overlay of 100 m of salt:

"The calculated times required for a vapor-liquid droplet to escape from the salt formation because of the natural thermal gradient of the earth are 10^6 and 10^3 years, respectively, for vapor-liquid droplets with and without foreign gas in their vapor phase. These calculated times, however, again ignore the ability of grain boundaries to trap migrating droplets."

With reference to the facts presented and discussed in section 3.6, it appears justified to ignore hypothetical statements such as those quoted from Anthony and Cline (1974) without further detailed discussion.

However, the results of careful investigations by Holdoway (1974) cannot, and should not, be ignored; Holdoway investigated the "behavior of fluid inclusions in salt during heating and irradiation" at the demonstration site referred to in the foregoing; "petrographic examination (of cores cut at an angle of 45° on the outside of the two holes of the main heating array, entering the holes 2.7 m below the mine floor) showed that irradiation resulted in coloration of the salt, producing colors ranging from blue-black nearest the radiation source, to pale blue and purple farther from the source."

"Bleached areas are common in the radiation-colored salt, many representing trails produced by the migration of fluid inclusions towards the heat source. During laboratory radiation studies, it was found that these bleached areas frequently colored much less readily than ordinary colorless salt which had not been irradiated previously. This observation, and the relationship between primary chevron structures and migrated fluid inclusions, suggest that the bleached trails represent the total amount of migration of the inclusions from the beginning of Project Salt Vault, until the ambient temperature was reached after the conclusion of the experiment."

For further details, the reader is referred to the reference publication; on p. 304, it is stated "that more water entered the array holes than had been anticipated. This water is thought to have been derived from fluid inclusions within the salt which migrated toward the radiation sources. If salt beds are to be used as a repository for radioactive wastes, such migration presents a possible hazard. On entering the array holes, water would become contaminated with dissolved radioactive wastes. Further migration of this water could lead to contamination of the water supply."

There is no indication of how "further migration" of water through 100 m of overlying salt *against* the temperature gradient could possibly occur. However, on p. 306, it is referred to Anthony and Cline (1972) as follows: "Inclusions with less than ten percent vapor migrate toward the heat source, and those with more than ten percent vapor migrate away from the heat source." This reference is repeated on p. 310, and it is concluded "that it is possible that gases produced during irradiation may have reduced the rate of migration of fluid inclusions toward the array hole and, in some cases, may have caused some migration away from the array hole". However, no evidence of such migration against the temperature gradients could be detected.

Besides, if indeed brine droplets should decide to reseal themselves at the surface by partial evaporation, and to turn around and migrate down the temperature gradient as proposed by Anthony and Cline (1974, see quotation), there would be a simple solution to stop such activities: just to prevent evaporation, e.g., by sealing the array holes with wax as done by the authors referred to in their experiment. There are, of course, other possibilities of preventing evaporation in array holes.

The "stress effects" referred to by Holdoway (1974, p. 310) apparently reflect hypotheses in which "an increase in stress around the array hole" is postulated; this is also indicated by the reference to "a large amount of stored energy (which) was released from samples 15 cm from the array hole".

These relationships are discussed in more detail in section 4.2. It appears most unlikely that energy could be stored where apparently the plasticity of the salt was increased, as shown by accelerated creep. In general, the above quotations contradict each other frequently; this appears indicative of attempts at drawing conclusions from data which are insufficient; some of the conclusions are definitely premature and require reconsideration on the basis of conflicting facts presented in this text.

3.8 DETERMINATION OF STRESSES IN SALT DEPOSITS

3.8.1 Introduction and literature review

Knowledge of the loads supported by mine pillars and/or abutment zones around mined-out areas is essential in mine design for adequate control of the deformations caused by extraction. In other words: the vertical stresses in support structures designed to carry the deadweight of the overburden should be known as precisely as possible. Unfortunately, stresses cannot be measured, neither in the laboratory on samples nor in situ in a rock massive.

However, in the case of elastic behavior both in situ and in the laboratory, it is possible to use known equations of elasticity for the conversion of measured strains to stresses; the required mechanical parameters of any particular rock can be determined in the laboratory under defined conditions of loading, temperature, etc.; various "stress meters" operate according to these principles. Such methods, if adequately executed, provide worthwhile information in cases where the prerequisite — elastic behavior — is met.

It should be apparent that such methods of indirect determination of stress cannot be applied to cases where the prerequite is non-existent, i.e., in rocks which exhibit inelastic behavior such as salt rocks. Additional problems are related to the strain-hardening of salt rocks in the laboratory, see section 3.4, and particularly to the fact that plastic deformation (creep) strongly depends on time.

On the other hand, the plasticity of salt rocks in situ, and their impermeability, provide for a very simple method of measuring the pressure in artificial fluid inclusions established at suitable depths into pillars and/or walls of single underground openings, see sub-section 3.4.3. Details of measurement results are given in sub-section 3.8.2.

As the recent literature has been inundated with claims of successful measurements of stresses in salt and potash mines, it appears necessary to review the various methods shortly, and to show the reasons why most methods cannot produce any reliable results on which mine design could be based.

(1) Stress-relief techniques. Stress-relief techniques are being used extensively in hardrock mines where they are feasible because of the elastic or near-elastic rock behavior. Strain measuring devices are bonded to the rock in situ at depths into pillars and/or walls at which the stresses are to be determined. The devices then are recovered by overcoring, i.e., a core of the rock to which the strain meter is bonded, is recovered by using a larger-diameter core bit. The deformations due to stress relief are measured and converted to stresses by the use of equations of elasticity.

Abel (1970, p. 200) reported on attempts at using such techniques in salt rocks: "Mohr (1956) measured the apparent stresses in salt pillars and demonstrated stress concentrations near their edges, such as would be predicted elastically. Potts (1964) reported similar results from other measurements. Mohr's and Potts's peak stress levels indicate the true short-term load carrying capability of the pillar."

Borchert and Muir (1964, p. 269) and Dreyer (1967, p. 77; 1972, p. 204) criticized Mohr's data for it "only indicates the strain released. It is exceedingly difficult to convert these values to absolute pressures in kp/cm^2. The main trouble is that we cannot yet tell to what extent the preliminary boring alters the stress conditions in the pillar." To overcome these difficulties in application of elasticity theories, the released strains were "compensated" in the laboratory; see the following sub-section.

Gimm et al. (1970) attempted to determine absolute stresses in rock salt pillars using "the" stress-relief technique; no details are reported; it was concluded that there must exist a plastic zone near the edge of the pillar, while the pillar interior remains in an elastic state.

Singhal (1971) reported similar conclusions drawn from "laboratory and field experiments to study the application of the stress-relief technique of absolute stress measurement in evaporites"; major principal stresses varying from 138 to 2582 psi (about 10—180 kp/cm^2) were calculated for depths of 4—5 ft. (1.2—1.5 m) into rock salt pillars. Most of the values were "less than expected" and "rejected as obviously erroneous". Only two values were "considered realistic" for they were "indicative of a stress concentration near the pillar edge". It is apparent that such arbitrary selection of values which match preassumed theoretical stress distributions over the pillar cross-section proves nothing but rather questionable methods in arriving at wished-for data.

The problems involved in applications of stress-relief techniques in salt rocks become apparent from Fig. 3-11, which shows measured expansions of salt cores during and after the process of overcoring (Mohr, 1955). The expansion is clearly time-dependent; the decay of the expansion rate resembles the exponential decay of stress-relief creep rates.

It appears justified to assume that time-dependent stress-relief deformation inevitably occurs prior to overcoring, during the time period required to bond the strain measuring device to the future core. Depending on the length of time between drilling the first hole and the overcoring hole, the original stresses are altered to various degrees; thus, at the best, strains measured during and after overcoring reflect the state of stress at the time of overcoring, which is not the original state of stress. Even if elasticity equations were applicable to clearly time-dependent deformations for conversion

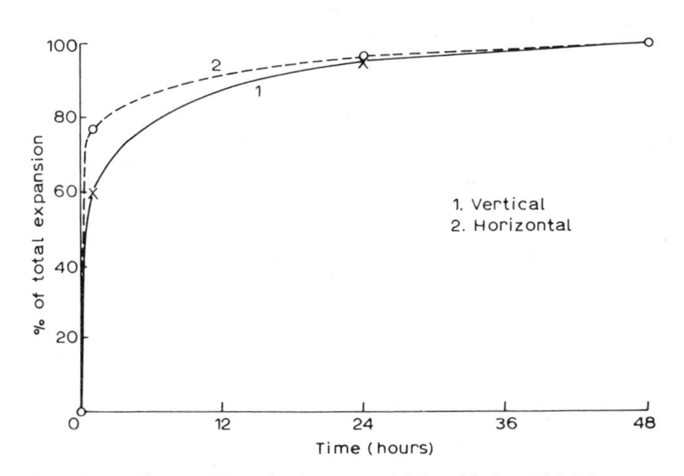

Fig. 3-11. Expansion of salt cores. (After Mohr, 1956.)

of strains to stresses, the stress changes prior to overcoring need to be considered, as emphasized in the quoted criticism.

(2) Compensation methods. The ideas behind strain-compensation methods to determine the stresses which exist at the location where a sample is taken, are described by Dreyer (1972, p. 204) as follows: "The pin method" (just another name for the "compensation method" described by Borchert and Muir, 1964, p. 269) "differs from other overcoring techniques. . . in that no knowledge of the elastic modulus or of Poisson's ratio is required. These stress-dependent rock properties are not needed, because the stresses in-situ are determined directly in the laboratory by reproducing the in-situ strain of a sample. No formulae are needed, which are based on homogeneous and isotropic rock properties and into which often erroneous moduli are inserted."

Such reasoning completely disregards the facts: (a) that salt rocks deform plastically, by creep, in response to stress differences which exceed the limit of elastic behavior; (b) that the reproducing of in-situ strain in the laboratory causes strain-hardening. The extremely dangerous consequences of such disregard of basic principles have been shown in section 3.4.

It is emphasized that the "yield point technique" (Dreyer, 1972, pp. 193—197), called "point of inflection method" by Borchert and Muir (1964, pp. 263—268), is based on similar erroneous assumptions; therefore, the stress values obtained by the "yield point technique" and the "pin method" may be of similar magnitude. However, in no way can such methods "provide, for the first time, a means by which stresses in-situ can be quantitatively assessed" (Dreyer, 1972, p. 197). Catastrophic consequences of such assessments are shown in detail in Volume 2.

(3) Other "stress meters". Changes to the diameter of a borehole between two points at its circumference can be measured very precisely with modern instruments. Scientists dealing with the interpretation of measured changes to borehole diameters realize the problems involved in cases where long-term borehole closures occur, and demonstrate inelastic behavior of the rock around the measuring points. Obviously, elasticity formulae cannot be applied for the calculation of stresses from measured inelastic strains.

Modern instruments for measurement of radial borehole deformation could be utilized for measurements of the initial stress-relief closure of boreholes drilled into salt rocks around mine openings; in this way, the magnitude of the stress-relief closure at various distances from the opening can be measured, and related to the existing stress field, provided the following conditions are met: as the initial stress-relief creep rates exhibit a rapid exponential decay, measurements must commence as soon as possible, i.e., within minutes after the hole has been drilled at any particular measuring point. Each measuring point must be chosen at exactly the same distance from the temporary face of the borehole; this distance must be sufficient to eliminate effects of the end restraint near the temporary face.

As the stress-relief creep curves immediately after drilling are regular curves — see Fig. 3-7 for comparable curves produced by pressure release in hydraulic cells — the curves for each measuring point can be extrapolated with sufficient accuracy from measurements over time periods of, say, 10 minutes. It is necessary, of course, to calibrate the resulting curves for any specific salt rock by absolute stress determinations with hydraulic cells; see the following sub-section.

The principles of such in-situ measurements have been known for about two decades; they are reported in a number of publications (Baar, 1958, 1959a, and subsequent publications in English, e.g., 1966a). It is not known to this writer whether more measurements according to these principles were conducted in situ. However, at least one in-situ experiment that shows the effects of stress-relief creep into boreholes is referred to repeatedly in various recent publications. The following report is translated and shortened from Gimm (1968, p. 216, figs. 4/58 and 4/59) who refers to an in-situ experiment conducted by Dreyer (1965); the reasons why Dreyer (1967, 1972, 1974a) fails to mention this in-situ experiment in his own publications are not known.

According to Gimm, three small-diameter "blasting holes" were drilled in normal direction into a rock salt face, as shown in Fig. 3-12. Exact dimen-

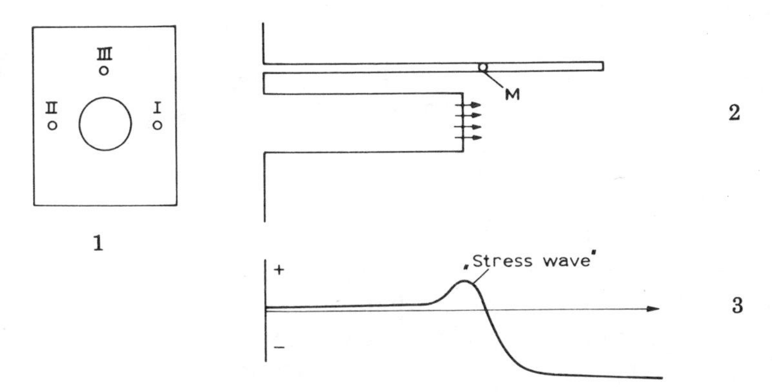

Fig. 3-12. Stress-relief deformations caused by drilling holes at different times. (From Gimm, 1968, after Dreyer, 1965.)

1. View of the rock-salt face into which the small holes *I—III* had been drilled prior to drilling the 34-cm diameter hole in the centre. 2. Section shows the location of the measuring point at which the radial borehole deformation caused by drilling the large hole was measured. 3. Results of measurements, and interpretation by Gimm, after Dreyer.

+ = "stress wave" travelling ahead of the advancing face of the large hole, assumed to be indicated by temporary widening of the small hole; — = creep convergence of the small hole after the face of the large hole had passed the measuring point.

sions are not given, except for the diameter of 34 cm of a larger hole that was subsequently drilled between the small-diameter holes; the time elapsed since drilling the small holes is not indicated. Assuming that the Figure is drawn to scale, and that all holes were drilled parallel to each other, the distances between the small holes and the large hole were approximately 17 cm; the effects of drilling the large hole were measured at the distance of approximately 1.2 m from the mouths of the small holes. The data is shown in the lower part of Fig. 3-12, and interpreted as indicative of a wave of elastic stress concentrations which travel ahead of the advancing face of the large borehole.

Fig. 3-12 clearly shows this interpretation inconclusive: the positive deformation at the time when the large hole approached the measuring point indicates widening of the small hole rather than shortening of its diameter, as must be expected if a travelling stress concentration affects the small holes. In fact, the temporary widening was caused by rapid stress-relief creep ahead of the advancing large hole; *after* the face had passed, stress-relief creep at very small rates into the small holes resumed.

Most writers realize the problems in interpreting measured borehole deformations, although increases in creep rates are usually attributed to increased stresses (McClain and Bradshaw, 1967, p. 254, with reference to extensive testing of various borehole deformation meters).

It must be pointed out, however, that borehole convergences in salt do not necessarily indicate loads; stress-relief creep has similar effects, as demonstrated by measurement data recently published by Dreyer (1972, pp. 338—344) who apparently dropped his earlier hypotheses quoted above after Gimm (1968).

In contrast, another "new stress meter to determine absolute stresses in rock-salt pillars" (Dreyer, 1972, pp. 205—215) is still being praised as before in various publications (e.g. Borchert and Muir, 1964, pp. 269, 270; Gimm, 1968, pp. 210, 211); of course, this "stress meter" too is just another strain meter, the strains being measured in boreholes to be drilled into the rock massive. Strains are measured in the vertical direction, and in two horizontal directions, as is the case with other modern strain meters. The particular advantage of the "new stress meter", called "measuring cartridge" by Borchert and Muir (1964, l.c.) and "salt plug" by Dreyer (1972, l.c.), is believed to be that this strain meter is made from the rock material itself, i.e., rock salt; "the strain gauges and the strain gauge lead wires are the only 'foreign' elements" (Dreyer, 1972, p. 209). In this way, "the danger that the measured deformation will relate less to the rock itself than to the characteristic properties of the material from which the measuring device is constructed", is overcome (Borchert and Muir, 1964, p. 270). Singhal (1971) gave the same reasons for the construction of a "low-modulus plug made of rock salt" which resembles the one referred to above.

It is evident that all these authors are thinking in terms of elasticity theories applied to elastic rocks, in which case it makes a great difference if a measuring device is either too stiff or too soft compared to the rock itself. As Dreyer (1972, p. 205) put it: "These stress meters will furnish absolute stresses only in exceptional cases, where the compressibility of the stress meter just equals the compressibility of the surrounding rock. Generally, measured stress values will be larger or smaller than the stresses in-situ depending on whether the compressibility of the stress meter is lower or higher than the compressibility of the rock."

Such reasoning may be adequate in cases where the rock exhibits elastic behavior, in which case elasticity formulae may be used to calculate stresses from measured strains; it is definitely inadequate to apply such principles to rocks which clearly show plastic behavior, as did the rock-salt pillar in which Dreyer (1972, pp. 210—215) installed his "new stress meter" to determine absolute stresses. The strain readings presented in four figures just confirm what was known from measurements employing hydraulic cells (Baar, 1959a, 1966a), see the following sub-section: after approximately 32 weeks, the initial disturbance due to the presence of the borehole was fully compensated by creep in the two radial measuring directions, i.e., the stress-relief creep zone around the borehole was eliminated, and "force in-situ equilibrium" (Dreyer, l.c., p. 214) was obtained; this is the equilibrium pressure in hydraulic cells referred to in sub-section 3.4.3; the time requirement for creep to establish such equilibrium depends on the delay time between drilling and installing the measuring device—see the next sub-section (on hydraulic measurements) for further details.

However, in axial direction, i.e., towards the pillar surface, creep continued after 32 weeks at virtually constant rates which decrease with increasing depths into the pillar; this also is a well-known fact established by deformation measurements—see Chapter 4.

To summarize the strain measurements with salt plugs referred to: known facts were confirmed, but the problem of converting measured strains into stresses was not resolved—to the contrary: as elasticity formulae were used to calculate stresses from apparent plastic deformations, the results cannot be correct; on the basis of stress values obtained in the laboratory using the "compensation method", the vertical and horizontal compressive stresses near the pillar surface are given as 213 and 188 kp/cm^2, respectively. These values certainly are in agreement with other values reported by the same author, see sub-section 3.3.1; however, "elastic analyses are bound to yield erroneous results" (Dreyer, 1972, p. 124), and it must be emphasized again that the quoted values of compressive stresses near the pillar surface cannot be reconciled with the known very low creep limits of salt rocks in situ, and with the measured creep rates in axial direction which became virtually constant at all locations up to 9 m into the pillar.

The calculated stresses at 9 m distance from the pillar surface are given as 268 and 225 kp/cm^2, respectively, for the vertical stress and the horizontal stress normal to the pillar surface. "Generally, it was observed that the pillar stress increased systematically towards the centre of the pillar" (Dreyer, 1972, p. 215). The quoted absolute stress values should be disregarded as apparently erroneous; the statement confirms previous results obtained by other investigators (Baar, 1959b, 1966a, 1972a,b; Mraz, 1973) who found regular stress gradients in situ, the stresses increasing from free surfaces to the centres of pillars, or to the boundaries of creep zones around isolated openings. The general statement is welcomed for it confirms regular stress gradients; it should be realized that this statement clearly contradicts the hypotheses which call for stress peaks near the surface of underground openings in salt rocks (high circumferential stresses or stress envelopes; see sub-section 3.3.1 for similar hypotheses promoted by other authors).

(4) "Calibrated stress meters". Various meters, including photo-elastic devices, have been developed to measure the loads applied to the metering instrument; after calibration in the laboratory by applying loads which are known, such instruments can be used in situ, e.g., in boreholes, for measurements of the loads applied to the instrument because of deformations of the rocks around the instrument.

These loads usually cannot be related to stresses in kp/cm^2 existing in the surrounding rock for the simple reason that the volume of rock contributing to measured loads is unknown, and cannot be determined. To elucidate these relationships by simple examples: the load on a single prop set underground can be measured, and so can the load acting on the bearing plate of a rock bolt; however, these loads have little, if any, relation to the stresses within the supported rock mass: the prop or the bolt may just support the dead-weight of a piece of rock completely detached from the rock mass; or these supports may be loaded by creep deformation in the surrounding rock mass to the point at which the load exceeds the supporting capacity of the element which fails, either due to excessive compressive load, or due to excessive tension in the case of the rock bolt.

These relationships cause difficulties in the evaluation of loads measured by such calibrated meters in rocks which exhibit elastic behavior; they definitely exclude the possibility of obtaining reliable stress values for rocks which deform inelastically by creep, particularly salt rocks. Actually, this is the reason given by Dreyer (1972, p. 205) for the development of the "salt plug", see quotation on p. 00, and for his criticism on stiff inclusions for stress measurements in salt rocks (l.c., pp. 230—232); referring to the borehole stress meter of Potts (1957), it is stated that "the use of Pott's stiff inclusion technique is problematic in that the properties of the inclusion may be fully reflected in the gauge readings".

It is puzzling to read the same author's appraisal of another calibrated stiff inclusion (l.c., p. 231): "The friction stress meter of Erasmus (1958) offers an equally simple means to monitor absolute stresses even over extended periods of time." "Concentration of stress with time near the free faces of a pillar, which had just been blocked out in a salt mine at a depth of 800 m", is shown in a diagram (l.c., p. 232); at the time zero, the stress also is zero; within about 20 months, "the stress finally approaches 212 kp/cm^2.

If the stress values shown in the diagram were correct, they would confirm rapid stress-relief creep after excavation, as the installation of the device requires some time during which the original stresses were lowered to zero as shown in the diagram; there is no possibility of raising such zero stresses to final values of over 200 kp/cm^2 in case the strain readings obtained with salt plugs (l.c., fig. 56, p. 212) are correct: the strain rates over the time period of 20 weeks, and beyond, are virtually constant, and nearly equal in both the vertical direction and normal to the free face. Such creep at virtually constant rates definitely rules out the possibility that zero stresses could have been raised to 212 kp/cm^2; strain-hardening as invoked from laboratory testing is ruled out by the constant creep rates as well as by the inevitable effects on the salt around the stiff inclusions that should have taken place, preventing the loading of the inclusions.

(5) Ultrasonic stress determination. The "very elegant technique" of measuring ultrasonic velocity profiles of mine pillars for stress determination (Borchert and Muir, 1964, p. 271) apparently did not live up to the expectations expressed in numerous publications (e.g., Buchheim, 1958; Dreyer, 1967, p. 78; Gimm, 1968, pp. 211, 212). It is not mentioned in Dreyer's (1972, pp. 193—240) review of methods of stress measurement in-situ.

Like other rock parameters which can be measured, "the measured velocity requires some standard which will enable it to be converted to the actual stress present in pillars" (Borchert and Muir, 1964, p. 269). Some of the problems in calibration of in-situ measurements of velocities were outlined when the first results obtained in salt and potash pillars were published (Baar, 1959b, pp. 134—138, discussion to Buchheim, 1958). The main problem apparently is caused by stress-relief creep around the boreholes which are required for the positioning of the instruments; as stress-relief creep changes the stress field rapidly, corresponding changes to the in-situ sound velocities occur (Baar, 1959b); these and other calibration problems apparently could not be resolved.

The in-situ experiment described by Fig. 3-12 drastically demonstrated the effects of stress-relief creep around boreholes; this experience may have contributed to Dreyer's decision to drop the "very elegant method" of ultrasonic measurements from his list of methods of stress measurement in salt rocks, and to include what is called the pressure-bag method (Dreyer, 1972, pp. 215—218) discussed in detail in the following sub-section.

3.8.2 Hydraulic measurements

The unique behavior of salt rocks in situ, particularly their low creep limits and their fast response to any differential stress exceeding the creep limits, enable the application of measuring methods which are not feasible in most other rocks. Basic principles of plasticity can be utilized to measure the local stress field at any point that can be reached by a borehole: the low creep limits of salt rocks do not allow any local stress difference exceeding the elastic limit to persist beyond the time period required to eliminate excess stress differences by creep.

Equations of plasticity under simplified and idealized conditions may be found in pertinent textbooks. The following quotation from Jaeger (1962, 1969, chapter 29) states the basic principle utilized in hydraulic stress measurements in salt rocks: "The maximum shear stress at every point in a region behaving plastically must have the same value of $\frac{1}{2}s_0$ (s_0 is defined as the difference between the maximum and the minimum principal stress). In most problems in which a body is subjected to prescribed stresses, it will be divided into elastic regions in which the magnitude of the maximum shear stress is less than $\frac{1}{2}s_0$ and in plastic regions in which it is equal to $\frac{1}{2}s_0$."

Applying these principles to salt rocks around underground openings created by Man, considering the proof of plastic behavior provided by creep at constant rates into such openings as measured at numerous locations, and assuming the creep limit of 10 kp/cm^2 valid for a specific salt rock, it follows that the maximum possible stress difference at any point in this specific salt rock is 20 kp/cm^2.

The questions to be answered when the feasibility of hydraulic stress measurements in situ was investigated (1955—1957), were the following:

(1) How much time is required for creep to re-establish the original stress field after it had been disturbed by drilling a borehole to the intended distance from an existing underground opening; (2) does strain-hardening due to creep raise the creep limit as observed in the laboratory.

The answer to the second question is "no" as shown in sub-section 3.4.3 and Fig. 3-7.

There is no clear-cut answer to the first question for various reasons which are outlined in the following.

(1) The drilling of a borehole into highly stressed salt rocks reduces the radial stresses at the periphery of the hole to the atmospheric pressure, i.e., to zero for all practical considerations; the surrounding salt rocks respond immediately to such removal of principal stress normal to their surface along the periphery of the hole; in other words, the minimum principal stress at all points of the surface becomes zero. Disregarding the resulting instantaneous elastic reaction, the surrounding salt rocks respond immediately by stress relief creep.

It is shown in detail in Chapter 4 that stress-relief creep rates are characterized by an exponential decay. This means: the more time is allowed for stress-relief creep to reduce the stresses in the stress-relief creep zone, the more time is required for creep to re-establish what was termed "equilibrium pressure" in sub-section 3.4.3. The equilibrium pressure built up in a closed hydraulic cell is the pressure which *terminates* the creep of the salt rock around the cell, as demonstrated in Fig. 3-7.

Contrary to Dreyer (1972, pp. 215—218, figs. 60 and 61), who postulates "asymptotical attenuation" at an "asymptotic value" given in kp/cm^2, it must be emphasized that the equilibrium pressures built up in hydraulic cells are defined final values which terminate the creep according to the following conditions:

(a) If an established equilibrium pressure is lowered, creep of the surrounding salt is terminated when the re-established equilibrium pressure reaches the minimum principal stress acting in the surrounding rock.

(b) If the same equilibrium pressure is increased from outside, the creep of the surrounding rock is terminated when the acting maximum principal stress is equalized in the hydraulic cell.

This is demonstrated in Fig. 3-7 which also shows that relatively little time is required for creep to re-establish equilibrium pressures, provided that changes to the equilibrium pressures are performed quickly to prevent major creep deformation around the cell.

It may be mentioned that the purpose of the measurements referred to was not to measure precisely the absolute maximum and minimum stresses around mine openings, and in mine pillars, but to obtain reliable approximate values of loads supported by pillars. According to hypotheses widely accepted at the time of the measurements, and still promoted presently by some theoreticians, stress peaks exceeding several times the original stresses were to be found near underground openings in salt rocks, particularly in mine pillars; this meant vertical stresses up to approximately 600 kp/cm^2, as still postulated by Dreyer (1972, pp. 359, 360, and fig. 109) for the case referred to. For this reason, the pressure gauges used during these investigations had to cover the whole range from zero to 1000 kp/cm^2; more precise instruments for the measurements reported in Fig. 3-7 were not available on short notice, as would have been advantageous; however, there was no time left as the mine was lost by flooding even before the planned measurements could be carried through with available gauges and recorders.

(2) Fig. 3-13 shows the build-up of equilibrium pressures in two hydraulic cells at depths of 4 and 5.5 m, respectively, installed in the wall of an entry in virgin ground at the depth of 820 m. The dash connections between the solid lines of pressure readings indicate intentional changes to the cell pressures; such changes were made to check the reaction of the cell pressure, and to aid the build-up of equilibrium pressures by increasing the cell pressures above the expected equilibrium pressures. The latter procedure proved extremely helpful in combination with the reduction of the time period

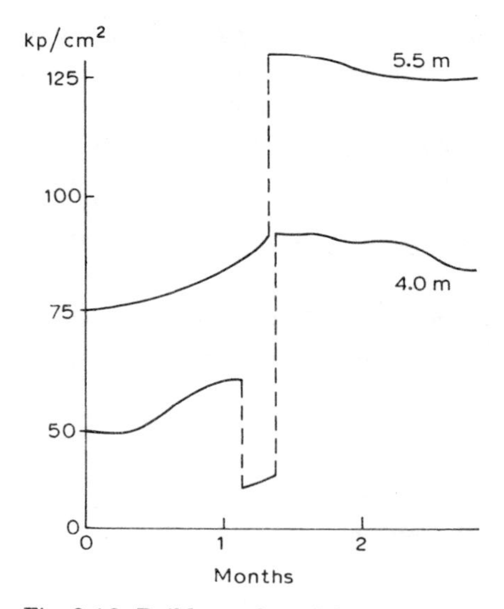

Fig. 3-13. Build-up of equilibrium pressures in hydraulic cells (Baar, 1966a).
The dashed lines indicate intentional changes to the cell pressures. Conditions as in Fig. 3-6.

between drilling and establishing the hydraulic cell in the deepest part (30 cm) of slightly inclined boreholes.

Some of the procedures adopted after extensive experimentation may be outlined shortly; further details can be found in previous publications (Baar, 1959b, 1961b; particularly 1972b).

(a) Immediately after drilling, the needed quantity of hydraulic oil was filled in, and pressurized immediately to pressures which could be taken by packer-type preliminary seals without slipping; the borehole then was sealed permanently over lengths deemed necessary. The details of the design at the deepest part of the borehole, and outside the borehole, are exactly those shown by Dreyer (1972, fig. 60); however, the boreholes were not filled with "fillmass" and cement to the mouth, as such extensive sealing proved unnecessary and impractical. Since neither Borchert (1959) nor Dreyer (1967) eliminated their identical incorrect description of this method in the respective English translations of their original text (Borchert and Muir, 1964, p. 268; Dreyer, 1972, p. 218), it is emphasized again that the use of gas in such measuring systems was never considered or attempted; the obvious reason is that gas would have required considerable compression by borehole convergence before any reliable indication by pressure gauges outside the borehole would have been achieved; it is not known how the identical references to considerable difficulties in the construction of really gas-tight seals originated. It appears typical that these writers, although repeatedly notified of their error, just didn't bother, but continued publishing their erroneous descriptions which lack common sense: in the first place, *if* gas should be used in such pressure cells, there would be no reason for not using just air; after all, air is a gas mixture that follows the same gas laws, and it is readily available everywhere in a mine. Secondly, without pressurizing an isolated gas inclusion from outside for many times, it would take decades or centuries for the salt around the inclusion to bring the gas pressure close to an "asymptotic value" as shown by Dreyer (1972, fig. 61).

Both Dreyer and Borchert knew at least since 1956 — see sub-section 3.4.3 — that hydraulic oil was being used in numerous pressure cells in which equilibrium pressures had been established; however, the conclusions were rejected, and extended testing for expected strain-hardening effects — see Fig. 3-7 — was requested. The identical references (Borchert, 1959; Borchert and Muir, 1964, p. 268; Dreyer, 1967, p. 77; 1972, p. 218) to a publication by Baar (1952b) also characterizes rather questionable methods used by some writers: the senseless idea of using gases in pressure cells for the determination of rock stresses is attributed to a writer who merely stated the facts: in natural fluid inclusions in salt, the pressure equals the lithostatic stress which is hydrostatic in undisturbed salt deposits. This applies to any fluid inclusion, be it gas, oil, or brine.

The incorrect reference to a 1952 publication becomes conspicuous indeed when, exactly 20 years later, another publication appears in which the author reports on his "efforts made to develop a technique to measure the magnitude of a hydrostatic pressure" in sealed borehole sections (Dreyer, 1972, p. 215), using "a fluid" identified as "hydraulic" rather than following the senseless idea of using gases, as allegedly proposed twenty years ago in a publication quoted again (l.c., p. 218). The idea behind such continued incorrect references appears rather obvious in view of the claim that, by using a hydraulic fluid, an "absolute stress" value was obtained; this value, although not asymptotic, but finally constant as shown in Figs. 3-7 and 3-13, is then used to arrive at "almost hydrostatic" stress values—exactly as described by the author who allegedly used gas which wouldn't work.

It is difficult to understand the apparent expectation of a writer to get away with such methods in a publication designed as a textbook; everybody interested in stress measurements in-situ has the possibility of checking the publications referred to, and everybody responsible for correct mine design will do so in view of the published controversies which have been going on for decades because of the disastrous consequences of application of erroneous hypotheses.

(b) It is disadvantageous to fill boreholes over their whole length with cement, and particularly to place a "cement stopper" at the mouth of the borehole as shown by Dreyer (1972, p. 216, fig. 60). The reasons are as follows. The axial expansion of the salt around the cemented borehole is considerable in cases of high original rock stress, and in cases of re-loading of pillars after initial stress-relief creep. The reader is referred to published measurement results (Baar, 1966a) some of which are included in Chapter 4. The forces developed by creep in cases where elastic elements such as steel rods in boreholes tend to prevent creep of the salt, are enormous: loads up to 20 metric tons developed quickly; within few days, steel rods were torn apart; in another case, heavy reloading of a pillar raised the load to 70 metric tons on a stronger steel rod that also failed in tension (Baar, l.c., p. 25). Evidently, a "cement stopper" at the mouth of a borehole is bound to produce similar effects on the high-pressure tubing when lateral expansion occurs: it will pull the instrumentation apart; this is the reason why the length of the seal at the pressure cell was kept small in the investigations reported above. The initial difficulties in constructing tight seals of limited lengths were resolved by adding suitable chemicals to the cement mixture; at that time, high-quality plastic resins, which are now commercially available for such sealing jobs, were not yet available.

There is another important reason which calls for seals of limited lengths only: at the locations where hydraulic stress determinations are of practical interest for the assessment of the loads supported by pillars, heavy slabbing off the pillar surface is common; when such slabs separate along shearing cracks, an open borehole provides time until the shearing displacement is bound to destroy the free floating high-pressure tubing. A cemented hole does not, as the shearing forces are immediately transferred to the tubing.

For these reasons, it is definitely advisable not to cement boreholes for hydraulic stress determinations over their whole lengths; advantage should be taken of the well-established fact that the lengthening of unit lengths normal to the surface of pillars decreases with increasing depths into pillars, as also confirmed by the above-mentioned measurements with salt plugs.

(c) In cases where it appears doubtful that a tight pressure cell can be established by simply packing off a section of a borehole, as Sellers (1970, p. 241) put it with reference to Baar (1966a), any type of hydraulic pressure cell "made along the lines of the cylindrical pressure cell developed by the U.S.B.M. for the measurement of the in-situ modulus of rigidity" can be used. It makes no difference whether copper or another suitable material such as rubber is used to build the cell. As such cells can be manufactured at little cost, and can save lots of troubles in sealing boreholes absolutely tight, their use is advisable, particularly near the surface of pillars where salt rocks may have lost their impermeability.

Measurement results obtained with U.S.B.M. copper cells are reported by Baar (1972b) and by Mraz (1973); the results "confirmed earlier conclusions (Baar, 1966a): in the creep zone around openings in salt rocks, local near-hydrostatic stresses develop which increase from the opening surface to the boundary of the creep zone". W. Dreyer (personal communication, 5.11. 1970) fully agreed to the quoted conclusion that a hydraulic cell, which is not too long, measures the quasi-hydrostatic pressure at its location; with increasing depths into the massive salt rock, the measured values increase regularly until the original pressure according to the depth below surface is reached. If the hydraulic cell is too long relative to the depth of the creep zone, it indicates an average value as it can indicate only one constant pressure. Dreyer emphasized that there is absolutely no disagreement with the quoted conclusion. It is left to everybody's guess why complete agreement is expressed in personal written communications, while complete disagreement based on misleading erroneous reference to gas-filled pressure cells is subsequently published, announcing the development of a new hydraulic pressure cell which happens to be identical with one described in various previous publications which are not even mentioned. As stated previously: such short-sighted activities are difficult to understand as unconditional belief in quite contradictory publications should not be expected from engineers who sign responsible for the adequate design of underground operations in salt deposits.

Chapter 4

DEFORMATIONS IN RESPONSE TO STRESS RELIEF BY EXCAVATION

4.1 GENERAL REMARKS—OUTLINE OF PROBLEMS

The pressure reductions in hydraulic cells shown in Fig. 3-7, and dealt with in more detail in sub-section 3.8.2, resemble excavation processes: the excavation of underground openings changes the existing stress field by the removal of rock which previously had provided for a balanced stress field of nearly equal principal stresses. At every point at the periphery of a new excavation, the minimum principal stress becomes zero for all practical considerations in conventional mines; in solution mining operations, the minimum principal stress is reduced to about one half of the original stress as the hydrostatic pressure of the brine is roughly one half of the lithostatic pressure originally existing in the salt deposit. In gas storage caverns which are being operated "dry" after removal of the brine, the minimum principal stresses may vary considerably, depending on the maximum and minimum storage pressures.

It must be borne in mind that the very first deformations around any new excavation underground are caused by the reduction of stresses, in contrast to most laboratory tests in which the stresses are increased.

There can be no doubt of instantaneous elastic response of the rock around any new opening to the reduction of the minimum principal stress. Compared to the subsequent plastic deformations by creep of salt rocks, instantaneous elastic reactions are negligible and insignificant, with the exception of cases where gases are occluded in the salt rocks around new openings.

The geological conditions of the formation of salt rocks with occluded gases have been outlined in sub-section 2.2.3; in section 3.5, the low tensile and shearing strengths of salt rocks are shown responsible for brittle fracture of salt with gas inclusions. Such brittle fracture, originating at the free surface, extends in a chain-reaction into the surrounding salt massive.

Outbursts of occluded gases after brittle failure of the occluding salt are of particular importance in some salt and potash mining districts; they require a detailed discussion which is presented in sub-section 4.5.1.

It is re-emphasized that explosion-like failure of carnallite without gas inclusions — see sub-section 3.3.2 — can be produced by sudden stress relief in the laboratory; however, it is not observed in situ, even in cases of mining depths exceeding 1000 m. Referring back to sub-section 3.8.1, the fast response by creep to excavation in salt rocks at great depths is re-emphasized: there is a stress-relief creep zone travelling ahead of advancing faces, reducing the effective stress differences at the face itself sufficiently to prevent brittle fracture of any practical significane; the subsequent elastic response to continuing creep in the creep zone around openings is limited to superficial slabbing of stress-relieved segments which tend to behave elastically, shearing off the ideal "active" section; this is shown in sub-section 4.2.4.

The "ticking noise" which "can be heard during the first few hours" after advancing the face with boring machines at the depth of about 950 m (Coolbaugh, 1967, p. 72), is indicative of superficial damage to the salt rock around openings: individual crystals at the newly cut surface respond elastically to further deformation forced by stress-relief creep at greater distances from the opening; typically, these noises cease after few hours, i.e., after the rapid initial stress relief creep has decelerated as shown in Fig. 4-5.

The buckling of separated salt beds is clearly time-dependent, although it reflects the elastic reaction of such completely stress-relieved salt beds; the time requirement of buckling indicates its cause, which is the continuing creep outside the "active" section of an opening.

Precise measurements of stress-relief creep strains do not require any sophisticated instrumentation; the strains to be measured in situ are relatively large compared to strains in cases of elastic behavior, or to strains observed in the laboratory on small samples and models. As a matter of fact, the less sophisticated the measuring system is, the more reliable are the results; in addition, simple measuring systems allow extensive data acquisition at low costs, eliminating the need for using questionable single strain values obtained in only one series of strain measurements over extended periods of time. Exceptional single values usually indicate some faulty equipment, or just human error; such values must not be used in arriving at averaged values, but must be disregarded completely. In other words: before any strain data obtained in situ is fed into a computer, common sense must be used to eliminate questionable values.

The best way of eliminating questionable single strain values prior to further evaluation is very simple and easy: to plot the measured strains above a time axis to obtain what has been termed "cumulative creep curves" or "cumulative closure curves" (e.g., Serata, 1968, pp. 305, 306, and fig. 16). There is no advantage in calculating percentage strains or closure rates prior to plotting the measured values; however, in further evaluation of measured changes to creep or closure rates, it is of great advantage if the cumulative strain curves are plotted above a time axis with the exact date of the measurements taken. For measurements of stress-relief strains immediately after excavation of openings, the time axis must be chosen such that strains during less than one hour can be plotted at the exact time of measurement.

The above statements are based on a re-evaluation of extensive strain data obtained at the IMC-K 1 potash mine near Esterhazy, Saskatchewan (Baar,

1972c). The measuring system is described in various publications (Barron and Toews, 1964; Zahary, 1965; Coolbaugh, 1967). Measuring data was obtained at 69 sites throughout the mine in 1965, three years after mine development had begun; many more stations were installed when the mine developed further, the data obtained representing the most comprehensive in-situ data collection to date. Similar data was obtained in other Canadian potash mines since they became operational in 1968—1970. The following interpretations of stress-relief deformations in situ are mainly based on the data evaluation by this writer during a research program carried out 1971—1973 (Baar, 1971a,b). Regrettably, some data is still considered confidential; it is possible, however, to derive conclusive evidence from recent publications which include in-situ strain data, trying to fit the data into preconceived hypotheses (Serata, 1972, 1973, 1974; McKinlay, 1972; Serata and Schultz, 1972; Duffield, 1972; Rininsland, 1972; Mackintosh, 1975).

The data collected in other Canadian potash mines confirm, in principle, the conclusions derived from IMC data (Baar, 1972c; Mraz, 1973); however, the different geological conditions in the Saskatoon district — the potash is bedded rather than homogeneous and isotropic as in the IMC mines — cause some differences in the deformations by stress-relief creep around individual openings.

It is essential to recognize the differences in stress-relief deformations as effective countermeasures must be based on the actual in-situ behavior of the respective salt rocks rather than on preconceived hypotheses based on inadequate laboratory experiments. The only reasonable way of arriving at unbiased interpretations of measured strains appears to be to refine the data by eliminating questionable results as outlined above, and secondly, to obtain meaningful data, i.e., to measure the strains which occur shortly after the excavation of an opening, or during the widening of an opening to its final width or height.

For obvious reasons, standard strain measuring devices — see Sellers (1970) for detailed descriptions of measuring systems suitable in salt mines — cannot be used in single openings for measurements of strains which occur immediately after the opening has been excavated: the installation of the measuring system requires time; so the strains of interest have already occurred when the measuring system becomes operational. There are two possibilities of getting around this problem:

(1) To install the measuring system in an existing opening prior to the excavation of a second opening; the strains caused by the second excavation can be recorded conveniently either by manual measurement or by continuous recording with suitable instrumentation. Such instrumentation is easily constructed as the strains to be measured amount to several centimeters during the first hours after excavation of the second opening: steel wires are mounted at various distances in boreholes drilled from the existing opening; the wires are kept under equal tension by weights and conducted in such a

way over a slowly moving chart that writing hands can be fixed to the wires to record any strains occurring at the anchor point of the respective wire. Numerous records obtained in this way are the basis of the interpretations of stress-relief creep strains at the IMC potash mine presented in the following.

Similar strain measurements were made previously in a German potash mine: the results, reported in various publications (Baar, 1959a,b, 1966a, 1970a) were confirmed by the more comprehensive Canadian measurements and records referred to in this text.

(2) To use convergence recorders which can be set up immediately after an opening has been made, and which are left in place while the opening is cut wider; again, such recorders are constructed easily as the strains to be recorded are rather high for the first hours and days—see Fig. 4-5.

The second measuring system is not suitable, of course, in mines where drilling and blasting is used; however, disregarding the known effects of blasting on initial deformations (Baar, 1970b), the basic pattern of stress-relief strains is similar in all conventional room-and-pillar salt mines.

It is emphasized that the design of stress-relief methods in Canadian potash mines is entirely based on information obtained at the IMC potash mine by extensive measurements using the two measuring systems described above. However, it is also acknowledged that useful data had been obtained incidentally by previous measurements according to the possibility (1) described above (e.g., Zahary, 1965). The correct interpretation of these data was jeopardized by preconceived hypotheses, and by erroneous assumptions of stress arches which make "the amount of strain in the vertical direction greatest at a point several feet above the room surface"; "in the horizontal direction the amount of movement decreases as the distance from the surface increases. However, in the vertical direction, the maximum movement does not occur until a point is reached 6 or 7 ft. (1.8 or 2.1 m, opening width 6.3 m) above the opening. This corresponds exactly with the stress arch observed in the picture of the photo-elastic plastic attached to a laboratory specimen" (Coolbaugh, 1967, p. 72, fig. 9; also Zahary, 1965, fig. 14).

The above quotation refers to photo-elastic analyses reported by Serata (1966, pp. 10—12, figs. 16—18; 1970, pp. 267—270, figs. 25, 26).

The postulated stress envelope at the above quoted distance into the roof, and at much smaller distances into the walls, is still shown in publications, e.g. Serata (1974, fig. 2). At the same time and in the same publication, the same author also claims that he invented the "Stress Relief Method" which takes advantage of the large extent of stress-relief creep zones (also Serata and Schultz, 1972; Serata, 1973). Such self-contradictory publications should be taken for what they are; i.e., commercial advertisements to promote methods which are just copied, but not understood. Such copying has repeatedly resulted in disaster.

Regrettably, such advertising articles are accepted by scientific publishers and/or editors; as a result, some inexperienced writers have been using rather questionable methods of manipulation of in-sity data in their attempts to prove certain hypotheses correct. This has been shown for certain cases in previous publications (e.g., Baar, 1970a).

In view of the serious consequences of practical applications of hypotheses, such as the stress arch and stress envelope concepts, in salt and potash mining, the significance of rapid and extensive stress-relief creep into newly excavated underground openings in salt deposits is re-emphasized. There is definitely no need for more disasters such as those caused in June 1975 by potash mining in Germany; however, as similar catastrophes have been caused repeatedly in the past by application of erroneous mine design principles (Baar, 1966a, 1973, 1974b), they are bound to occur again unless erroneous hypotheses are eliminated from practical application. For these reasons, the significance of stress-relief creep in salt deposits is given particular attention in the following.

It is emphasized that stress-relief creep occurs in salt rocks, but not in other evaporites, such as anhydrite beds, which often are found near underground openings in salt deposits. In cases where stress-relief creep zones extend to the boundary of interbedded formations which exhibit elastic behavior, corresponding changes to the following statements regarding stress-relief creep are necessary. This is shown in detail in Chapter 5.

4.2 STRESS-RELIEF CREEP AROUND SINGLE, ISOLATED OPENINGS

4.2.1 Rapid decay of initial stress-relief creep rates

The rapid decrease of creep rates immediately after the excavation of underground openings in salt rocks has been known for some time. In earlier publications (Kampf-Emden, 1956; Höfer, 1958a,b), such decelerating creep is related to strain hardening; this explanation is still given in most recent publications dealt with in section 3.4.

Contradictory evidence was published by Wilkening (1958, 1959) and by Baar (1959b, 1966a, 1970a). Some rather puzzling results of in-situ measurements are explained by the apparent rapid decrease of creep rates during the first hours or days after the excavation of an opening; for example, the closure measurements published by Zahary (1965, fig. 6) and by Coolbaugh (1967, fig. 5) show the vertical convergence of an opening (width 6.3 m, height 2.25 m) smaller than the horizontal one, see Fig. 4-1. The opposite relation would be expected from the opening dimensions. However, as shown by Baar (1970a, fig. 2), the vertical stress relief convergence decreased rapidly over the first four days during which time no measurements were taken; several centimetres of vertical convergence are missing in the respective figures. Apparently, stress-relief creep during the first four days had resulted in nearly complete elimination of stress differences in the immediate

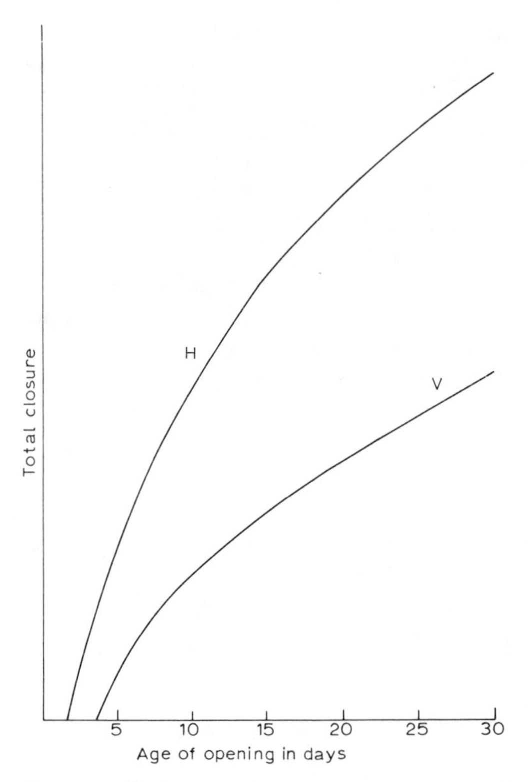

Fig. 4-1. Closure of the first test room at IMC. (After Zahary, 1965, and Coolbaugh, 1967.)

Measurements commenced two days after excavation in the horizontal direction, and four days after excavation in the vertical direction. Reference points anchored in boreholes 0.5 ft. (0.15 m) deep. Opening dimensions given in Table 4-I.

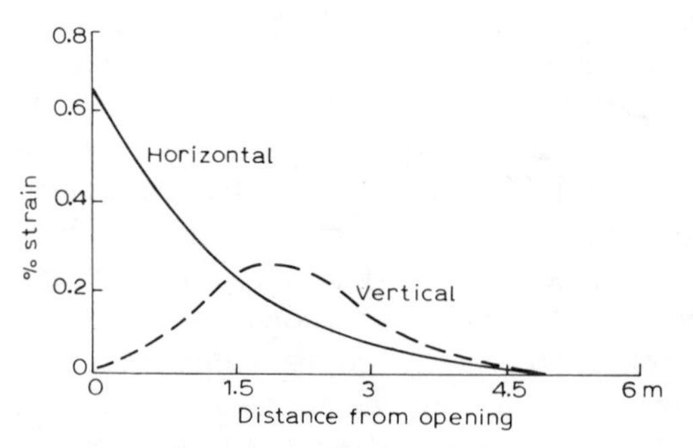

Fig. 4-2. Comparison of calculated strains around the first test room at IMC. (After Coolbaugh, 1967.)

Strains calculated from the data shown in Fig. 4-4.

roof and floor of the openings; there, the rock was behaving elastically during the subsequent measurements. For this reason, the measured strain in the vertical direction was greatest at the distance of about 2 m from the room as shown in Zahary's (1965) detailed evaluation. Fig. 4-2 shows the strains (converted to per cent strain), measured over the time period of 200 days after excavation, in sections at increasing depths from the opening (Coolbaugh, 1967, fig. 9); the lengths of the reference sections reflect the distances at which the respective measuring bolts were anchored; these distances are 0.5, 5, 10, and 20 ft. (0.15, 1.5, 3, and 6 m) from the opening in both the vertical direction and the horizontal direction.

Disregard the questionable procedure of calculating the average strains for individual sections—this is definitely incorrect in cases where bed separations or shear cracks develop: the total movement between two reference points, or any percentage of it, may be caused by movement at the point of separation, while no or little strain occurs in the rest of the section. In the case of Fig. 4-2, the calculation of average strains may be justified by the homogeneous structure of the salt rocks involved in creep: at the centreline of the opening, shear or tension cracks develop only during the very last stage of separation of roof slabs, see Fig. 4-3 (Mraz, 1973, fig. 13).

The development of shear or tension cracks at the centreline of the opening shown in Figs. 4-1 and 4-2 can be safely excluded—after 8 years of continued measurements, there was no visible indication of such cracks which start developing along the walls. The conclusion that several centimetres of vertical convergence occurred during the first four days, appears inescapable. Fig. 4-2 clearly indicates where the missing creep strains occurred: beyond the distance of about 2 m from the opening, the measured vertical strains are greater than those of corresponding horizontal sections; obviously, the mis-

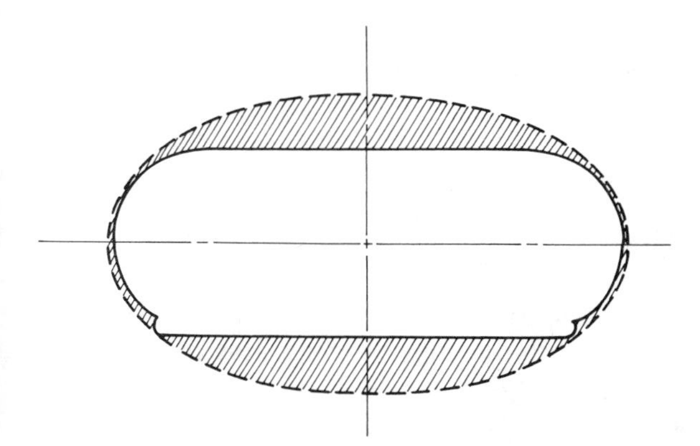

Fig. 4-3. "Active" cross-section of openings at IMC (Mraz, 1973, after Baar, 1970b).
The indicated segments inside the elliptical section are eventually sheared off as they exhibit elastic behavior after stress-relief creep.

sing creep strains occurred within the first two metres, i.e., in the segments identified in Fig. 4-3: these segments show hardly any strain during the period of measurements. As Fig. 4-2 shows the total convergences measured between reference points in roof and floor, one half of the missing creep strains is attributable to the roof segment, the other half occurred in the floor segment.

The inconsistency in the evaluation of the horizontal movement shown in Fig. 4-2 (Zahary, 1965; Coolbaugh, 1967) becomes apparent from a comparison with the quoted statement regarding the postulated stress arch at the depth of about 2 m into the roof: if maximum strains were indicative of maximum stresses — Zahary (1965, p. 7) emphasizes "the consistency of loading and deformation" with reference to Fig. 4-1 — the maximum stresses in the walls, i.e., the envelope of increased stresses, would be located at the walls; the stresses would decrease with increasing depths into the walls. As any plastic deformation must take place in the direction away from stress concentrations, a stress envelope at the walls would cause shortening of sections deeper in the walls.

Figs. 4-1 and 4-2 show that such shortening did not occur. To the contrary, the strains shown in these Figures demonstrate very regular decreases for the horizontal strains from the surface to the point of zero strain, and for the vertical strains from about 2 m distance to the point of zero strain. Such regular decreases in strains necessitate regular stress gradients; in addition, according to the basic principle of plastic behavior quoted in sub-section 3.8.2, the maximum possible stress difference, that could persist around the opening, equals the creep limit of the salt rocks whose expansion in the stress-relief creep zone caused the measured strains and opening convergences shown in Figs. 4-1 and 4-2.

The stress gradients in the walls — apparently regular gradients — can be calculated if the horizontal extent of the stress-relief creep zone is known; this also holds for the vertical stress gradients beyond the 2-m distances in roof and floor which apparently indicate the inner boundary of the creep zone in the vertical direction.

Unexpected creep occurred outside the area covered by the measuring system in both the horizontal and the vertical directions. However, the regularity of the measured strains made it possible to extrapolate the distances to the outer boundaries of the creep zones as shown by Zahary (1965).

The development and the extent of stress-relief creep zones is dealt with in more detail in the following sub-section 4.2.2. Here, it appears indicated to show the further development of the convergence creep shown in Fig. 4-1 for the first 30 days after mining the opening.

Fig. 4-4 (Zahary, 1965, fig. 7) demonstrates "both the capability of the instrumentation and the regularity of the creep deformation for the first 450 days after mining". However, some contradictory interpretations by Zahary (1965, p. 7) and by Coolbaugh (1967), see Fig. 3-7, must be corrected.

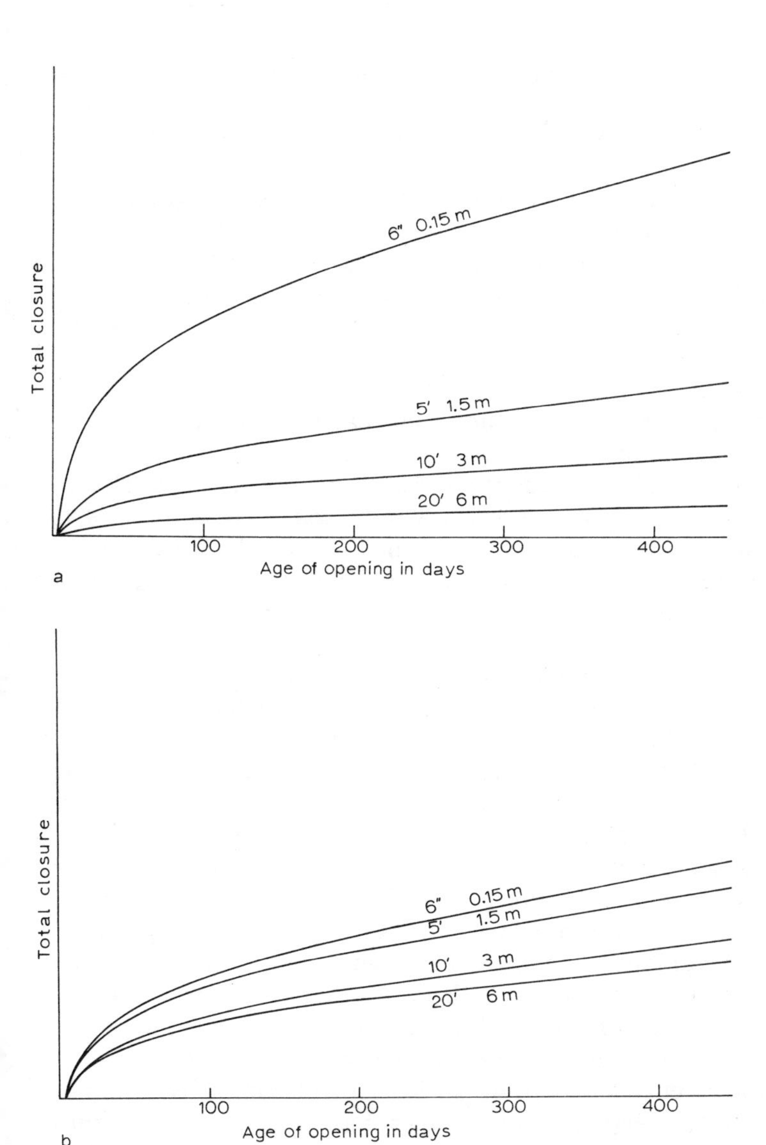

Fig. 4-4. Closure of the first test room at IMC, measured between points at the indicated distances from the opening. (After Zahary, 1965.)
a. Horizontal convergences; b. vertical convergences.

Zahary stated that "the rate of closure (horizontal convergence measured between the anchor points at the depths indicated in the figure) decreases with the distance into the wall and with the increasing age of the opening". The latter statement is then contradicted by saying that "15 ft. (4.5 m) from the edge of the opening, the strain rate is constant over the full 450-day period".

The vertical convergences as shown in Fig. 4-4,b exhibit a similar, very regular development over the 450-day period; however, the measurements on the deepest pins (20 ft., 6 m) show that about one half of the total vertical convergence resulted from creep beyond these reference points, comparing with only negligible creep beyond the corresponding points in the horizontal direction, see Fig. 4.4. This is an extremely important observation: it demonstrates that the depth into the massive salt rock to which stress-relief creep occurs, depends on the dimensions of the area from which the original stresses are removed by excavation.

As the evaluations of the convergence data by Zahary (1965) and Coolbaugh (1967) are given much attention in recent attempts to prove the "asymptotic" decrease of convergence rates to ultimately zero (Dreyer, 1974a, pp. 127, 131—132, 162), it appears indicated to re-emphasize the actual development shown by Zahary's published figures; the following quotations from Baar (1972c, pp. 36—39) may serve this purpose. They refer to convergence measurements at the IMC potash mine to which the other quoted publications also refer.

"In order to learn more about the amount of initial vertical relief creep, and its time requirement, closure recorders were constructed which can be installed immediately after an opening is cut. Some recorded figures may demonstrate the rapidness of initial vertical stress relief creep. Within the first few hours, the vertical relief creep into 14 ft. (4.2 m) wide openings 7.5—8 ft. (2.25—2.4 m) high amounts to 1 inch (2.5 cm) or more, the creep rates decreasing continuously."

"Each step of widening an opening by cutting additional 7 ft. (2.1 m) wide passes is reflected by an immediate increase in vertical creep into the original opening. The amount of immediate creep is less at each subsequent pass. Total initial vertical closures up to 3 inches (7.5 cm) were recorded in openings widened to 35 ft. (10.5 m). After any increase in creep caused by widening, the creep rate decreases, the cumulative creep curves tending to develop into 'normal' curves according to local conditions. Near the faces and walls of openings, creep rates are smaller, indicating the restriction to creep caused by non-excavated ground. These measurements demonstrate how rapidly the initial vertical stress relief creep extends into roof and floor."

"The development of the horizontal creep into an opening differs greatly. Due to less opening height compared to width, and due to the circular cross-section of the opening at the walls, the initial horizontal creep is much more restricted. However, the rapid initial vertical creep partially removes the restrictions, lowering the horizontal stresses above and below the opening, and enabling horizontal creep above and below the opening walls into the stress relieved zone. Such horizontal creep, in turn, facilitates the horizontal creep in the walls."

TABLE 4-I

Time between mining and beginning of creep at constant rates at various distances from an opening in sylvinitic potash; depth 3140 ft. (approx. 950 m); opening width 21 ft. (6.3 m), height 7.5 ft. (2.25 m)

Distance from opening		Time (days)	
(feet)	(m)	horizontal	vertical
20	6	90	150
10	3	100	175
5	1.5	150	200
0.5	0.15	200	225

"After the initial rapid vertical relief creep, the further development is dominated by the tendency to establish more uniform stress conditions around the opening by horizontal creep. Horizontal creep is converted into vertical creep above and below the opening. These interrelationships are indicated by the different courses of total closure curves. Vertical cumulative closure curves demonstrate the highest initial creep rates. After a few days, they flatten rather sharply, and the vertical creep rates become smaller than the horizontal ones. Horizontal cumulative creep curves develop smoothly over a period of weeks into straight lines (see Baar, 1970b, fig. 2)."

"The interrelationships between horizontal and vertical creep are also indicated by the different times at which the creep rates become fairly constant at various locations around the opening. In Table 4-I, it can be seen that con-

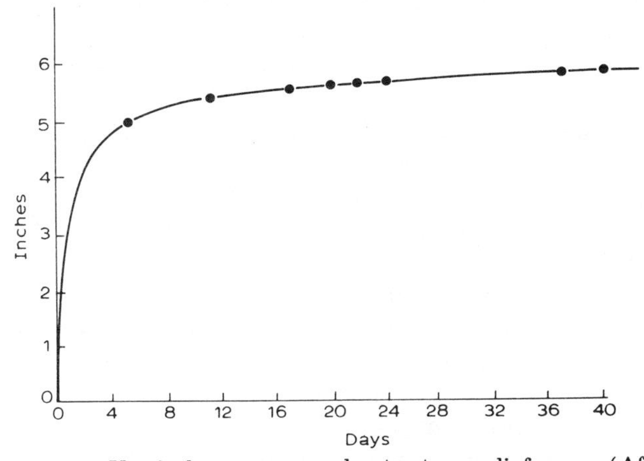

Fig. 4-5. Vertical convergence due to stress-relief creep. (After Mraz, 1973.)

Convergences recorded at IMC during and after the widening of the first cut (4.2 m) by subsequent cutting to the final opening width of 18.3 m.

stant creep rates first develop at points along the horizontal axis of the zone of creep, beginning at 20 ft. (6 m) and proceeding towards the opening. At points along the vertical axis, the creep rates develop correspondingly, but with a time delay of several weeks."

Fig. 4-5 (Mraz, 1973, fig. 14) is an example of vertical closure recorded in the first cut of an opening during its widening by subsequent passes of the borer to the final width of 61 ft. (18.3 m). It should be noticed that the vertical convergence had exceeded the amount of 10 cm within 4 days, indicating the considerable effect of stress-relief creep in such a relatively wide room; after only 3 weeks, the convergence rate became virtually constant as the cumulative closure curve became a straight line.

The further development of vertical convergence rates in response to reloading, if and when overburden subsidence is initiated, is dealt with in Chapter 5.

4.2.2 The development of stress-relief creep zones around single openings

Earlier attempts at measuring the development of stress-relief creep zones around single, isolated openings in salt rocks were jeopardized by the unexpected rapidness of stress-relief creep; as a matter of fact, the first known program to establish the development of stress-relief creep zones in situ by measurement (Wilkening 1958, 1959; Baar 1959b, 1966a) had to be abandoned as the mine in which the comprehensive program was in progress, was lost by flooding—due to extensive stress-relief deformations around an isolated mined-out area, see sub-section 5.3.2.

Nevertheless, the basic principles of stress-relief creep into single roadways with "standard" dimensions — see Fig. 3-6 — were established: the extent to which stress-relief creep occurs in the horizontal direction, was measured in openings approximately 5 m wide and 2.5 m high; the near-hydrostatic stress field around such openings was determined by hydraulic measurements described in sub-section 3.8.2.

The measuring systems which were used extensively at the IMC potash mine — see Fig. 4-4 for results of convergence measurements — made precise extrapolations of the horizontal extent of stress-relief creep possible: as the opening convergence itself, and the convergence between points at given distances from the opening are measured, the expansion by creep between the points at the greatest distance is known; in relation to the measured opening convergence, the measured expansion indicates the depth beyond the deepest points affected by stress-relief creep.

After it had been realized that the original systems with 20-ft. (6-m) pins did not cover the total extent of horizontal creep, 30-ft. (9-m) pins were used to establish precisely the horizontal extent of creep around isolated single openings. These measurements confirmed the reliability of extrapolations, and they also confirmed the horizontal extent of stress-relief creep shown in Fig. 3-6 for comparable conditions.

With respect to vertical creep, extrapolations cannot be made with similar precision because the amount of vertical convergence attributable to expansion between the deepest measuring points is usually only a small fraction of the total opening convergence, depending on the width of any particular opening. The wider an opening is made, i.e., the larger the area of complete removal of vertical stresses is made, the deeper into roof and floor stress-relief creep extends. However, exact measurements apparently have not yet been made; such measurements cannot be made in mines only one level of openings, e.g., in Canadian potash mines where the mine development takes place in the potash bed to be extracted. Such measurements require the existence of an opening at a higher or lower level in which the measuring system is to be installed. The wide-spread belief in erroneous hypotheses, which ignore stress-relief creep, has to date prevented the exact measurement of the vertical extension of stress-relief creep, although it is the most important parameter in cases where vertical stress-relief creep may result in gas or water discharge into a mine. This is shown in Volume 2.

The extrapolated dimensions of the stress-relief creep zone around the isolated opening, the closure of which is shown in Fig. 4-4, are listed in Table 4-II. The opening dimensions are given in Table 4-I. For this particular opening section, the measurement data indicate the development of a near-circular section of the stress-relief creep zone over a relatively long period of time; the initial vertical dimension is reduced with time, while the horizontal dimension is increasing slightly. After 2600 days since excavation of the opening, the section of the creep zone had become almost circular.

This development is in agreement with the measured convergence rates which, after few days, become higher in the horizontal direction than in the vertical direction—see Fig. 4-1. Curve *1* in Fig. 5-2 shows the vertical cumulative convergences of the opening referred to; it is emphasized that the mea-

TABLE 4-II

Time-dependent development of the vertical and horizontal dimensions of the stress-relief creep zone around an isolated opening (conditions as in Table 4-I).

Time since mining (days)	Extrapolated dimensions of stress-relief creep zone			
	vertical		horizontal	
	(ft.)	(m)	(ft.)	(m)
450	94.5	28.4	73	21.9
1000	90.5	27.2	76	22.8
2000	88.5	26.6	81	24.3
2600	86.5	25.9	83	25.0

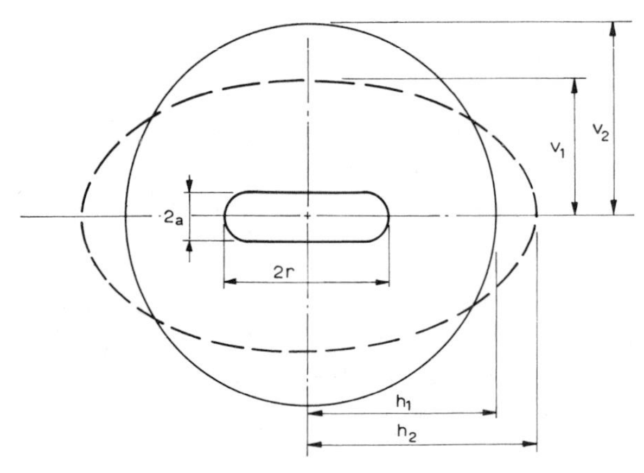

Fig. 4-6. Development of the creep zone around an opening at IMC (Mraz, 1973, after Baar, 1970b).

Measurement data listed in Table 4-III. Note: the initial extent of the stress-relief creep zone is not shown.

surements commenced four days after mining, so the most rapid initial stress-relief convergence is missing as the curve shows the measured convergences.

With reference to Fig. 3-3, and to Table 4-I, it is re-emphasized that both the vertical and the horizontal convergence rates developed into constant rates which remained constant over the following years.

The tendency for the initial section of the stress-relief creep zone around

TABLE 4-III

Time-dependent development of the stress-relief creep zone around an isolated opening in carnallitic salt rocks (carnallite content up to 10%; opening width 28 ft. (8.4 m); other conditions as in Table 4-I)

Time since mining (days)	Dimensions of creep zone			
	vertical		horizontal	
	(ft.)	(m)	(ft.)	(m)
200	108	32.4	90	27.0
500	92	27.6	92	27.6
1000	86	25.8	94	28.2
1500	78	23.4	98	29.4
2000	62 *	18.6	102	30.6

* Creep was restricted to the rock between the anchor points at 30-ft. (9-m) depths into roof and floor.

Note: the creep zone reached a circular cross-section after 500 days, see Fig. 4-6.

an oblong opening — this initial section resembles an ellipse with the longer axis in the vertical direction, see Fig. 3-6 — to develop into a circular section, becomes more apparent with the increase in widths of openings with equal height. This is shown in Fig. 4-6; some of the basic measurement data for this figure are listed in Table 4-III. The opening was excavated in carnallitic salt rocks, with the potash bed as well as roof and floor containing up to 10% carnallite.

Apparently, carnallite as a component of salt rocks facilitates creep deformations, in particular if carnallite occurs as a bond between halite and sylvite crystals as shown in Fig. 2-12. In addition, the measuring site was slightly affected by reloading, see curve *10* in Fig. 5-2. Reloading may have accelerated the development shown in Fig. 4-6, in which the initial ellipse with the longer axis in the vertical direction is not shown.

The development of the elliptical section with the longer axis in the horizontal direction is based on measured data given in Table 4-III. The measured data are in line with the extrapolated data used in the construction of the initial sections of stress-relief creep zones.

In Table 4-IV, some more data are listed to show both the effects of greater opening widths and the effects of carnallite contents on the development of stress relief creep zones; reloading also is involved, so the data should be considered representative of the variability of stress-relief creep zones under various conditions.

It is particularly emphasized that the initial dimensions of the stress-relief creep zones listed in Table 4-IV were over 100 ft. (30 m) in the vertical direction; the changes towards smaller vertical dimensions, with the horizontal dimensions increasing as shown in Tables 4-II and 4-III, may occur rather

TABLE 4-IV

Time-dependent development of vertical axes of creep zones under various conditions of loading (other conditions as in Table 4-I, with the exception of opening widths and potash ore types as indicated)

Room width		Type of salt rock	Vertical dimension of creep zone					
			after 100 days		after 300 days		after 1000 days	
(ft.)	(m)		(ft.)	(m)	(ft.)	(m)	(ft.)	(m)
21	6.3	sylvinitic	105	31.5	90	27.0	80	24.0
21	6.3	carnallitic	70	21.0	50	15.0	45	13.5
28	8.4	sylvinitic	95	28.5	90	27.0	85	25.5
28	8.4	carnallitic	50	15.0	47.5	14.25	45	13.5
35	10.5	sylvinitic	95	28.5	85	25.5	67.5	20.25
35	10.5	carnallitic	105	31.5	55	16.5	not available	

Note: values exceeding 55 ft. (16.5 m) are extrapolated values.

rapidly or relatively slowly, depending on the local conditions at any particular measuring site. These conditions are dealt with in more detail in Chapter 5. Here, the following statement regarding the random values listed in Table 4-IV may be quoted from the original publication (Baar, 1972a, p. 48):

"Attempts to quantitatively relate any particular development of the shape of the creep zone to the influencing factors would be merely speculative as long as the pillars loads at any time, and the mineralogical composition of the rocks involved in creep are not exactly known. It must be stressed that the effects of horizontal creep may be entirely different in bedded salt deposits, in particular if clay or anhydrite layers act as parting planes above or below openings."

4.2.3 Interrelationships between vertical and horizontal stress-relief creep

Before dealing with mine openings in tabular deposits which normally are larger in width than in height, having an oblong cross-section, it is of interest to consider radial convergences of roadways with "ideal" circular cross-sections. Such measurements resemble a welcome link between borehole convergence measurements, and closure measurements in large storage caverns which normally are made with circular cross-sections in horizontal planes.

Since the early 1940's, many thousands of kilometres of roadways were cut with circular sections in German and Russian potash mines; the diameter cut by most of the machines with rotating cutting heads is approximately 3 m. Difficulties caused by rapid stress-relief convergences in long exploration headings were repeatedly reported in publications (e.g., Baar, 1953a, 1971a).

Fig. 4-7 shows measured convergences of a circular roadway in rock salt

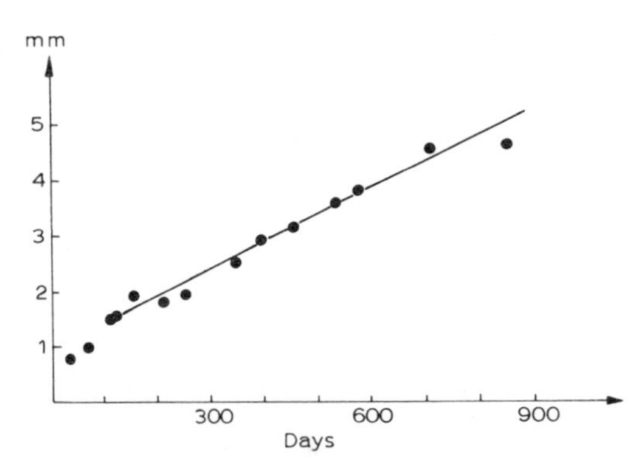

Fig. 4-7. Radial closure of a circular roadway in rock salt. (After Höfer, 1964.)

Roadway drilled at the depth of 1000 m, diameter 3 m. Initial creep closure missing. The measured values are represented by a straight line after 100 days, indicating constant closure rates.

at the depth of about 1000 m (after Höfer, 1964, fig. 9); according to the author, the 3-m diameter roadways were driven in virgin ground hundreds of metres from other mine workings. This classifies the measured convergences as stress-relief creep convergences. The measurements were made between measuring points set vertically, horizontally, and at angles of 45°; the author (l.c., p. 66) emphasizes that "convergences in the four directions of measurement were almost identical; this means that the hypothesis of a hydrostatic state of stress. . . can be regarded as largely assured".

The interpretation of Fig. 4-7 by its author cannot be accepted (l.c., p. 68) because the measurements commenced too late, "with the result that what should have been measured had, in fact, already taken place", i.e., the rapid initial stress-relief creep had decelerated. The creep rate is virtually constant over the time period from 100 to 900 days after mining. Erratic measured values indicate difficulties in exact measurement, see Fig. 3-3, or displacements of individual measuring points; for these reasons, the very last measurement shown at day 850 should be disregarded completely as it shows no convergence for the preceding 150-day period, while all other 150-day periods since day 100 show convergences of approximately 0.75 mm, i.e., constant convergence rates. Typically, the measured convergence at day 100 is about twice as high; in addition, most of the initial stress-relief convergence obviously is missing.

The point to be emphasized with regard to these measurement results is the following: apparently, in openings with a circular cross-section, the creep in radial directions at an angle of 45° to the horizontal direction includes equal amounts of displacements in both the vertical and the horizontal directions. This is a very significant difference compared to the creep around openings with oblong cross-sections: as the vertical stress-relief creep into oblong openings occurs much faster, for some hours to few days, than does the horizontal stress-relief creep during the same periods of time, the rapid vertical development of the stress-relief creep zones allows more horizontal creep above and below the walls of oblong openings than is the case around circular openings with equal heights.

These inter-relationships between vertical and horizontal stress-relief creep around openings with an oblong cross-section are of great practical significance; they must be recognized as they play the key role in the prevention of roof failures in both homogeneous and stratified salt rocks.

The following interpretation, based on the comprehensive IMC data referred to above, is applicable only to homogeneous and isotropic salt rocks in which no discontinuity planes of any significance, such as bedding planes or fracture zones, exist. The description follows largely that presented in an earlier publication (Baar, 1972a, pp. 49—52).

The inter-relationships between vertical and horizontal stress-relief creep into oblong horizontal openings with infinite lengths at depths around 1000 m may be summarized as follows: Three stages can be distinguished, the *initial* stage, the *intermediate* stage, and the *final* stage of stress-relief creep; any stage may last for variable periods of time, depending on the local conditions of reloading dealt with in detail in Chapter 5.

The initial stage is characterized by rapid stress-relief creep which extends faster in the vertical direction than in the horizontal direction; the reasons apparently are related to the fact that the original vertical stresses are removed from larger areas of roof and floor, compared to the smaller areas at the walls from which the horizontal stresses are removed. Depending on the ratio of opening width to opening height, the initial stage of faster vertical creep — and greater vertical convergence — may last for hours or few days; there is no such initial stage in openings with a circular cross-section as the radial creep is equal in all directions. The wider an opening is made in relation to its height, the longer becomes the time period during which the vertical stress-relief creep rate exceeds the horizontal stress-relief creep rate; in practice, this is partly due to the time requirement for widening an opening to its final width.

The point in time at which the vertical closure rate becomes equal to the horizontal closure rate may be taken as the end of the initial stage of stress-relief creep. Admittedly, such a definition is in no way justified by the development of the creep processes as the transitions from one stage to the subsequent one occur at different times throughout the stress-relief creep zones.

The intermediate stage is characterized by the increasing significance of the horizontal component of creep deformations above and below the walls of openings. As the rapid vertical stress-relief creep above and below the opening results in reduction of the vertical stresses, the horizontal stresses too are reduced according to the basic law of plasticity quoted in sub-section 3.8.2, such reductions of the horizontal stresses above and below the opening allow corresponding stress reductions above and below the walls by stress relief creep which, by necessity, has much larger horizontal than vertical components.

The horizontal creep above and below the walls of an opening is converted into vertical creep above and below the opening; such vertical creep may be termed conversion creep. It results in the changes to the vertical dimensions of creep zones shown in Tables 4-I to 4-IV, and Fig. 4-6: conversion creep tends to establish circular sections of the stress relief creep zones around oblong openings.

As the rapid initial stress relief creep reduces the stress differences throughout the stress-relief creep zone of an opening, the final adjustment of local stress differences to the creep limit of the salt rocks involved in conversion creep takes much longer periods of time, as indicated by the data presented. In most practical cases, it is extremely difficult to exclude any possibility of reloading effects due to extraction mining; the transfer of abutment loads over relatively large distances has been verified repeatedly, see Chapter 5.

The relationship between the initial rapid vertical stress-relief creep and the resulting conversion creep can be, and has been, utilized in the following way: as the volume of rock above and below an opening, which is involved in stress-relief creep and subsequent conversion creep, is a function of the width of the opening, the amount of conversion creep also reflects the width of the opening; the wider the opening, the greater the volume of rocks which perform the conversion of a given amount of horizontal creep above and below the walls of the opening.

In other words: the wider an opening is made, the smaller is the ratio of vertical to horizontal closure rates after the rapid initial stress-relief creep. Widening of openings to sufficient widths eliminates the hazard of roof and/or floor failure due to shearing of the segments shown in Fig. 4-3 off the "active" cross-section. It has been shown that these segments are behaving elastically because of complete elimination of stress differences exceeding the creep limit; so the segments tend to maintain their shape and separate from the "active" cross-section which is continuously converging due to creep deformation. The less conversion creep occurs, i.e., the smaller the ratio of vertical to horizontal closure rates can be made, the safer is the opening against shear failure.

The conclusion that openings in homogeneous salt rocks must be made wider if shear failure is to be prevented, cannot be reconciled with hypotheses which call for stress increases at the walls of oblong openings, as may be justified in the case of elastic behavior; these hypotheses assume increased "excavation stresses" or circumferential stresses at or near the openings if they are made wider relative to their height; in order to avoid such assumed stress accumulation which in turn is assumed to cause excess shear stresses (e.g., Hedley, 1972), these hypotheses call for narrower openings to prevent roof failure. For this reason, "the first measure taken to correct this failure problem (at IMC-K 1) was to reduce the room width; this did not eliminate the problem but intensified it" (Serata and Schultz, 1972, p. 39).

The re-evaluation of the measurement data, and other observations (Baar, 1972a) confirmed that the ratio of vertical to horizontal closure is indeed drastically reduced simply by excavating the openings wider, as the conversion creep is reduced due to the greater rock volume involved above and below wider openings, as outlined above. Some data to show this effect of wider openings is compiled in Table 4-V; due to the faster creep of carnallitic salt rocks, the positive effects of widening openings show up earlier in carnallitic salt. It is emphasized that the remaining horizontal stresses prevent any intergranular loosening of homogeneous salt rocks, and secure selfsupport of the roof; eventually, a homogeneous thick salt bed may bend or buckle if the opening is made too wide under given conditions.

Similar conditions develop in mined-out panels with narrow pillars which allow too much overlapping of stress-relief creep zones; in such cases, the mined-out area resembles a single wide opening, and extensive conversion

TABLE 4-V

Ratio of vertical to horizontal convergence rates after constant convergence rates had been attained in openings of different widths (conditions as in Table 4-I; loading conditions not comparable as both original loads and reloading history were different)

Site No.	Opening width (ft.)	Opening width (m)	Type of salt rock	Constant convergence rates attained at (days since mining)	Ratio of vertical/ horizontal convergence rates
1	21	6.3	sylvinitic	300	0.75
2	21	6.3	sylvinitic	550	0.69
3	28	8.4	sylvinitic	600	0.65
4	35	10.5	sylvinitic	400	0.6
5	21	6.3	carnallitic	400	0.6
6	35	10.5	carnallitic	200	0.3

creep and/or buckling may occur in the stress-relieved areas above and below the mined-out areas. These relationships are shown in detail in Chapter 5.

The final stage of stress-relief creep into isolated single openings is characterized by constant convergence rates in both the vertical and horizontal directions. As conversion creep dictates the vertical convergence rates, see Table 4-V, both the opening width and the opening height, and in particular the ratio of opening width to opening height, are to be considered carefully in the design of mine openings for long-term service.

Frequently, the final stage of constant convergence rates by stress-relief creep is not attained in mine openings, as the widths of abutment zones are grossly underestimated. As the result of reloading, the creep rates accelerate, and aggravate the failure problems in openings designed for long-term service. Details of methods to take advantage of the relatively small creep rates during the final stage of stress relief creep are given in Volume 2.

In the strongest way possible must be re-emphasized that, in bedded salt deposits where individual salt beds separate easily from each other, buckling of separated salt beds rather than conversion creep may occur in response to horizontal creep deformations above and below the walls of openings. Such elastic buckling is time-dependent as the causing forces depend on time; buckling causes increased vertical convergences and may result in tensile failure. These relationships are extremely important in the practical application of stress-relief methods as shown in Volume 2.

The sections of man-made openings for long-term service in salt deposits are usually either circular or oblong, i.e., the horizontal dimensions of near-horizontal roadways and exploration drifts exceed the vertical dimensions. The sections of roadways driven in the Alpine salt deposits of Austria represent an interesting exception: for centuries, these openings were excavated with trapezoidal sections, the height being approximately 1.8 m; the average width was 1.2 m, with the roof approximately 1 m wide, and the floor approximately 1.4 m wide. The development of such original sections into circular or elliptical sections is shown in numerous photographs published by Schauberger (1950). These observations confirm extensive creep deformations behind the "ideal" cross-section which intersects the four corners of the original section, resembling in this regard the situation shown in Fig. 4-3.

Due to extensive flowage during the Alpine orogeny, the original structure of the evaporite deposits can hardly be recognized; some geologists believe in purification of the salt during such flowage, the less mobile components of evaporite sequences having been left behind. Also, blocks and small fragments of non-evaporites may have been incorporated in the salt deposits. The resulting mixtures of salt and other rocks resemble similar mixtures observed in salt diapirs and salt domes. The percentage of salt in such mixtures exhibits considerable variability which seldom reflects the original percentage of salt. The Austrian name of this type of salt rocks is "Haselgebirge"; for unknown reasons, German writers (e.g., Dreyer, 1974, and Röhr, 1974) are using the Austrian word in their English publications when referring to this type of heavily brecciated rocks with varying amounts of salt, the salt exhibiting the characteristics of extensive and repeated recrystallization. The "Hasel mountain" referred to by Dreyer (1972, p. 402) represents the result of literal translation of the local Austrian term; it should not be considered to represent an abnormal type of salt rock as very similar salt rocks are encountered in salt diapirs all over the world.

Schauberger's (1950) instructive photographs clearly demonstrate the long-term reaction of such salt rocks to the excavation of near-horizontal roadways with non-ideal cross-sections: inside the "ideal" section shown in Fig. 4-3, the rock mass desintegrates with continuing creep deformation of the rock mass outside the "ideal" section. Some of the photographs show nearly circular "final" sections established by deformations over several decades. Most of the photographs show elliptical "final" sections, the longer axes of the ellipses exhibiting near-horizontal directions, reflecting the fact that the areas of the walls are larger than those exposed in roof and floor of the initial cross-section. This observation is in agreement with Figs. 3-6 and 4-13 for openings with heights smaller than widths which exhibit stress-relief creep zones with the longer axes in the vertical direction.

It is emphasized that the denomination "final" circular or elliptical section, respectively, does not imply complete stabilization with no further closure; the meaning is that the shape of the sections is expected to remain circular or elliptical, provided that no directional stresses — be it residual tectonic stress or stress developing from extraction mining — develop. Schauberger (1950, p. 138) emphasizes the fact mentioned in sub-section 2.6.2: old mine openings, worked centuries or millenniums ago, are completely closed; only tools and, in a few cases, the conserved bodies of trapped miners have shown long ago that the "pseudo-conglomeratic mixture of salt, shale and anhydrite" is deforming continuously until the original status is re-established; complete recrystallization of the salt makes it impossible to recognize any difference between the original salt rock and salt rock which flowed into man-made openings.

Schauberger (pers. comm., 1976) provided some interesting results of convergence measurements carried out in roadways with the dimensions listed above; the original section of such openings calculates as approximately 2.2 m^2, i.e., the openings are relatively small in comparison to openings excavated for the purpose of salt extraction from the Alpine deposits as well as from other salt deposits. The measurement results are compiled in Table 4-VI. It is emphasized that the convergence rates were virtually constant over measuring periods of 12—18 years, so the calculation of the time required for complete closure at any particular location appears justified.

As the previous mining history around each of the measuring sites is not known, and the question of whether or not residual tectonic stresses actively influence the deformations is still under discussion among geologists, the calculated lengths of time periods to complete closure are not important. The fact that constant convergence rates over decades are established by measurement should be borne in mind.

4.2.4 Stress-relief creep into vertical openings (shafts, storage caverns)

4.2.4.1 Corrections to misinterpretations of data. As the original stress field in salt deposits can safely be assumed as hydrostatic, there are no differences in the principles of stress-relief creep into openings with the longer axis in the vertical direction, or in other directions.

TABLE 4-VI

Convergence of roadways, Hallstatt diapir, Austria [1]

Site No.	Salt content (wt.% NaCl)	Closure rate (mm/year)		Time to complete closure (years)	Remarks
		vertical	horizontal		
1	80	3.6	2.6	460	
2	80—90	7.9	6.4	190	rich in $Mg/Na/KSO_4$
3	80	14.7	15.8	75	
4	60	10.2	10.5	114	
5	35	30.7	22.6	53	heavy spalling
6	35	42.0	30.0	40	heavy spalling

[1] Constant closure rates measured over 12—18 years in development entries in "Haselgebirge" (salt with various amounts of shale and anhydrite). The initial cross-sections are trapezoidal with heights of 1.8 m and average widths of 1.2 m. The times to complete closure are calculated on the basis of the measured constant closure rates, disregarding the initial closure rates which were probably much higher, exhibiting the exponential decay shown in Fig. 3-3.

Qualitative evidence of continuing horizontal creep deformations into mine shafts at great depths has been known for many decades, e.g., from deep potash mines in France and Germany: at depths of about 1000 m, or more, it is nearly impossible to construct shaft linings which can withstand the forces generated by continuing horizontal creep of salt rocks behind elastic shaft linings. As near-hydrostatic stress fields build up if horizontal shaft convergences are prevented, shaft linings must be constructed to withstand forces equal to the deadweight of the overburden; although this may not be impossible, it is definitely expensive.

In order to avoid frequent repair work in deep salt shafts with inadequate lining, such shafts may be left unlined, allowing for "natural" convergence as it occurs. Or the stress field behind linings may be controlled by stress-relief methods dealt with in Volume 2.

Unfortunately, no attempts to measure the continuing convergences of deep salt mine shafts have been undertaken to date. However, many hypotheses were developed on the basis of theoretical assumptions, and/or on the basis of laboratory testing of models which were loaded to represent the in-situ conditions. As most theoreticians assume hydrostatic stress fields, the hypotheses for vertical shafts resemble those developed for horizontal roadways, and they are equally inapplicable to in-situ conditions if derived from elasticity theories which assume tangential stress concentrations at or near the periphery of shafts or other openings with similar cross-sections in horizontal planes.

One series of measurements of radial shaft convergences over a time period of 36 days shortly after sinking was made in the No. 1 shaft at IMC, Esterhazy, Canada. These measurements are of particular interest as they can be compared with the data presented in sub-section 4.2.3. The results received much publicity in recent publications (Dreyer, 1972, 1974), in which they are misinterpreted, and misused to prove the validity of creep laws which predict an exponential decay of convergence rates in deep vertical openings in salt. As a matter of fact, the measured creep rates indeed exhibit an exponential decay over 36 days; in view of the publicity given these measurements also in other publications (e.g., Coolbaugh, 1967), it appears indicated to discuss the measurements and their significance on the basis of the detailed data published by Barron and Toews (1964).

Fig. 4-8 shows mean displacements relative to the shaft axis for points at the indicated distances from the shaft; the measurements were taken in two directions across the shaft diameter; the measuring method was the one also used in the measurements reported in the previous sub-section. For further details, the reader is referred to the original publication.

The following conclusions, drawn by Barron and Toews (1964, p. 124) in their evaluation of the data, may be quoted: "there is insufficient data available to draw anything but the most tentative conclusions and inferences from the results. These may be summarized as follows:

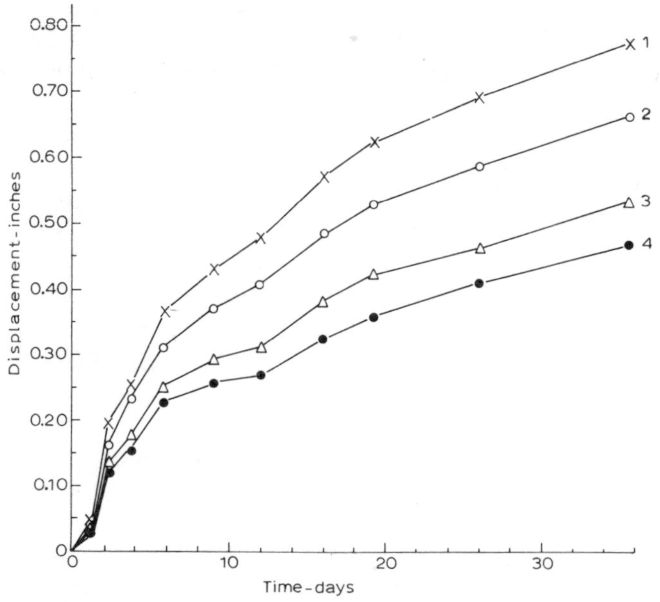

Fig. 4-8. Radial closure of a circular mine shaft in rock salt. (After Barron and Toews, 1964.)

Depth approximately 900 m, diameter 5.4 m. Initial creep closure missing. The measured values indicate the mean displacement relative to the shaft axis of points at the indicated distances from the shaft wall: *1* = 0.15 m; *2* = 1.2 m; *3* = 2.1 m; *4* = 3 m.

(1) All anchors converge towards the shaft axis and the rate of displacement diminishes with time and with increasing radius from the shaft axis. . .

(13) The agreement obtained between measured data and assumptions made in theory suggests a possible method of using deformation measurements to estimate field stress. In view of the tentative nature of many of the above points, it is thought that much could be learned from further measurements under similar conditions. . . The measurements should be continued until it can be safely assumed that the displacement is approaching an asymptote. . . Note should be made of the shaft-sinking timetable in an effort to define the time between the creation of the opening and the start of measurment. . ."

These statements indicate why only "most tentative conclusions" must be drawn in serious attempts to interpret the measurement data:

(1) The time elapsed since the shaft had been excavated to its temporary bottom is not known; on the basis of subsequent measurements detailed in sub-sections 4.2.1—4.2.2, it is safe to say that rather large deformations immediately after the creation of the opening are missing in Fig. 4-8.

(2) Much has been learned indeed from measurements which continued until assumptions regarding a postulated asymptotic end-value of the creep

deformations were no longer required: at least since 1964, it has been evident that constant convergence rates had developed in isolated single openings throughout the mine referred to. For this reason, publications by writers who had access to the data (Serata, 1966, 1968, 1970; Coolbaugh, 1967; Dreyer, 1969, 1972, 1974), and knew of the fact that constant convergence rates had developed, but continued publishing premature data to prove decreasing rates, cannot be regarded as scientific attempts to resolve vital problems in the design of openings in deep salt deposits.

The data shown in Fig. 4-8, and the evaluation by Barron and Toews (1964), can be supplemented as follows:

(1) To date, the displacements in the shaft have not approached an asymptote; to the contrary, the expectation (Coolbaugh, 1967, p. 73) that structures in the shaft would not be damaged by closure of the shaft, proved erroneous. Measurement data are not available.

(2) It is not known at which elevation above the temporary bottom of the shaft the measuring system was installed; there is reason to assume that the measuring system was installed at a convenient level, possibly only 1—2 m above the temporary shaft bottom. In such a case, the shaft bottom would have been acting as restraint against horizontal convergences, and the measured convergences would be too small in comparison with openings of much greater axial extent.

(3) All four curves in Fig. 4-8 are clearly levelling off until day 12, i.e., the convergence rates decreased rather rapidly; the course of the curves resembles the courses of the stress-relief creep curves in Fig. 4-1. Between days 12 and 16, all four curves show an acceleration in convergence rates, the cumulative convergence curves levelling off again over the following 10 days or so, possibly over a longer period of time for which no data are available. There can be little doubt that, between days 12 and 16, shaft sinking continued, and caused the convergence rates to accelerate temporarily as the shaft bottom restraint was removed. These relationships must be considered in any attempt of drawing conclusions from the data.

(4) The data shown in Fig. 4-8 demonstrate that more than 50% of the total measured convergence of the shaft was caused by creep deformation outside the deepest measuring points at the distance of 10 ft. (3 m) from the shaft. This is indicative of a creep zone with dimensions similar to the vertical dimensions of the creep zone around the first horizontal test room, see Table 4-II; however, extrapolations are equally difficult in both cases and should be considered "most tentative", if attempted.

For obvious reasons, the dimensions of creep zones around vertical openings can be measured only in cases where the openings are accessible, i.e., in mine shafts. However, the opening convergences can be determined in openings which are not accessible to men, be it for the reason that the openings are too narrow such as boreholes, or be it for other reasons which apply, for example, to storage caverns in salt deposits.

4.2.4.2 Corrections to misleading publications. Before dealing with published results of convergence measurements in boreholes and storage caverns, some remarks regarding controversial statements about measured convergences in deep salt shafts appear indicated.

Gussow (1970, p. 145), referring to statements about the plasticity of halite and carnallite by Richter-Bernburg (1970, p. 145), expressed the opinion that "in open mine workings, at a depth of 1500—1800 m, halite would begin to creep but this would decrease with time on account of work hardening. Actually, the creep only occurs because there is no confining pressure. . . Creep rate increases with greater depth of burial and at 12,000 ft. (3658 m) becomes uncontrollable in an open drill hole. . . Unless the temperature of the salt was at approximately 300° C, even then no flow would occur unless there was a great enough pressure differential to cause plastic flow."

Richter-Bernburg (1970, pp. 145, 146) disagreed, stating that "experiments have value, but experience has more value. We have observed a difference between the creep in the halite of Zechstein 2 and 3. These two gray halites are indistinguishable by grain size, color, petrology, or chemical composition. . . Yet a shaft in Zechstein 2 begins to close immediately and if equipment is installed, it becomes inoperative in about two months. If the shaft is left open for about one year, the creep stops and it can then be used. Yet in Zechstein 3 the equipment can be installed the next day after it is completed and no significant creep occurs. We cannot explain this; we cannot find any difference between the two salts except that one is in a younger sequence than the other."

Gussow (1970, p. 146) pointed at possible differences in the degree of work hardening. Richter-Bernburg (l.c., p. 146), referring to the salt glaciers of Iran, invoked "great orogenetic force" as required for salt extrusion in diapirs. Gussow strongly objected, emphasizing that "the salt glaciers, if attributed to tectonic forces, would come up like the solid spine of a volcano and go straight into the air. But if the rock salt is plastic, then it moves out under gravity like a lava flow. In fact not all lava flows are liquid, some are just plastic and exhibit plastic flow. I feel that the salt, when extruded, had to be about 300° C."

Such high temperatures, of course, cannot be invoked for the presently observed flow of salt glaciers, so Kinsman (1970, p. 146) suggested that "solution and reprecipitation" may be the mechanism which enables the salt to flow down in the glacier; "there is plenty of ground water around to help out".

Such "ivory-tower" discussions, of course, are of little help in resolving the creep problems in deep mine shafts as there is no water around to help out, and the temperatures at depths around 1000 m may be somewhat higher than comfortable surface temperatures, depending on the effectiveness of the mine ventilation system, but they enable men to work for 8 hours at such depths.

Therefore, many critical questions were raised particularly regarding the creep strain rates in deep mines; Bradshaw (1970, pp. 170, 171) stated that "in the Hutchinson, Kansas, mines (650 ft., 195 m, deep), the convergence rate continues to decrease for some time between 10 and 20 years, after which it apparently becomes constant." Richter-Bernburg insisted in complete termination of creep in mines to depths of 1200 m: "when we measure (the convergence) we find that after two years the (cumulative convergence) curve drops to horizontal" (i.e., it runs parallel to the time axis).

It is apparent from these published statements that the gap between practical experience over many decades in deep salt mine shafts, which require frequent repairs because of continuing creep, and between convergence measurements which allegedly show that the creep terminates after one year or two, cannot be closed by statements which are not supported by published results of measurements over sufficient periods of time, as postulated by Barron and Toews (1964).

4.2.4.3 Measurements of borehole creep closure. Convergence measurements in single horizontal openings apparently are not accepted, by some theoreticians, as indicative of the development of constant convergence rates in-situ; another possibility might be that in-situ measurement data such as shown in Figs. 4-4, 4-7, and 4-8 remain unnoticed; or premature conclusions drawn in the respective publications are welcomed as "proof" of the exponential decay of in-situ creep rates. For these reasons, more in-situ measurements in vertical openings may be presented, and discussed in the following.

Serata (1966, pp. 13—17) announced the development of an electronic device to measure the creep rates (borehole closure rates) in deep boreholes in salt deposits for verification of creep rates which plot as straight lines on log-log diagrams, see Fig. 3-3. The results of an announced test at the depth of 3780 ft. (1134 m) in the Prairie Evaporites near Saskatoon, Saskatchewan, were reported by Serata (1970, pp. 262—264, fig. 14): the convergences of the borehole drilled on May 24, 1965, with the nominal diameter of 6 inches (15 cm) were measured over approximately a month, starting at the end of September of the same year: "the recording of the closure disclosed a significant irregularity of the curves". Only four of the supposedly recorded six curves are shown in the figure referred to; two curves show virtually constant convergence rates over the total measuring period of one month. One curve shows decreasing creep rates for about 10 days, and constant closure rates thereafter. The remaining curve shows increasing closure rates for about 4 days, and no convergence at all for the rest of the measuring time. "Judging from the nature of the irregularity and magnitude of the cavity closure, we assumed that both the viscoelastic creep and probable crystal growth on the exposed surface contributed to the excess closure."

Disregarding the apparent failure of the electronic borehole creep meter in producing meaningful results for the intended proof of exponentially decreasing creep closure rates, the fact that at least two curves with virtually constant closure rates were recorded appears to be noteworthy.

In the same publication (Serata, 1970, p. 264 and fig. 15), "good agreement with the theoretical prediction" of the closure of an experimental solution cavity at the depth of about 10,300 ft. (about 3100 m) is claimed for caliper logs taken at different times: 1, 78, 182, and 347 days. The diameter of the experimental solution cavity — probably just a borehole — is not indicated, neither is the time elapsed since creation of the cavity known. The logs show about equal convergences during the first period of 78 days and the last period of 165 days, but no convergence during the intermediate period of 104 days. The theory which predicts such rather unusual development of convergences, is not indicated. It appears safe to assume calibration problems rather than actual convergences which would terminate for three months and then start again at about half the rate of the first measuring period.

"Another method of cavity creep closure measurement is to record the overflow of the filling liquid from the test solution cavity. As the theory indicates, the overflow rate continues to decrease with time due to the establishment of the structural equilibrium around the cavity. A typical overflow characteristic from a test well made in a salt dome is illustrated" in a figure (Serata, 1970, p. 265, fig. 16). This figure resembles a previously published figure (Serata, 1966, fig. 21) which shows the closure rate of a model cavity in rock salt confined by hydrostatic pressure of 3000 psi; however, some significant changes should be noticed: the viscoplastic flow of the older figure became viscoelastic, and vice versa, in the figure which supposedly shows the overflow rate; both the closure rate and the overflow rate are plotted on a logarithmic scale; the time scale is non-logarithmic, but differs greatly: the total flow plots as a straight line after 15 days for the closure rate, but only after 150 days for the overflow rate, which represents the decrease in volume.

Assuming that the change to the time scale by exactly the factor of 10 — i.e., by adding a zero which makes 10 days 100 days, and so on — is not just due to an error, it would indicate considerably slower decay of in-situ creep rates compared to laboratory rates. Considering earlier decay rates calculated by the same author to plot as straight lines on log-log diagrams — see Fig. 3-3 — there is evidently "insufficient data available to draw anything but the most tentative conclusions", as Barron and Toews (1964) put it.

However, one point is to be stressed regarding the convergence rates of boreholes and experimental in-situ solution cavities referred to in the foregoing: the decay of creep closure rates is definitely not related to loading, but reflects stress-relief creep over measuring periods which began at unknown points in time after the openings had been created. Regarding the further development of the creep closure rates after relatively short time periods of measurement, another statement by Serata (1970, p. 266) should be borne in mind: "When a cavity is created in a medium under the visco-plastic stress state, the cavity will never reach a stable equilibrium. Instead, the cavity continues to deform linearly."

This statement certainly is in agreement with the measurement results presented in sub-section 4.2.2; however, it cannot be reconciled with the opposite concept of "ultimate opening closure" values to be achieved by strain hardening (Serata, 1968, p. 305) which does not occur in situ, see sub-section 3.4.3.

4.2.4.4 Measurements of storage cavern creep closure. It is not surprising to see Serata (1973) contradicting the above quoted statement regarding linear long-term cavity closure by exactly the opposite statement that cavity instability is due to the non-linear long-term creep properties of salt rocks; apparently, such renewed confusion was brought about by "entirely unexpected anomalies" in the closure of a series of storage caverns in the Eminence salt dome, about 20 miles north of Hattiesburg, Mississippi, U.S.A. Allen (1971) gives technical details of the construction of the first two caverns which are in operation since 1970. The unexpected anomalies in the closure of the first cavern included a rise of the cavity bottom by 120 ft. (36 m) and a cavity storage space loss possibly up to 40%.

Serata (1973) gives the following account supported by numerous figures: Cavity No. 1 between the depths of 5750 and 6550 ft. (1725 and 1965 m) had been washed out with an original diameter of about 100 ft. (30 m), i.e., it resembled a cylinder with slightly larger diameters in the bottom half of its vertical height of about 800 ft. (240 m). The "cavity loading" and the corresponding viscoplastic radial closures are described in three stages:

I. Solutioning and dewatering. Around the top of the cavity, the original hydrostatic stresses of 5700 psi (approximately 400 bar) were changed to more than 8000 psi (560 bar) tangential stress, and 1000 psi radial stress. The latter value apparently refers to the minimum gas pressure maintained after dewatering, as the radial stress in the brine-filled cavity cannot drop below about one half of the original stresses because of the specific gravity of the brine compared to that of the overburden. "During Stage I, the stress distribution pattern changes quickly with time" as shown in a figure which resembles, in principle, the changes shown by Nair et al. (1974, p. 138), Chao (1974, pp. 121, 122), and previously by numerous others, particularly Salustowicz (1958): the assumed stress concentration at the walls of the cavity is assumed to move away from the cavity, resulting in the "time-dependent expansion of the plastic radius from zero to 145 ft." (43.5 m).

"By the end of the one-year period of Stage I, the stress distribution pattern stabilizes"; this statement is not consistent with fig. 2 of the publication under discussion, according to which the calculated 2.2% radial closure occurred during the first two months of the one-year period, while virtually no closure is shown for the following ten months.

At the bottom of the cavity, the calculated radial closure at the end of Stage I, prior to the shown reduction of the internal cavity pressure from over 3000 psi to 1000 psi, amounts to 4%. This hypothetical closure also is assumed to have occurred during the first two months of the one-year period.

II. First cycle of gas storage. Disregard the obvious error in relating the calculated expansion of the plastic radius by 145 ft. (43.5 m) during Stage I to the first two months of solutioning: fig. 2 of the publication under discussion clearly defines Stage II as the time period of about 3 months during which the gas pressure was reduced to about 1000 psi, kept at this value for more than 2 months, before it was increased to about 4000 psi.

The calculated expansion from zero to 145 ft. evidently must have occurred during Stage II if the figures are correct; if the text is correct, the plastic radius expanded from 145 to 400 ft. (43.5 to 120 m) during Stage II.

"It should be noted here that the plastic zone expands to about 800 ft. (240 m) in the radial direction at the end of Stage II. This makes the diameter of the plastic zone 1700 ft." (510 m). This statement (Serata, 1973, p. 3) makes the confusion quite obvious: the reader has the choice between 3 values of the expansion of the plastic radius at the end of Stage II: 145 ft., 400 ft., and 800 ft. (43.5, 120, 240 m).

"A large increase in the closure took place during a relatively short period of Stage II, reaching 13.5 and 31.0% radial closures respectively" (at the top and the bottom of the cavity; the "total cavity closure" is given as 33%). Apparently, the large increase in closure is related to the cavity pressure drop to 1000 psi at the beginning of Stage II; at the end of Stage II, the closure is shown as terminated at the levels which had been obtained according to the respective figures.

III. Repeated cycles of gas storage. "No significant change of the closure was found during State III. The change in the stress condition was found to be rather limited during Stage III."

It must be emphasized that the quoted descriptions are incorrect; so is the reported data; the available correct data is summarized in Fig. 4-9, and supplemented as follows:

Stage I. Leaching was completed on 21.12.69, dewatering was finished on 8.10.70. On 25.05.70, the cavity bottom was at the depth of 6560 ft. Sonar logs are shown in the publication by Allen (1971, fig. 4). Due to the known difficulties in measuring exactly the volumes of brines and gases under the varying conditions of pressure and temperature during the leaching and dewatering operations, exact closure data for Stage I is not available.

Stage II. No measurements of cavity volume and bottom elevation were made after the first cycle of gas storage. After the second cycle, on 28.04.72, it was found that the cavity bottom was at 6408 ft., which means a loss of 152 ft. (45.6 m) in about two years. On 23.6.72, the cavity had been refilled with brine; the sonar log shown by Serata (1973, fig. 5) was taken after refilling.

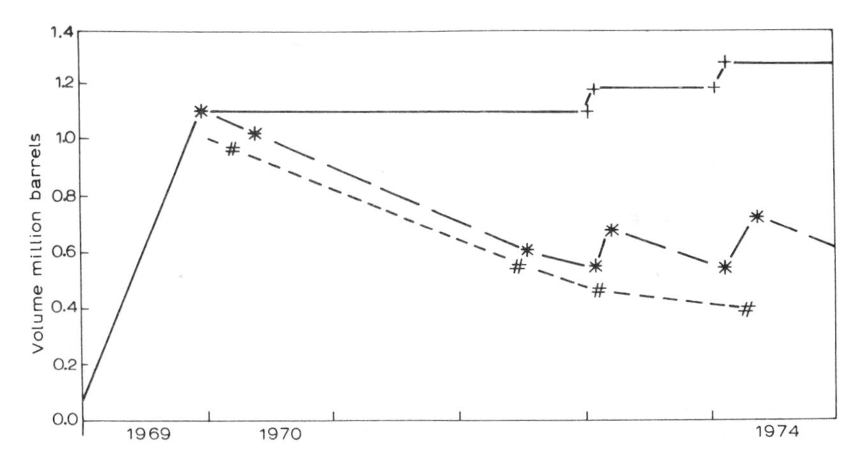

Fig. 4-9. Closure of the Eminence cavern No. 1. (Modified after Allen, 1974.)

Initial depths of cavern top and bottom approximately 1725 m and 1965 m, respectively; initial diameter of the cylindrical cavern approximately 30 m (December 1969).

+ = leached volume, calculated from the amount of dissolved salt: * = displacement volume, calculated from the volume of brine displaced when the cavern was filled with gas; # = sonar volume, calculated from sonar surveys. The displacement volume shown after the last survey in 1974 is the volume projected on the basis of previous measurements.

Stage III. "Considerable wall sloughing is indicated in cavern No. 1 from the first sonar survey on March 12, 1970, to the second survey on June 27, 1972. . . The last two sonar surveys were run after filling the cavern with fresh water and the largest cavern volume loss occurs above the depth of 6200 ft. This indicates more salt was dissolved from the lower part of the cavern" (Allen, 1974).

"This extensive wall failure appeared rather abnormal. . . the salt which flowed into the cavity by viscoplasticity failed to remain in the wall. Instead, it dropped and piled up on the bottom of the cavity" (Serata, 1973, p. 4).

It goes without saying that the volume of the cavity is not changed no matter how much salt drops to the bottom; the usable cavity volume, of course, is reduced as the brine cannot be removed from the rubble pile.

To explain the apparent true loss in cavity volume, Dreyer (1974b, pp. 90, 91; 1974a, pp. 137–139, figs. 57 and 58) publicized another rather inconclusive hypothesis: ". . .asked to compare the 40%-value (of cavity closure) with predictions" by his convergence formula, according to which "the volume convergence was calculated to be 4%", Dreyer stressed "that pure and very coarse-textured rock salt shows an increasing tendency to creep", contradicting Serata's (1973, p. 6) opposite view of brittle behavior that resulted in rather abnormal and extensive wall failure. Dreyer called for floor pressure which resulted in buckling of supposedly existing rock salt and anhydrite layers within the range of the cavern floor: "buckling starts whereby tensile cracks at the upper surface of the anhydrite strata and shear fractures at the lower boundary developed simultaneously, leading to a push-out of the broken layers and to a better understanding of the observed floor lift". Such theorizing makes no sense for the following reasons:

(1) Horizontally bedded sequences of rock salt and anhydrite layers must not be expected in salt domes which rose vertically from great depths.

(2) If buckling of horizontal layers occurs, it is caused by horizontal forces rather than by floor pressure.

(3) If vertical pressure were the cause of floor lift as shown by Dreyer (1974a, fig. 58) for the Eminence No. 1 cavern, no voids could possibly form between separating layers for the simple reason that vertical forces cannot be transmitted through voids from one horizontal layer to the next one in the vertical direction.

(4) Push-out of broken layers, resembling elastic reaction to the reduction of the cavern volume by creep deformation, does not change the cavern volume.

On the basis of a careful evaluation of all available data from the Eminence storage caverns, Allen (1974) arrived at the following realistic conclusions regarding the closure of the caverns: "The closure rate appears to be a function of pressure differential and time. The least indicated closure was in Cavern No. 1, at a rate of about 12 percent per year in 1973 and early 1974. This cavern was operated during this time at pressures above that equivalent to a hydrostatic head of brine water at the reference depth of 6200 ft., except for about 45 days during the filling of the cavern with fresh water."

"The closure rate in Cavern No. 1 has been equivalent to a storage volume loss of about 2 percent of the remaining storage volumes per month since being placed in service. Cavern No. 1 has the longest operating history of reasonably consistent periods of low pressure operations and a constant percentage storage volume loss per month; therefore, the loss rate of 2 percent per month is probably the most representative figure to use for projecting future storage volumes in all of the present caverns at Eminence."

"The initial closure rates in Caverns No. 1 and No. 2 were about 18—20 percent per year when the low cavern pressure was about 3 months per year. In Cavern No. 3, the closure rate was about 26 percent per year and the low pressure time was 3½ months during a 9-month period or at a rate of 4.9 months per year. The largest closure rate of about 40 percent per year was in Cavern No. 4, and the low pressure time was at a rate of about 6½ months per year."

The apparent relationship between the lengths of the low-pressure times and the amounts of creep closure during these times characterizes the creep as stress-relief creep: during high-pressure times, the creep closure may have been affected in a way similar to Fig. 3-7. It is emphasized that the closure rates for Caverns No. 3 and No. 4 apply to the first year of operation.

It is evident from the convergence data of the caverns at Eminence that convergence formulae which predict 4% where up to 40% actually occurred, are equally inapplicable to openings with long axes in the vertical direction and in the horizontal direction.

Some theoreticians, particularly Dreyer (1972, p. 418), predict only half the convergence of shaft-like openings for spherical cavities according to their formulas; "consequently, in the planning of the cavern dimensions it is desirable to strive for the least possible height and for a possibly spherical shape" (l.c., p. 428). Fortunately, in-situ measurement data to assess the validity of these statements is available; as a matter of fact, the data referred to in the following is published in a number of recent publications by Dreyer (1972, 1974b) and by Röhr (1973, 1974). The measurements were taken in the gas storage cavern Kiel 101 in 1967; the cavern had been washed with an inversed-pear shape between the depths of 1305 and 1400 m. Due to the high content of insolubles, 28,000 m^3 of the calculated initial cavity volume of 68,000 m^3 were filled with residues when the leaching was terminated; according to sonar measurements, the usable remaining cavern volume was 39,600 m^3. The brine-filled cavity had a nearly spherical shape with the horizontal diameter of about 40 m and the vertical diameter of about 30 m.

Starting about 1.11.67, the pressure at the roof of the cavity was lowered from 156 kp/cm^2 to practically zero by pumping the brine out of the access well; "following this inadmissible pressure relaxation, about 3 m of the roof fell in; further breakage occurred when the (brine) level was lowered again" (Dreyer, 1972, p. 405). Fig. 4-10 (Dreyer, 1972, fig. 128) supposedly shows the "internal pressure in the area of the cavern roof during the *first* lowering" of the brine level, and the pressure build-up thereafter when the brine level was allowed to rise in the access well. If this is correct, the above quoted pressure drop is incorrect: Fig. 4-10 shows the drop from only 131 kp/cm^2. However, this is of minor significance; it just shows once more that data on which some theoreticians are basing their hypotheses, require careful checking.

Fig. 4-11 shows the rise of the brine level in the access well after it had been lowered at the indicated times to the indicated levels. These curves are extremely significant as they clearly reflect the development of stress-relief creep curves shown in Fig. 4-1, 4-4, 4-5, 4-7, and 4-9. It must be borne in mind that an increase in brine levels means a corresponding increase in pressure against the cavity walls, so the curves seen in Fig. 4-11 reflect a continuing decrease of the differential pressure between the content of the cavity, and the original stresses outside the creep zone of the cavity. The other stress-relief creep curves referred to are not affected in such a way as the air in mine openings is under the atmospheric pressure, and the differential pressures remain constant for all practical considerations.

Accordingly, the convergence rates reflected by the curves in Fig. 4-11 — the slope of the curves is indicative of the convergence rate — decrease faster

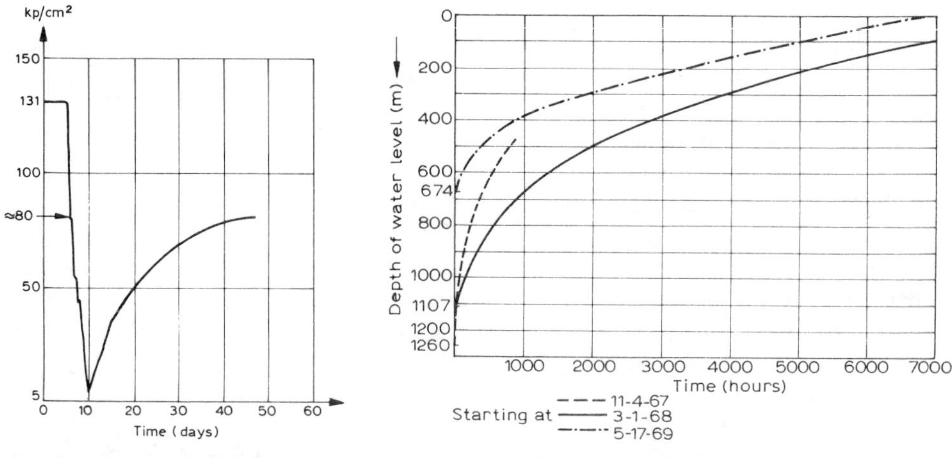

Fig. 4-10. Internal pressure at the roof of the Kiel 101 cavern. (After Dreyer, 1972.)
Pressure drop due to dewatering of the access well. Initial depths of cavern top and bottom approximately 1305 m and 1400 m, respectively; maximum diameter about 40 m. The lower part of the cavern was filled with insoluble residues; the vertical diameter of the brine-filled cavern is given as about 30 m.

Fig. 4-11. Rise of brine levels, Kiel 101 cavern. (After Röhr, 1974.)
Brine levels in the access well were lowered at the indicated times to the indicated levels. The curves reflect the cavern closure due to stress relief rather than "cavern loading".

than in the other curves referred to. However, there is no indication that the convergence rates approach an asymptotic value; to the contrary: the upper curve, measurements of which began on 17.5.69, became a virtually straight line after about 1000 hours or 42 days, indicating constant convergence rates for the following 250-day period of measuring time, although the differential pressure continued to decrease slightly.

Röhr (1974, p. 96) emphasized that "in evaluating the outflow rate, the influence of temperature must be taken into account. From the end of the leaching, the temperature in the cavern at a depth of 1310 m rose within 282 days from 32°C to 46°C but was still well below the stationary wall rock temperature of 67.5°C". "The measured rates of outflow are many times greater than the quantities corresponding to the rate of temperature rise. These rates become negligible if the measurements of rise and fall of the water table (level) are considered." This statement may be correct; however, the known influences of temperatures on the plasticity of salt rocks — see sub-section 3.7 — probably contributed a great deal to the offsetting of the influence of increasing pressures on the cavity closure rates to the effect that the vitually constant closure rates shown in Fig. 4-11 developed.

There are other possible effects on the usable cavity volume, such as changes to the volume of the residue due to dewatering, swelling of some of its components due to hydration, and uplift of its surface in response to horizontal cavity convergence, particularly below 1351.5 m where apparently much cleaner salt surrounded the cavity.

These and other uncertainties make any theoretical calculation merely speculative, particularly if it is based on laboratory experiments over short periods of time which cause strain hardening and related changes towards elastic behavior, as exhibited by "ideal" creep curves found in loading tests.

However, the sonar volumes referred to in the quoted publications definitely demonstrate the rather rapid loss in usable cavity volume: at the beginning of the first lowering of the brine level, the sonar volume was 39,600 m^3. The following survey after about 45 days, i.e., apparently at the end of the time period shown in Fig. 4-10, the sonar volume of 32,100 m^3 is indicated. This means an apparent loss of 7,500 m^3, approximately 18.5% of the original volume. If related to the calculated initial capacity of 68,000 m^3 — Röhr (1974, p. 94) felt that "the total capacity must be taken into account for rock-mechanical considerations" — the loss in volume amounts to 11% in 45 days; according to Fig. 4-10, this period of time includes more than one month during which the pressure rose to about 80 kp/cm^2; there can be little doubt that such an increase in pressure decelerated the convergence rate considerably.

For these reasons, the calculated apparent volume convergences of 18.5 and 11%, respectively, during 45 days have little, if any, significance with respect to the subsequent development; these rapid initial convergences just confirm the rapidness of stress-relief creep shown in Fig. 4-5. As the two sonar surveys were taken at points in time which were 45 days apart, there is no clear evidence of such a rapid decrease in convergence rates over few days, as verified in deep conventional mines, see sub-section 4.2.1. However, the relatively fast rise of the brine level shown for the first few days after 4.11.67, at which time the pressure was practically zero, see Fig. 4-10, strongly suggests similar relationships.

The third sonar survey, carried out approximately 5 months after the second one (Röhr, 1974, fig. 4), showed an additional loss of 1900 m^3 in usable cavern volume, comparing with the previous loss of 7500 m^3 over 45 days. Relating the additional loss to the volume at the beginning of the 5-month period, it amounts to another 17% of loss in usable cavern volume. Converting both the loss for the first time period and for the second period to losses per month results in volume closure rates of 12.2 and 3.4%, respectively, for the two periods of time. The total volume convergence for the 6-month period between the first and the third sonar survey is approximately 23.5%, averaging 4% per month, comparing to the average monthly closure of 3.4% for the period between the second and the third survey.

These figures also suggest a rather rapid decrease in convergence rates during the first few days after the first sonar survey, coinciding in all probability with the low-pressure time. There appears no justification for further discussions of apparent errors in some of the figures shown by Röhr (1973, 1974), particularly his fig. 5 which shows the "inital load" of the cavern in terms of stress differences between the brine pressure and the theoretical lithostatic stresses at the walls of the cavern. In his text, the writer, like Dreyer (1972, 1974b), notes 156 kp/cm^2 as the initial load; in the figure referred to, it is shown at 170 kp/cm^2 compared to Dreyer's 131 kp/cm^2. Such inaccuracies characterize the value of publications, but they are negligible in view of fundamental errors: whichever of these values may be correct, it represents the assumed stress difference at the roof of the cavern; this assumption can only be correct if no creep had occurred during the leaching operation, i.e., if the creep limit of the salt rock is higher than the quoted values; as a matter of fact, Röhr (1974, p. 98) refers to 232 kp/cm^2 as the "brittle strength" obtained by Dreyer for laboratory samples and cavern models which were loaded until "failure began at a stress difference of 232 kp/cm^2" (l.c., p. 96).

In the cavern in situ, the stress differences were created by lowering the brine level to practically zero, i.e., by unloading the cavern walls, or by stress relief, which is the term preferred in this text. There is no loading of the cavern involved; to the contrary: complete unloading took place over few days. The appearance of fracturing indeed "can be explained by too rapid a pressure release" (Röhr, 1974, p. 94); however, such stress relief does not lead "to a build-up of stress too rapidly to be compensated by the plastic flow of salt. In this way, stresses were obtained exceeding the brittle strength; a relatively narrow rock zone around the cavern was subjected for an extended time to an excessive stress difference." Such reasoning applies to laboratory testing which causes near-elastic behavior of salt rocks under conditions of rapid loading; it has been shown previously that the hypotheses based on such tests are not applicable to in-situ openings in deep salt deposits; therefore, further discussion of the calculations presented by Röhr (1973, 1974) and by Dreyer (1972, 1974a) makes no sense: these calculations are based on erroneous assumptions right from the outset.

However, it appears indicated to correct another apparent error in Röhr's publications: after the third sonar survey in May 1968, which showed 30,200 m^3 of usable cavern volume, an increase in cavern volume to about 32,000 m^3 early in 1970 is shown; this is in sharp contrast to Fig. 3-9 which shows a constant rate of volume decrease after the initial stress-relief convergence at higher rates, and similar developments after each additional leaching for increased usable volumes; however, in Röhr's publications, neither additional leaching after the third sonar survey nor further sonar surveys are mentioned. Apparently, the increase in cavern volume is assumed on the basis of theoretical calculations; it is emphasized that "only 32,000 m^3 of brine could be forced out by the initial filling of the cavern with town gas" (Röhr, 1974, p. 99). If the calculated volumes were correct, they would indicate that the cavern was behaving near-elastically after the second sonar survey taken at the end of the 45-day period during which 18.5% loss in volume due to creep deformation occurred.

In fact, elastic behavior due to strain hardening, and asymptotic end values of cavern closures are predicted by the hypotheses to which Röhr refers, see Dreyer (1972, 1974a), and particularly Serata (1968, 1972). It is acknowledged that Serata (1973) changed his hypotheses entirely, possibly on account of the totally unexpected continuing creep closures of the Eminence caverns; as a matter of fact, Serata overshot his target considerably by postulating the diameter of 510 m for the plastic zone around Cavern No. 1 with the horizontal diameter of 30 m, and the cavern roof at the depth of 1725 m.

For comparison, the theoretical diameters of plastic zones around single caverns may be listed according to Dreyer (1972, pp. 474—476); the figures apply to a theoretical cavern at the average depth of 1400 m with the volume of 200,000 m^3; this volume requires either a cylindrical cavern with the height of 200 m and the diameter of 50 m, or a spherical cavern with the diameter of 73 m. According to Dreyer's table 85 and his fig. 136, the creep shear stress (flow limit) of 5 kp/cm^2 is reached at distances of 75 m from the spherical cavity, and at distances of 100 m from the cylindrical cavity. The resulting diameters of the plastic zones are 223 m for the spherical cavity, and 250 m for the cylindrical cavity. Outside these plastic zones, the creep shear stress of 5 kp/cm^2 is not reached, the salt rock remains in an elastic state; this is in complete agreement with the basic principle of plastic behavior, see sub-section 3.8.2. Inside the plastic zones, an increase in shear stresses towards the cavities is shown, as postulated in elasticity theories. Increased shear strengths which allow the salt to withstand increased shear stresses are related to strain hardening which is supposed to provide for considerable ultimate strengths, and stable conditions, i.e., elastic behavior, see section 3.4.

As the diameters of plastic zones around single caverns must be known as precisely as possible in the design of multiple-cavern storage facilities, more evidence of actually measured dimensions of plastic zones around single openings in salt deposits is presented in the following; this appears also indicated in view of other publications in which elastic behavior of storage caverns in salt deposits is postulated on the basis of theoretical considerations and laboratory testing which indeed causes elastic behavior due to strain hardening.

4.3 STRESS-RELIEF CREEP AROUND TWO PARALLEL OPENINGS

4.3.1 Introduction and outline of design problems

Stress-relief creep in the horizontal direction reduces the vertical stresses in the stress-relief creep zone around any single opening. Overburden support is removed in addition to that already removed by the excavation. The rapid extension of stress-relief creep zones causes particular design problems in developing conventional mines, and in storage facilities with numerous caverns: if the stress-relief creep zones of neighboring openings overlap in the salt rocks left between the openings for overburden support, such support elements, usually called pillars, do not provide effective support until overburden formations subside, and reload the pillars.

The time requirements for stress-relief creep and for overburden subsidence differ greatly. In addition, the time requirement for overburden subsidence is changing with the development of the area in which multiple openings are excavated. This is shown in detail in Chapter 5: reloading of pillars after the rapid initial stress-relief creep may commence immediately at a rate which depends on the mechanical properties of the overburden, or it may never occur in cases where the overburden load is fully transferred to abutment zones. The mechanism of load transfer also depends on the mechanical properties of the overburden; not only is this mechanism different under different geological conditions, it also can be changed by suitable mine development and application of various methods of overburden subsidence control.

For the purpose of discussion of stress-relief creep around two parallel openings in this sub-section, the time period of stress-relief creep is defined as the time period between the excavation of any particular opening, and the point in time at which an acceleration in convergence rates indicates the beginning of pillar reloading, i.e., the beginning of overburden subsidence. By necessity, overburden subsidence commences in the formations immediately above stress-relieved zones in the salt formation, and it proceeds towards the surface, again depending on geological conditions: elastic overburden formations may delay the subsidence of overlying formations until certain spans are reached; inadequate mine design may result in sudden subsidence due to tensile failure of the bridging overburden formation; this in turn causes sudden reloading of stress-relieved mine pillars which may be destroyed, as shown in detail in Volume 2.

Under geological conditions where elastic roof formations delay the subsidence of the overburden, limiting the vertical extension of stress-relief creep around multiple mine openings in this direction, the presence of gas or water bearing formations below the mining level poses particular problems: while pillar reloading is delayed as the initiation of subsidence is delayed, the extension of stress-relief creep towards the floor formations is not delayed, and proceeds as the span of the mined-out area becomes larger. This mechanism has caused numerous mine catastrophes, see Volume 2. Similar mechanisms may become effective above mined-out areas in cases where water-bearing formations above the salt are reached by stress-relief creep into mine openings. These statements also apply to multiple-cavern systems.

These relationships between stress-relief creep and overburden subsidence make it imperative to know the vertical dimensions of stress-relief creep zones around multiple openings as precisely as possible. Regrettably, there is no quantitative information based on in-situ measurements available; theoretical calculations are of little, if any, use as they can be based only on assumptions regarding the initiation of overburden subsidence and its continuation. As the development of overburden subsidence until the point in time at which full support by pillars is provided depends on the development of the mined-out area, it is unpredictable as the mine development usually cannot be predicted; this statement applies to all potash mines, to most salt mines, and to storage facilities in salt deposits where the geological conditions encountered in the solution well determine whether or not the leaching of a storage cavern is feasible at that particular location; in homogeneous and isotropic salt bodies where any designed mining or solutioning pattern can be executed according to plan, the development of the excavations may be predictable, but the problems in predicting the overburden subsidence remain.

The data presented in sub-section 4.2.2 for the vertical dimensions of stress-relief creep zones around single openings may be regarded as reasonably accurate as the development of these dimensions is consistent with the subsequent development after initiation of overburden subsidence. Measurement data to show the relation between overburden subsidence and reduction of the vertical dimensions of stress-relief creep zones is presented in the following.

Qualitatively, the extent of stress-relief creep zones into the floor of multiple mine openings has been known for many decades; stress-relief creep below mined-out areas of limited extent has caused numerous mine disasters as mentioned above and examplified in Volume 2, Chapter 6.

4.3.2 Stress reduction between two parallel openings

Fig. 4-12 shows schematically the redution of vertical stresses in a potash mine pillar at the depth of 820 m (Baar, 1966a, fig. 11). The Figure is based on measurements of the expansion of a 6-m pillar, one side of which was increased in height to approximately 7 m from its original height of 2.5 m. Due to geological conditions, all panels in the reference mine were relatively small; overburden subsidence of significance regarding pillar loads could not take place in such small mined-out areas. Numerous attempts to measure postulated stress accumulations in such pillars were frustrating: all the hydraulic cells installed in or below pillars indicated stresses less than 50 kp/cm^2; it took some time, and measurements in isolated openings, see Fig. 3-10, until it was realized that the indicated pressures indeed were indicative of drastically reduced pillar stresses (Baar, 1959b).

As the theoretical pillar load calculated from the extraction ratio is not changed by increasing the height of the pillar, it became evident that the

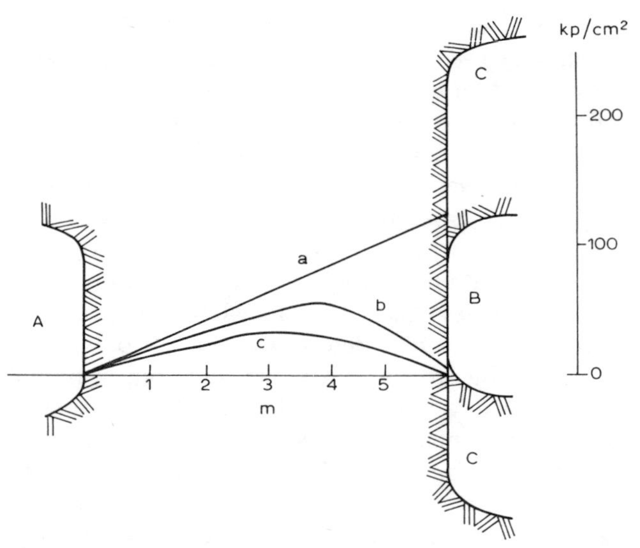

Fig. 4-12. Schematic representation of stress development in a potash pillar (Baar, 1966a).

Depth approximately 830 m. Excavation sequence as follows: A = pre-existing roadway with dimensions as shown in Fig. 3-6; a = vertical stresses in the future pillar; B = room excavated parallel to the roadway A; b = reduction of pillar stresses due to stress relief by excavation of room B, based on measurements of horizontal pillar deformations; C = final section of room B after mining roof and floor as indicated; c = further reduction of pillar stresses due to the increase in pillar height by excavation C.

measured horizontal pillar deformations were not related to increased pillar loads, but rather to increased pillar heights which enabled additional stress-relief deformations. These conclusions were confirmed by measurements in, and above, a larger panel (Wilkening, 1958, 1959) dealt with in detail in Chapter 5. As a matter of fact, extensive stress-relief deformations below the panel referred to resulted in flooding of the mine.

Extensive horizontal pillar deformation by creep immediately after the pillar had been created by the excavation of an opening parallel to an existing one, were reported and discussed by Zahary (1965, p. 8, figs. 15, 16). The measuring system which enabled the measurements of creep deformations related to the excavation of the second opening, is shown in Fig. 4-13 (see also Fig. 4-14). Zahary's evaluation may be quoted as it shows "the approach adopted in arriving at the basic mining plan (which) was somewhat unique in that a rational basis for designing the openings was formulated prior to the start of production mining. Although practical mining judgement played a significant part in the final decisions, the theoretical work was of prime importance. Laboratory studies were carried out and general design criteria were formulated. These were applied to the assumed mining conditions and basic dimensions of rooms and pillars were established. . . Labora-

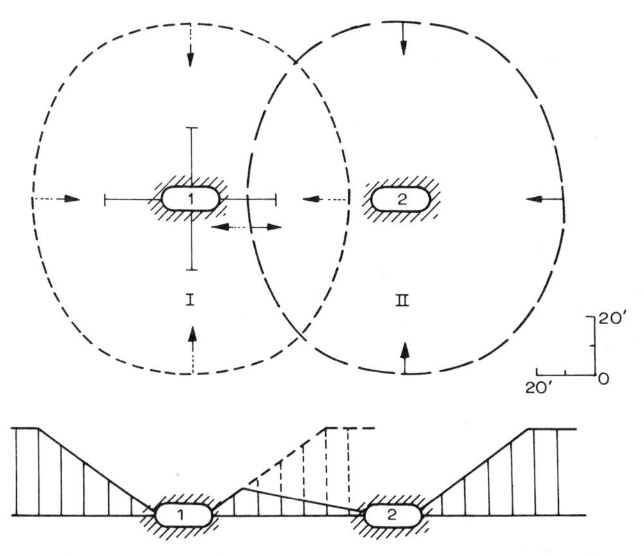

Fig. 4-13. Development of stress-relief creep zones around two parallel rooms, IMC-K 1 potash mine, first panels, extracted in 1963, see Fig. 5-1 (Baar, 1972b).

1 = first room with instrumentation; measuring pins anchored in boreholes at depths of 0.15, 1.5, 3.0, and 6.0 m; opening dimensions as listed in Table 4-1; design width of the pillar: 16 m; *2* = second room, mined one week after the excavation of room *1*.

Upper part of the figure: *I* = extrapolated extent of the stress-relief creep zone around room *1* at the time of mining room *2*; *II* = stress-relief creep zone around room *2*; movements of the 3-m and the 6-m pins in the pillar, in response to mining room *2*, are shown in Fig. 4-14.

The lower part of the figure indicates the reduction of vertical pillar stresses by stress-relief creep towards room *2*.

tory work and analyses were carried out by a consultant whose views on the stability of openings in salt have been published" (Serata, 1964). Similar statements were published by Coolbaugh (1967).

Zahary's (l.c., p. 8) evaluation of the results of measurements at the first measuring site located in the first mining panel reads as follows: "Horizontal movement has been plotted against the distance from the edge of the pillar. Ten days after mining the movement on the 20-ft. pin is slightly negative, that is, movement is toward the centre of the pillar. Since this anomalous behavior is not due to reading error, it is assumed to be some fault in the measuring system being used. The curve showing horizontal movement for 100 days after mining has been extended to the centre line of the pillar on the assumption that the centre line of the pillar does not move. Horizontal strain in the pillar has been calculated from the curves of horizontal movement. The results indicate a minimum value of strain about 15 ft. from the edge of the pillar."

The "basic dimensions of rooms and pillars" in the first extraction mining

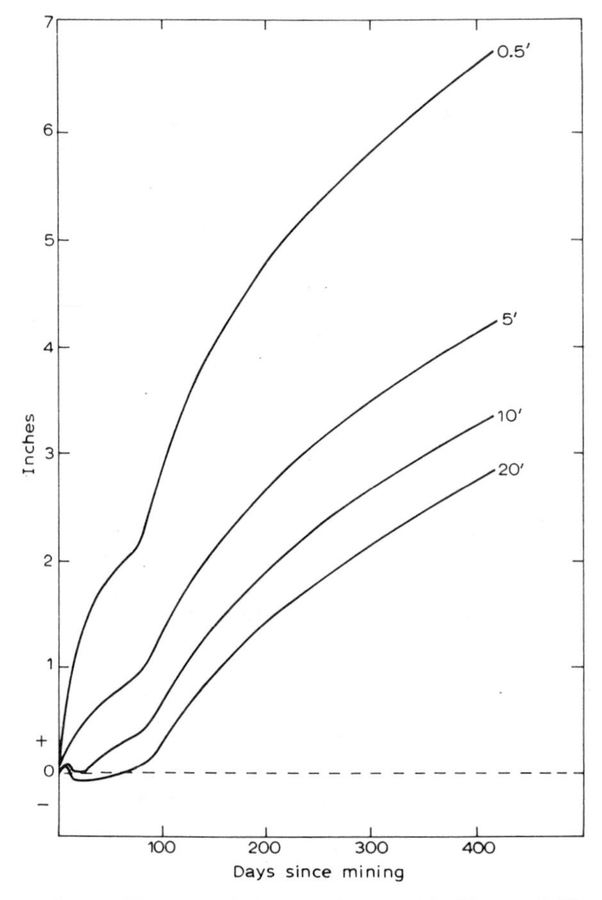

Fig. 4-14. Horizontal closure of room *1* in Fig. 4-13 (Baar, 1972b).

Temporary "negative" convergences in response to the excavation of the second room indicated by the pins anchored at pillar depths of 3 m and 6 m. Accelerated creep, caused by pillar reloading when overburden subsidence began, is indicated by all measuring pins.

panel are shown by Serata and Schultz (1972, fig. 3) as follows: room dimensions 20 × 7.5 ft. (6 × 2.25 m) as in Table 4-1 (disregard the 1-ft. difference in room width); pillar dimensions: width 50 ft. (correct is 52 ft. = 15.6 m), length 140 ft. (42 m).

The "high-stress concentration zone formed immediately above the roof" and near the walls of rooms with the quoted dimensions is shown by Serata (1974, fig. 2); the "stress envelope", with tangential stresses 33—100% higher than the original stress, is shown between the opening and the above quoted "minimum value of strain about 15 ft. from the edge of the pillar". As the tangential stress is the vertical stress in a plane thought to intersect the pillar horizontally at one half of its height, it appears evident that Zahary's "assumption that the centre line of the pillar does not move", is based on identical hypotheses in which stress envelopes are thought to prevent horizontal creep deformations outside the stress envelope. This is also apparent from Coolbaugh's (1967, p. 73) statement which reads: "Mine pillars are designed so that there is no internal overlap of zones of plasticity within them, and they are kept as nearly square as possible in order to obtain the maximum effective confinement with the minimum amount of surface area."

The measurement data, and the conclusions quoted above, receive particular attention in the recently published textbook by Dreyer (1974a, pp. 127 ff.), who claims that the data confirm his hypotheses, particularly his convergence formulas which call for decreasing creep rates due to strain hardening in situ.

In view of the significance attributed to these data by the authors referred to, it appears indicated to emphasize again some quite contradictory data ascertained by subsequent measurements throughout the mine (Baar, 1972b). With particular reference to the above quoted publications by Zahary (1965) and Coolbaugh (1967), the following corrections to some conclusions were published under the heading" Horizontal extent of relief creep in pillars" (Baar, 1972b, pp. 44, 45; the numbering of figures referred to was changed to fit the numbering in this text):

"Measuring and recording the horizontal extent of relief creep caused by excavation can be done conveniently by conventional instrumentation installed in a future pillar."

"The movement (reported by Zahary, 1965) was toward the centre of the pillar. However, the centre of the pillar also moved toward a new opening excavated parallel to the one in which the measuring site was installed. One day after the measurements had begun in the existing opening, the parallel opening was excavated, the width of the pillar being 52 ft. (15.6 m). Stress-relief creep into this new opening caused the negative movement of the 20-ft. (6 m) pin. The anchor point of that pin at 32 ft. (9.6 m) distance from the new opening moved slightly toward the new opening. The 10-ft. (3 m) and the 5-ft. (1.5 m) pins continued moving toward the older opening."

"After one week, the negative movement of the 20-ft. pin reached its maximum value of 0.15 in. (3.75 mm) and remained unchanged during the following 50 days. Then panel retreat mining caused subsidence to begin, and to reload the pillars, which in turn caused increasing horizontal pillar creep. The creep of the 20-ft. pin was no longer negative, the pillar centreline became the neutral line between movements in opposite directions toward both openings."

"Similar measurement results were obtained at various measuring sites. Hence there can be no doubt as to the reality of 'negative' movements in pillars which are not wide enough to cover the two zones of creep around two parallel openings."

"Typical closure curves demonstrating the horizontal creep around two parallel openings excavated at different times are shown in Fig. 4-14. Pillar and room dimensions are identical to those at the site referred to by Zahary (1965). The measuring site was installed in a room in the centre of the adjacent panel during panel entry development (see Fig. 5-1)."

"The development of stress-relief creep, and the resulting vertical pillar stresses are schematically indicated in Fig. 4-13, which is in excellent agreement with earlier measurements (Baar, 1966a)", see Fig. 4-12.

The hypotheses allegedly confirmed by the in-situ measurements at IMC-K 1 during the early mine development are still promoted in publications (Serata and Schultz, 1972, fig. 1; Serata, 1974, figs. 2, 4); these figures show the "high-stress concentration zone formed immediately above the roof" which is assumed to cause "roof failure regardless of the competence", i.e., according to the former publication (p. 39):

(1) in homogeneous salt rocks, "the stressed mass yields and then expands to failure; at the last stage in this failure process, superficial slabbing may or may not appear" (see sub-section 4.2.1, particularly Fig. 4-3, for correct interpretation of this failure process which is initiated in openings which are too narrow, regardless of the competence of the roof);

(2) in salt rocks with discontinuity planes or clay seams, "the roof fails by buckling rather than mass failure; usually, numerous slabs are formed along the seams immediately following the excavation of the ground and the slabs fail quickly by buckling one after another" (such buckling is clearly time-dependent and indicative of the elastic response of separated salt beds to horizontal creep deformation above the walls, i.e., in the expanding creep zone of an opening; it makes no difference whether the creep is caused by stress relief or by reloading; if reloading is prevented, buckling failure can be prevented by correct design of the opening width, see Volume 2 for details of stress-relief methods).

It must be emphasized strongly that the width of an opening has relatively little influence on the horizontal extent of stress-relief creep zones if the height of compared openings is the same; the opening width is significant with respect to the reaction to horizontal creep above and below the opening walls: if horizontal creep is converted into vertical creep, it results in shear failure as shown in Fig. 4-3; if it is converted into buckling of separating beds, it results in buckling failure, the sooner, the wider an opening is excavated. The principles which govern these interrelationships are outlined in sub-section 4.2.1, the practical application of these principles in stress relief methods is shown in Volume 2.

Regarding Fig. 4-14, which shows horizontal closures greater than those reported by Zahary (1965), it should be noticed particularly that the measuring site was located in the second panel of a developing extraction mining area, see Fig. 5-1. Fig. 5-2 demonstrates the effects of abutment loads on the initial stress-relief creep rates. The convergence curves of Fig. 4-14 show the effects on the initial stress-relief creep caused by the abutment loads transferred from the first panel, and the effects on reloading creep when retreat mining of the second panel caused an increase in loads due to the development of additional abutment loads. These important relationships are dealt with in detail in Chapter 5.

Referring back to Fig. 4-13, and to the apparent reduction in vertical stresses due to horizontal stress-relief creep in the 52-ft. (15.6 m) pillar, the following relationship between reloading creep as shown in Fig. 4-14, and the development of mined-out panels with the original IMC design may be emphasized with Baar (1972b, p. 46): "In Fig. 4-14, the creep rates begin to increase approximately 70 days after mining. Similar increasing creep has been measured at numerous measuring sites in the IMC mines. It begins as soon as a mined-out panel reaches a span of approximately 400 ft. (120 m), regardless of the time elapsed since mining a particular opening. This means that, under the geological conditions of that mine, pure stress-relief creep is restricted to mining areas which have a span less than 400 ft. (120 m). If openings are excavated at the boundary of an area which has exceeded that critical span, the effects of stress-relief creep and of reloading creep are superimposed, resulting in more uniform cumulative creep curves which become straight lines as soon as no more changes in pillar loads occur (see Fig. 5-2)."

It appears logical that reloading of 52-ft. pillars, as demonstrated by accelerated creep convergence in Fig. 5-2, requires previous reduction of vertical stresses as shown schematically in Fig. 4-13. The degree of stress reduction by rapid initial stress-relief creep at exponentially decreasing rates apparently depends on the length of time between the creation of the pillar by the excavation of the second opening, and the initiation of overburden subsidence. As the initiation of subsidence is not governed by time, but rather by mine development and geological conditions, it also appears logical that there is no possibility of calculating theoretically the degree of reduction of vertical stresses in a pillar between two parallel openings unless the governing factors are known. This underlines the need for reliable stress determinations in mine pillars prior to attempting to develop theoretical solutions on the basis of assumptions.

Even in cases where the span of mined-out areas, at which overburden subsidence begins, is established by measurement, the further development of the stress gradients in a pillar remains dependent on mine development and behavior of overburden formations at higher levels, until full pillar loads according to the extraction ratio are established. This is demonstrated by the development of the convergence curves in Fig. 5-2. Further evidence is presented in Chapter 5.

Prior to dealing in more detail with the stress redistribution around multiple openings in room-and-pillar mines, the development of pillar stress gradients during initial stress-relief creep, and subsequent reloading creep, is to be clarified; this is still a highly controversial issue between laboratory researchers and practicioners, and it requires special attention in the following section.

4.4. DEVELOPMENT OF STRESS GRADIENTS IN STRESS-RELIEF CREEP ZONES

4.4.1 Stress gradients around individual openings

There are two possibilities of establishing the stress gradients which develop in stress-relief creep zones in salt rocks: (1) to determine the local stresses at various distances from an opening by suitable measurements, see sub-section 3.8.2, and to extrapolate the stress gradients, as shown in Fig. 3-10; (2) to determine the extent of stress-relief creep zones under conditions which allow safe assumptions regarding the original stress field, and to calculate the stress gradients.

The basic principle of plasticity must be borne in mind: throughout the creep zone around an opening in salt rocks, the maximum possible shear stress equals one half of the difference between the maximum and the minimum local principal stresses; any greater stress difference causes creep deformation until the limit of elastic behavior is re-established. The creep limits of salt rocks are extremely low, see section 3.3. Rapid stress-relief creep immediately after the excavation of an opening in highly stressed salt rocks in-situ creates near-hydrostatic stresses throughout the creep zone. In conventional mines, the horizontal stress difference at the walls, between salt rock and opening, is reduced to practically zero; provided the original stresses are known, the extent of stress-relief creep into the surrounding salt rock allows the calculation of stress gradients. As the maximum possible stress difference, depending on the creep limit of the respective salt rocks, allows only slightly higher vertical stresses, the overburden support provided in creep zones can be determined with sufficient accuracy from stress gradients which have been determined by hydraulic cells, or by calculation from the depths of creep zones.

In an assumed horizontal plane at one half of the height of an opening, the vertical local stresses in a creep zone are, by necessity, slightly higher than the horizontal stresses normal to the opening walls. The opposite relation holds for the vertical stresses in a corresponding vertical plane through the centre of an opening: due to the horizontal component of creep that originates above and below the walls of the opening, the vertical stresses above and below openings are slightly lower than the horizontal stresses.

These statements apply to homogeneous and isotropic salt rocks, but not necessarily to salt rocks with bedding planes or equivalent discontinuities: if bed separation or equivalent void formation occurs in the creep zone due to elastic response of stress-relieved beds to further creep behind the "active" surface (see Fig. 4-3), the creep stress gradients build up from the active surface; however, highly pressurized fluids (brines and/or gases), if present, may migrate into the voids and load the separating beds excessively. This mechanism is extremely dangerous and requires special attention if the geological conditions are such that the mechanism could become effective.

There is one major difficulty in establishing stress gradients in stress-relief creep zones from the depths to which stress-relief creep extends into surrounding salt rocks: the original stress field can be calculated with sufficient accuracy for virgin ground conditions, but not for the conditions in abutment zones around mined-out areas; this holds particularly in new mining districts where the widths of abutment zones under various mining situations have not yet been established, and in cases where two or more abutment zones overlap at the location where an opening is excavated. For these reasons, only stress gradients established under virgin ground conditions is dealt with in the following.

It is very unfortunate indeed that, to date, only few reliable data on stress gradients in situ have been established, either by stress determination or by measurement of the extent of stress-relief creep under known conditions of original stress. Apparently, the investigations summarized in Figs. 3-10 and 4-13, and some of those to be dealt with in Volume 2, provide the most reliable information on stress gradients established by stress-relief creep in situ. It is emphasized that this information is being utilized extensively in Canadian potash mines; as a matter of fact, economic potash mining at depths of about 1000 m, or more, in Saskatchewan depends on application of this information on the development of stress gradients in stress-relief creep zones.

The data shown in Fig. 4-4 (Zahary 1965, fig. 7), supplemented by those compiled in Table 4-2, allows the calculation of stress gradients established by stress-relief creep in virgin ground; the original stress field is assumed hydrostatic at 3200 psi (225 kp/cm^2) according to the depth. Assuming the creep limit of 100 psi, the horizontal stress at the outer boundary of the stress-relief creep zone calculates as 3000 psi (210 kp/cm^2); at the wall, zero horizontal stress is assumed. The resulting stress gradients calculate as 115 psi/ft. (27 kp/cm^2/m) at day 450 since mining, and approximately 100 psi/ft. (23 kp/cm^2/m) at day 2600 since mining.

These gradients are in reasonable agreement with those shown in Fig. 3-10 for an opening with similar dimensions excavated under similar conditions: two years after excavation, the stress gradient of 20 kp/cm^2/m was determined by hydraulic measurements in boreholes.

The following stress gradients result from the data compiled in Table 4-3 for an opening in carnallitic salt rocks; the opening width is 28 ft. (8.4 m) compared to 21 ft. (6.3 m), the opening height is the same for both openings: 200 days after excavation, the stress gradient was equal to the above calculated gradient of 115 psi/ft. (27 kp/cm^2/m) at day 450; after 2000 days and some reloading (see Fig. 5-2, curve *10*) the horizontal stress gradient had decreased to approximately 80 psi/ft. (19 kp/cm^2/m).

The vertical stress gradients for the openings referred to above cannot be given with comparable accuracy as the extrapolations of vertical extents of stress-relief creep zones include over 50% of the measured convergences. Using the dimensions listed in Tables 4-2 and 4-3, and assuming the "active" opening surface at the distance of 7 ft. (2.1 m) into roof and floor, the vertical stress gradients calculate as follows:

21-ft. (6.3-m) opening at day 450: approximately 82 psi/ft. (19 kp/cm^2/m)
at day 2 600: approximately 92 psi/ft. (22 kp/cm^2/m)
28-ft. (8.4-m) opening at day 200: approximately 69 psi/ft. (16 kp/cm^2/m)
at day 2 000: approximately 150 psi/ft. (35 kp/cm^2/m).

It should be noticed that there is not too much of a difference between the final horizontal stress gradients for both openings (100 and 180 psi/ft., 23 and 19 kp/cm^2/m, respectively), and the final vertical gradient for the 21-ft. opening (92 psi/ft., 22 kp/cm^2/m); this appears attributable to the vertical maximum diameter of the "active" surface ellipse which nearly equals the width of the 21-ft. (6.3-m) opening.

For the 28-ft. (8.4-m) opening, the vertical stress gradient is considerably smaller after the rapid initial stress-relief creep (69 psi/ft., 16 kp/cm^2/m), reflecting the greater width; the increase by more than 100% at the end of the measuring period is related to reloading which resulted in recompression of the more distant portions of the creep zone above and below the opening, see Fig. 4-6.

The stress gradients calculated above cannot be reconciled with those shown for the same geological conditions in various recent publications by Serata and Schultz (1972, figs. 1, 4) and by Serata (1974, figs. 2, 3). For openings with the same height, but with different widths, the following distances between openings and so-called primary stress envelopes are shown (apparently, the edge of the stress arches is assumed at the point where the "measured" tangential stresses equal the original stress of 3000 psi (210 kp/cm^2), i.e., the boundary of the stress-relieved zone as defined above): For openings 16—24 ft. (5.8—7.4 m) wide, the vertical distance between openings and primary stress envelopes is approximately 6 ft. (1.8 m); the calculated stess gradient would amount to approximately 500 psi/ft. (117 kp/cm^2/m). In the walls, the stress gradients would be even greater; however, no exact data are shown in the figure referred to (Serata, 1974, fig. 2), which apparently reflects previous hypotheses of strain hardening and stabilization (Zahary, 1965; Coolbaugh, 1967; Serata, 1972, 1974, figs. 4 and 5).

"The stress distribution pattern developed around a 65-ft. (19.5-m) room" shown by Serata and Schultz (1972, fig. 4) is quite contrasting: the original stress level is reached at the horizontal distance of about 80 ft. (24 m) from the opening; the calculated stress gradient would be only about 37.5 psi/ft.

(8.7 kp/cm^2/m), i.e., about one half to one third of the gradients established by measurement at IMC for openings with the same height, and less than one tenth of the gradients shown by Serata (1974, fig. 2). Evidently, sound mine design cannot be based on such contrasting parameters. This statement equally applies to the "measured" stress gradients above the 65-ft. (19.5-m) room: in curve A (Serata and Schultz, 1972, fig. 4), the horizontal stresses to the depth of about 40 ft. (12 m) into the roof are shown as negative, i.e., tensile stresses; this is definitely unacceptable for homogeneous and isotropic salt rocks which behave plastically until the limit of elastic behavior is reached by creep that eliminates excess stress differences. Because of continuing horizontal creep above the walls, tensile stresses cannot develop unless bed separation, or separation of segments due to shearing occur.

Since separation of segments does not occur if the openings are made sufficiently wide — for this reason, the opening referred to was excavated wider, after these relationships had been proved at IMC — the inconclusive "measurements" of tensile stresses appear to be indicative of changes in hypotheses rather than of actual stresses.

In the figure under discussion, the "measured" horizontal roof stresses increase from zero at the depth of about 40 ft. (12 m) to the original stress at about 100 ft. (30 m) into the roof; the calculated stress gradient would be 50 psi/ft. (approximately 12 kp/cm^2/m). Calculating the stress gradient for the total distance between roof and a point 100 ft. (30 m) above it results in 30 psi/ft. (7 kp/cm^2/m). Such values are in line with the expectation of smaller stress gradients if the room width is increased to achieve more rapid vertical stress-relief creep in wider rooms.

Reduction of the room width at IMC "intensified the problem" of roof failure by shearing (Serata and Schultz, 1972, p. 39), and this experience was made during the early mine development in 1963. However, the room width was not increased to safe widths at that time, but only after the need for fundamental changes in applied design principles had been shown (Baar, 1970b, 1972b).

Apparently, Serata and Schultz (1972, fig. 4, curve B) adopted the concept of conversion creep during the intermediate stage of stress-relief creep, as outlined in subsection 4.2.3. In their figure, curve B supposedly shows the stress condition "a year after initial excavation"; in fact, the curve C shows the "measured" tangential stresses at that point in time; this becomes apparent from the identification of the dashed curves representing calculated stresses. Accounting for this apparent error, and taking the values of curve C, the horizontal roof stresses increase from about 200 psi (14 kp/cm^2) at the opening to the original stress at the distance of slightly over 100 ft. (30 m) into the roof; compared to the tensile stresses shown in curve A for the first 40 ft. (12 m) into the roof, this means a remarkable increase in horizontal stresses over one year. Apparently, horizontal creep above the opening walls must have occurred during this period of time—if the "measured" horizontal stresses have any meaning; this appears questionable as the stresses after "surrounding excavation" are shown further increased in the lower half, but decreased in the upper half of the roof section 80 ft. (24 m) in length. Fig. 5 of the publication under discussion shows the "changes in stress distribution" differently: first of all, it confirms the error in identification of the curves B and C noted above. Secondly, it shows the 200-psi (14-kp/cm^2) value of horizontal stress referred to above at the distance of 30 ft. (9 m) into the roof rather than at the opening. Thirdly, it shows increased horizontal stresses "after surrounding excavation" over the full length of the 80-ft. (24-m) roof section, with the exception of curve C_1 which intersects curve B about 10 ft. (3 m) below the 80-ft. (24-m) depth, and curve A at about the same distance above this point; horizontal stresses of 400 psi (28 kp/cm^2) at the opening, and 650 psi (45 kp/cm^2) at the distance of 80 ft. (24 m) into the roof, result in stress gradients of only 3.3 psi/ft. (0.7 kp/cm^2/m).

If these "measured" horizontal stresses were correct, the vertical stresses at 80 ft. (24 m) above the opening would calculate as 450 psi (32 kp/cm^2)— with 200 psi (14 kp/cm^2) being the maximum possible stress difference in a creep zone in salt rocks. This means: the support provided for the water-bearing shale 30 ft. (9 m) above the measuring point would be insufficient, as the water pressure is known to exceed 1200 psi (84 kp/cm^2); "surrounding excavation" would reduce the support, as shown in Fig. 4-13, rather than increase it as shown in the figures under discussion. Apparently, the two figures are based on only 13 "measured" stress values, five of which show tensile stresses which cannot occur in creep zones; the remaining eight values are rather questionable. The hypotheses derived from such and similar values require more corrections as shown in more detail in Chapter 5.

4.4.2 Stress gradients around two parallel openings

In the context of this sub-section, the effects of excavating a room "immediately next to an existing room" (Serata, 1974, p. 53, fig. 3) on the vertical stress gradients are in need of further clarification; Serata's hypotheses, which are also described in the publication by Serata and Schultz (1972), cannot be reconciled with the development of vertical stress gradients shown in Figs. 4-6 and 4-13. The "Stress Relief Method" as introduced into the technical literature by Serata (1973), and described in more detail in the above quoted publication, is claimed to differ from the so-called "Stress Control Method" (Serata, 1972) in which "more than two stress envelopes" are regulated in a certain time sequence; "the Stress Control Method is different in the manner of forming the final protective envelope" by pre-stressing for strain hardening; "inside of the protective envelope lies a large mass of stabilized ground which acts as a protective lining; the protective zone is designed to absorb the future increase of the ground stress" (Serata, 1974, pp. 51—55).

In order to avoid confusion, it may be emphasized that the different names — stress-control and stress-relief methods — were introduced by Serata in the quoted publications. Serata and Schultz (1972) do not mention such a difference, but refer to stress-control methods only, apparently including stress-relief methods; with reference to the measurements and hypotheses discussed above, these authors emphasize (l.c., p. 39): "When a water-bearing formation overlies the mine, two factors must be carefully balanced; one is room width. . . the other factor is *pillar width which determines the amount of reloading.*" This is the very first time that Serata is using the term "reloading" in publications he authored or co-authored, apparently admitting to the fact that postulated protective stress envelopes do not prevent reloading of so-called yield pillars if the room width is too large.

The "Stress Relief Method" (Serata, 1974, pp. 51, 52, fig. 3) is described as follows: "The roof stability (in the IMC area where the rooms had to be made wider to prevent shear failure) was further improved. . . when two of the 67-ft. (20.1-m) rooms were placed parallel, close together, by reducing the separation pillar width" to about 15—60 ft. (4.5—18 m). Such "narrow yield pillars" (Serata and Schultz, 1972, p. 40) are thought to eliminate the so-called "Primary Stress Envelopes" around each individual room, the two parallel rooms, with the narrow separation pillar yielded, behaving like one wide room. "The location of the high-stress concentration point is moved upward, away from the exposed roof, resulting in a smaller stress gradient." The location of the Secondary Stress Envelope above the yielded pillar is shown at about twice the distance of the Primary Stress Envelopes at the centreline of the two openings.

Referring back to the previously discussed measurements above a 65-ft. (19.5-m) single opening (pp. 4—59, 61), the reader is reminded that the Primary Stress Envelope for such a room is shown to be located at the minimum distance of about 100 ft. (30 m) into the roof; the "high-stress concentration points" are shown at distances of 140—160 ft. (42—48 m) into the roof. It follows from these figures that the Secondary Stress Envelope around two parallel rooms with a narrow yielded pillar is supposed to be located at the minimum distance of about 200 ft. (60 m) into the roof, with the high-stress concentration at about 300 ft. (90 m).

The implication of "smaller stress gradients" to be achieved by the Stress Relief Method described above, is to be considered on the basis of the fact that "a water-bearing formation overlies the mine" at the normal distance of only 110 ft. (33 m); it consists of non-evaporitic shales and dolomitic limestone; salt beds which may have existed above this formation, were removed by solutioning. The water-bearing formation above the mine is known to exhibit elastic behavior until the span of mined-out panels, at the neighboring IMC mines, has reached 400 ft. (120 m).

At IMC, the pillars in such mined-out panels are 52 ft. (15.6 m) wide, according to original design parameters. It goes without saying that the stress-relief creep zone above mine openings does not extend into formations which apparently behave elastically until a certain span of mined-out areas with 52-ft. pillars between the rooms is reached. There is no reason for the assumption of different behavior of the same roof formations if two parallel 67-ft. (20.1-m) rooms with a 60-ft. (18-m) separation pillar are excavated, resulting in the span of 194 ft. (68.2 m); this is only one half of the span at which 52-ft. pillars allow the initiation of measurable subsidence.

However, in contrast to the original design principles in which 52-ft. (15.6-m) pillars are assumed "competent", and stabilized by strain hardening (Serata, 1968, fig. 20), 60-ft. (18-m) pillars are "yielded" pillars after elimination of the anticipated Primary Stress Envelopes (Serata, 1974, fig. 3). This is a remarkable change in hypotheses; if the change were justified, and if indeed smaller stress gradients were created by the postulated yielding of 60-ft. (18-m) pillars, the consequences would be extremely dangerous for the following reasons: as the roof formations are to be expected to exhibit plate-like behavior, high-pressure water (1200 psi, 84 kp/cm^2) may gain access to the top of the salt; the postulated small stress gradients would be raised until the full load resulting from the water pressure would be imposed on the yielded pillar; this load would be considerably higher than the water pressure because of the wide openings which provide no support.

It is evident that the situation would be worse if the width of the "yield" pillars is only 15 ft. (4.5 m) as published (Serata and Schultz, 1972, p. 40; Serata, 1974, p. 53, fig. 3). However, it must be emphasized that these design data are incorrect; as soon as an attempt was made to excavate rooms according to these parameters, the design proved inapplicable because of excessive creep deformations.

The actual design adopted after apparent failure of the quoted principles is as follows: room width 67 ft. (20.1 m), "yield" pillar width 50 ft. (15 m), resulting in a room-pillar-room unit 184 ft. (55.2 m) wide; such units are separated by 180-ft. (54-m) so-called abutment pillars. Apparently, the basic idea behind this design is still the belief that stable abutments must be provided for the anticipated stress arches across the room-pillar-room units. Data on the reaction of this supposedly stable design — it resembles the "barrier"-pillar concept described by Höfer (1958); this concept proved inapplicable and dangerous, see Chapter 5 — to reloading when overburden subsidence becomes fully effective, are not available for publication; with reference to Figs. 5-1 and 5-2, it is emphasized that 50-ft. (15-m) pillars are satisfactory when fully loaded according to the extraction ratio; pillar failure due to excess loading has not occurred even under most adverse conditions. Undoubtedly, the 180-ft. (54-m) so-called abutment pillars of the present design support most of the overburden load until large-scale creep deformations have produced sufficient conversion creep to fully reload the 50-ft. (15-m) pillars of the room-pillar-room units. These relationships are dealt with in more detail in Chapter 5.

Apparently, even attempts to reduce the "yield" pillar width to 15 ft. (4.5 m) in room-pillar-room units with 67 ft. (20.1 m) wide rooms have not resulted in the mobilization of the 1200 psi (84 kp/cm^2) water pressure at the distance of about 110 ft. (33 m) above the units, i.e., in the development of a planar pressure of that magnitude at the top of the salt. In this case, it is possible to calculate the minimum stress gradient between openings and the top of the salt: it amounts to approximately 11 psi/ft. (2.6 kp/cm^2/m). Such minimum stress gradients are in line with the calculated vertical stress gradients in stress-relief creep zones of narrower openings at IMC, reflecting the effects of greater openings widths. However, these minimum stress gradients cannot be reconciled with the extremely small "measured" stress gradients discussed above.

The effects of planar vertical pressures on wide openings, around which relatively small stress gradients had been created by stress-relief creep, is shown in more detail in Chapter 5. It may be emphasized here that it makes not too much of a difference if relatively small stress gradients develop because of too wide an excavation of a single opening, or because of leaving too narrow pillars which provide hardly any noteworthy support after stress-relief creep.

4.5 ELASTIC DEFORMATIONS AROUND SINGLE OPENINGS IN SALT

4.5.1 Outbursts of salt and gas

Referring back to previous chapters, the following relationships are re-emphasized:

(1) Occlusion of gases in salt deposits occurred under special geological conditions, see sub-section 2.2.3.; many salt deposits, particularly deposits which do not include potash beds, are free from dangerous amounts of gases.

(2) Creep deformations in situ do not affect the impermeability of salt deposits; this is particularly important regarding stress-relief creep outside of the "active" surface of the salt rocks around an underground opening shown in Fig. 4-3. As a consequence, gases cannot be drained out of the salt massive by drilling advanced boreholes, or by similar measures which are applicable in permeable rocks, such as coal seams.

(3) The above statements do not apply to gases trapped in fissure zones in salt deposits, or between individual salt crystals if the voids filled with gases are interconnected; in such cases, gases bleed off in time until the pressure within the interconnected voids equals the pressure in artificial large openings. It is emphasized that this process is reversible, i.e., the pressure in interconnected voids is increased by application of pressures which exceed the existing pressures; these relationships are outlined in detail in Chapter 5 and in Volume 2.

TABLE 4-VII

Statistics on gas outbursts (1960—1965) in three potash mines, Werra district, East Germany (after data from Gimm, 1968, p. 565)

Mine No.	Number of outbursts			Average quantity [1]	Maximum quantity [1]
	(10^2—10^3 t)	(10^3—10^4 t)	(over 10^4 t)		
1	141	100	3	1700 t	20,000 t
2	107	34	6	2470 t	100,000 t
3	35	7	1	1000 t	15,000 t

[1] The quantities of fractured salt rocks per outburst are given in metric tons (t). Outbursts with less than 100 t are not included.

The following point is stressed particularly: because of the very low creep limits of salt rocks, the in-situ pressure in voids filled with gases equals the lithostatic stress field according to depths below surface and specific gravities of overburden formations. It follows from these relationships that the gas pressure is increased correspondingly in abutment zones around mined-out areas. In cases where abutment zones overlap, e.g., in so-called remnant pillars, the original in-situ gas pressure may have been increased to values several times the theoretical lithostatic stress in virgin ground.

Some statistical data on the amounts of salt rocks destroyed by brittle fracture due to gas outbursts, is compiled in Table 4-VII. It is not possible to relate the volumes of gases observed after such outbursts to the volumes of fractured salt rocks, particularly in cases of large outbursts; this is due to the relationships emphasized under (3) above: large outbursts usually open up existing fissure zones from which unknown amounts of gases bleed off in time, contributing to what has been termed the "exhalation phase" of a gas outburst in older publications.

In cases of larger outbursts, the resulting volume of gases may exceed the volume of the mine openings from which the air is displaced, the gases blowing out of the mine shafts for considerable lengths of time; the largest outburst listed in Table 4-VII caused gases to blow out of the two 520 m deep shafts for about 25 minutes; three people were killed as the CO_2-gases, being heavier than air, were flowing downhill from the shafts onto the surrounding area (Gimm, 1968, p. 570). It is emphasized in the publication referred to that similar accidents did not occur during the preceding ten years as the mines are cleared prior to the blasting that causes the gas outbursts; the blasts are set off electrically from surface, and in the case of a dangerous large gas outflow, the population is evacuated prior to being affected by the advancing clouds of gas.

Extensive efforts were made, and are still being made, to reduce the damages caused by gas outbursts near the locations underground where the

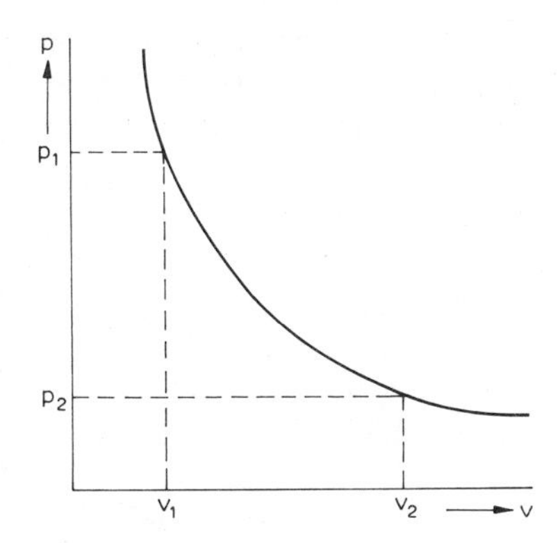

Fig. 4-15. Graph showing the isothermal expansion of gas with decreasing pressure; the product of pressure times volume of a given gas mass remains constant (Baar, 1961a).

outbursts occur; one principal goal is to limit the intensity of individual outbursts, i.e., to control the amounts of fractured salt carried away by the expanding gases after the chain reaction of a gas outburst is initiated. The chain reaction is initiated at the surface of the salt rocks around a mine opening if and when the gas pressure exceeds the tensile strength of the occluding salt which is destroyed by brittle fracture.

The basic gas law which governs the volume expansion of gases after the initiation of an outburst, is demonstrated in Fig. 4-15. This Figure shows the isothermal expansion of a gas volume v_1 under the pressure p_1 if the pressure is reduced to the value p_2; at that pressure the gas volume equals v_2: the product of pressure times volume of the gas mass remains constant. This means that 200 m^3 of gases, blown out of a mine as reported above, had the volume of only 1 m^3 prior to the outburst, if it occurred at the depth of about 800 m in virgin ground, the original gas pressure being 200 kp/cm^2 at that depth.

It appears conceivable that stress-relief creep ahead of an advancing underground opening is affected by high gas pressures; the creep rates are expected to be considerably increased. Measurement data to prove this point are not available. Apparently, the peculiar disking of cores from boreholes drilled into gas-bearing salt rocks (Gimm and Pforr, 1964) is also related to the expansion of occluded gases; in such cases, intergranular loosening may result in the emanation of unknown amounts of gases.

The fact that small-diameter exploration boreholes do not initiate gas outbursts, demonstrates another important relationship: the surface of the salt rocks, from which the original stress is removed unilaterally by excavation, must reach a certain areal extent to initiate a gas outburst; in addition, the shape of the cross-section of an opening is important: apparently, a circular section is "ideal" as it provides equal restraint against internal tension due to gas pressure; the smaller the radius, the more restraint must be overcome at every point of the exposed salt rocks to initiate the chain reaction of a gas outburst. After initiation due to brittle fracture by internal gas pressure, the gases "explode", increasing their volume according to Fig. 4-15, i.e., to volumes possibly 200 times, or even more, the original volume.

The following method of limiting the extent of gas outbursts is based on these relationships (Baar, 1954d, pp. 441—446); apparently, the method is now being used extensively (Gimm, 1968, p. 590 and fig. 10/30) in the mines referred to in Table 4-VII, after extensive underground measurements had confirmed the basic relationships. Earlier contradictory hypotheses were adapted over a time period of about 10 years prior to 1968; as this adaptation process is reflected in numerous publications by Gimm and co-workers, it is recommended to observe caution when studying obsolete publications by these writers.

Different geological conditions may require different methods of limiting the extent of gas outbursts; therefore, the following data must not be generalized.

The chain reaction of a gas outburst is terminated for one of two possible reasons (Baar, 1954d, p. 444): (1) when the outburst cavity reaches the natural, geologically controlled boundaries of gas-bearing salt rocks; (2) when the volume expansion of the "exploding" gases is too slow compared to the amount of gases set free in an outburst cavity; as the result, gas pressure builds up in the cavity to the degree that prevents further brittle fracture at the surface of the surrounding salt, terminating the outburst process prematurely.

As the brittle fracture is caused by excess tensile stresses due to gas pressure in the salt compared to the gas pressure in the cavity, it follows that an outburst which was terminated prematurely by pressure build-up in the cavity, can be re-activated when the cavity pressure is sharply reduced by corresponding gas flow out of the cavity. Exactly this process is demonstrated by a much publicized seismo-acoustic record of a 3-phase gas outburst which finally terminated after 145 seconds, after having been interrupted twice for periods of 10 and 25 seconds, respectively; the corresponding aero-dynamic record (Gimm, 1968, fig. 10/19, p. 576) clearly indicates the reduction of the flow velocities during the periods of interruption, i.e., the reduction of gas pressures in the outburst cavity. The reduced gas pressure apparently restored excessive stress differences which caused the resumptions of the outburst process.

The rates of pressure build-up and pressure reductions in an outburst cavity depend on the width of the passage way through which expanding gases must travel before reaching large mine openings; it should be borne in mind that fractured salt is carried out of the rapidly developing outburst cavities, the salt being sedimented in larger mine openings due to reduction of the gas flow velocities. The amount of fractured salt carried through the pipe which connects the outburst cavity to the mine openings, reflects the resistance to be overcome by the expanding gases; if the duration of an outburst is known, the amount of ejected salt allows the calculation of ejection per unit of time.

Gimm (1968, fig. 10/20, p. 577) has shown this in a diagram of amounts of fractured salt (metric tons) ejected per second plotted versus the idealized diameter of the pipe through which the salt is ejected; ignore the curve shown in the reference figure as it is obviously drawn rather arbitrarily to suit undisclosed hypotheses. Apparently, the relationship is linear. According to the data presented in the figure referred to, the minimum diameter of a pipe or borehole to initiate a gas outburst under the given conditions is approximately 0.5 m. In boreholes with smaller diameters, the "ideal" circular section apparently provides for sufficient natural restraint against initiation of brittle failure due to internal tensile stresses. Any larger circular section, and any "equivalent" section, allow the initiation of the chain-reaction type of tensile brittle failure. The only "activation" required for the occluded gases to "explode" is the removal of restraining salt rocks sufficiently close to gas inclusions; it makes no difference in principle whether the restraining salt rocks are removed by drilling, cutting, blasting, or solutioning; gas outbursts have occurred with any type of excavation (Gimm, p. 572), demonstrating that no special "impuls" is necessary to "activate" occluded gases.

The process of brittle failure due to excess gas pressure is reported to penetrate into the salt massive at rates of approximately 0.7—2.4 m/sec., the smaller rates probably applying to outburst cavities with smaller sections. Apparently, the resulting amounts of fractured salt and "exploding" gases cannot escape through boreholes at a corresponding rate if the holes are too small in diameter; as a consequence, sufficient gas pressure to terminate the chain reaction builds up. According to Gimm, the above-mentioned diagram is used to determine the diameter of the pipe between mine opening and gas-bearing salt; depending on the amount of fractured salt that can be handled at any particular location, the diameter is chosen accordingly (Gimm, p. 590).

In cases where roadways are to be advanced through salt rocks with gas concentrations sufficient to cause gas outbursts, the method of limiting the extent of gas outburst cavities is applied in a more effective way as follows (Baar, 1959d, Patent Claim Berlin WP 5 d/64 430): a borehole with the "safe" diameter of, say, 25 cm is drilled into the gas-bearing salt from a safe distance of the advancing face; the gas outburst is triggered by blasting in surrounding small holes, with the explosives positioned only in the gas-bearing salt. The outburst terminates prematurely due to pressure build-up; it results in partial stress relief ahead of the face, and in release of gases out of the surrounding salt to the effect that the face can be advanced safely to repeat the procedure if required. Gimm (1968, p. 591, fig. 10/34) shows the photo of a face after removal of 10 metric tons of salt by a gas outburst that had been triggered in this way; the publication is appreciated as it constitutes proof of the feasibility of the method which is based on the publication by Baar (1954d, pp. 444, 446); it is acknowledged that detailed data to refine the procedure was gathered by Gimm and co-workers (Gimm, 1968).

It is also acknowledged that Gimm (1968, pp. 569, 570) published definite proof of the danger involved in using cutting machines in cases where roadways are to be advanced through gas-bearing salt rocks; some theoreticians (see Baar, 1954d, for references) had proposed to use cutting machines, which advance 3-m diameter roadways, claiming that gas outbursts cannot be triggered by cutting the salt. These proposals resulted in heated published controversies after a small gas outburst had occurred in 1953 ahead of an advancing cutting machine (Baar, 1954a, p. 136). The issue was resolved in 1959 when an attempt to cut 3-m diameter roadways into gas-bearing salt rocks, ended in failure: after two outbursts had been caused, the dangerous hypothesis was declared erroneous; fortunately, the outbursts were only small, so no serious damage was caused.

Such and similar events forced considerable changes to hypotheses promoted previously by Gimm and co-workers; extreme caution in applying obsolete hypotheses is recommended again with reference to the published discussions to Gimm and Pforr (1964, pp. 446—449). Although a remarkable adaptation of previous hypotheses to the views summarized by Baar (1962) cannot be overlooked, some details in Gimm's (1968) textbook remained questionable; however, as large outbursts of gas and salt are not likely to occur in most salt deposits, no further attempts to clarify minor disagreements will be made in this text.

4.5.2 Other sudden failure processes caused by gas pressure

A gas outburst, like any other excavation process that creates a cavity, removes the vertical support of the roof strata. Under geological conditions where the immediate roof is made up of individual salt beds separated by bedding planes, some of the salt beds may locally consist of secondary salts which occluded gases as shown in sub-section 2.4.3. Similar conditions may exist at any other location where gases could be included in secondary salts which formed in voids, e.g., in fissures and cracks which opened up temporarily in salt deposits as outlined in section 2.4.

The gas pressure in individual roof layers with occluded gases results in abnormal stress-relief deformations into new excavations, no matter in which way the openings are created. The increase in volume of the salt substances, which deform by creep, is negligible in comparison to the increase in volume required for corresponding depressurization of occluded gases. The vertical forces acting on salt beds between an opening and roof layers with occluded gases remain nearly unchanged, resulting in large roof falls if and when other prerequisites are met. Roof falls of this type are enhanced by existing fissure systems, particularly in cases where two such systems intersect each other; this has been shown in some detail in previous discussions (Baar, 1952b, 1954c).

It goes without saying that roof layers with occluded gases most frequently exist at locations where the potash beds also contain large amounts of gases, i.e., at locations where gas outbursts create large cavities. This is the reason why extensive roof falls are frequently associated with large gas outbursts, the roof strata without occluded gases breaking down in relatively large pieces and blocks rather than suffering brittle fracture, as is the case with gasbearing salt rocks.

This type of roof falls may occur while the gas outburst is still progressing into the massive salt, i.e., at distances up to 100 m, or more. The larger blocks of fallen roof strata are not carried out of the outburst cavity, but may be piled up at locations where smaller sections of the cavity exist, or where such smaller sections are created by a roof fall; the flow of expanding gases and smaller particles of fractured salt is soon restricted by such barriers to the degree that terminates the gas outburst because of temporary build-up of gas pressure behind the barrier; the outburst may be re-activated when the pressure is reduced sufficiently, as shown in sub-section 4.5.1.

Outbursts, which cause roof failures as described above, have been classified as a special type of so-called block-type outbursts (Gimm, 1968, p. 568, fig. 10/14). There is no justification for such a classification from a rock-mechanical point of view; it depends on local geological conditions whether or not roof failures occur after an outburst cavity has been created. The time lapse between outburst and roof failure may be short, possibly less than one minute; it must be borne in mind that the outburst process may advance approximately 42—144 m into the salt massive in one minute, depending on the penetration rate applicable to any particular outburst. If roof failures occur while an outburst is in progress, they may result in multi-phase outbursts as described above; if roof failures occur after the final termination of an outburst, they force concentrated gases out of the outburst cavity, and this may result in dangerous situations similar to those dealt with in more detail in Volume 2.

The conspicuous spalling (slabbing, sloughing) around gas-outburst cavities is another manifestation of elastic reaction of salt rocks after stress relief due to the creation of a cavity; it resembles the spalling that develops around cavities in gas-free salt, see sub-section 3.3.2. The separation of spalls occurs inside the ideal "active" cross-section of the outburst cavity, see Fig. 4-3. Gas pressure inside the separating spalls apparently accelerates the spalling process considerably, as is also the case in conspicuously "disking" cores.

It must be emphasized that the disking of cores takes place at the face of boreholes in the direction normal to the axis of the advancing hole; in contrast, the spalling in outburst cavities takes place in the direction parallel to the axis of the cavities, i.e., normal to the face of the advancing cavities. After an outburst has terminated, spalling also develops at the face; this, however, is not indicative of a fissure system which runs ahead of the face of an outburst cavity (Gimm, 1968, fig. 10/16, p. 569), and which is believed to represent "the remnants of the rock-mechanical process that prepares the activation of the energy of occluded gases, and the release of the gases" (l.c., pp. 574, 575), whatever these phrases are supposed to mean. Evidently, spalling parallel to the axis, as shown in Gimm's fig. 10/15 for gas-outburst cavities, and in fig. 10/18 for openings in gas-free salt rocks, cannot be caused by a hypothetical process which travels ahead of the face to cause spalling normal to the axes of the advancing openings. In this regard, obsolete hypotheses which called for a special impuls to trigger gas outbursts apparently have not yet been adapted to contradicting facts summarized by Baar (1962). As shown in sub-section 4.5.1, a special impuls to initiate the chain-reaction of a gas outburst is not required. However, spalling around gas-outburst cavities is greatly accelerated by gas pressure within individual spalls, in which the gas pressure is insufficient to cause brittle fracture of salt crystals.

Another type of sudden failure caused by excess gas pressure on separating salt beds is interpreted in previous publications (Baar, 1954c, fig. 1, p. 340). Related observations are reported and discussed in other publications (Baar, 1952b, 1953a, 1961a, 1962; Gimm, 1968, with references to previous controversial publications). Similar failure processes occur in cases where excess gas pressure forces brines into voids between separating salt beds which remain impervious until fracture occurs. Separated salt beds behave like "clamped plates"—see Chapter 5 for details. Sudden fracture due to excess tensile stresses may occur.

Such sudden failure allows the explosion-like expansion of gases which had migrated into separation voids, or into porous interbedded strata, no matter whether such potential gas reservoirs occur in the roof or in the floor of mine openings. Fig. 4-16 shows a crater in the floor of a French potash mine; it was caused by explosion-like failure due to excessive gas pressure, the gases having migrated into the stress-relieved zone underneath the opening.

There is a simple method available to prevent this type of failure, and the sudden release of large amounts of gases: to drill relief holes at suitable distances; fluids which migrate towards mine openings are continuously drained out of potential reservoirs through such relief holes (Baar, 1953a, 1954d; Fine et al., 1964, with discussion). Contradictory statements by Gimm (1968, p. 593) are erroneous.

Relief holes into roof and/or floor should also be drilled in cases where bed separation is caused by buckling due to lateral forces which are acting on individual salt beds, if the presence of gases cannot be excluded with absolute certainty. This also holds true for geological conditions under which the

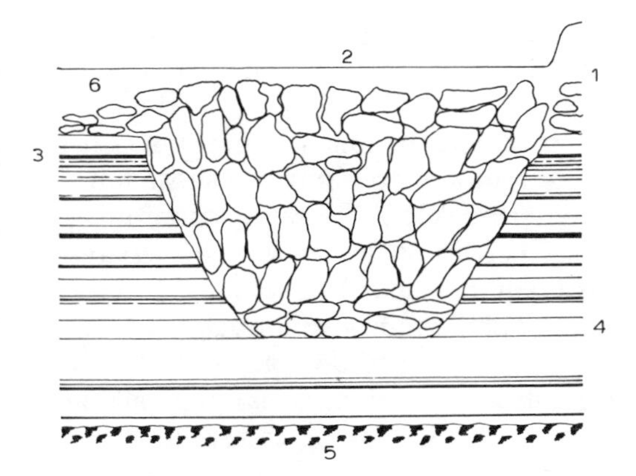

Fig. 4-16. Cross-section of a crater caused by gas blow-out in a potash mine. (After Fine et al., 1964.)

1 = mined-out area, caved-in by blasting of pillar remnants; *2* = roof after extraction of the potash seam; *3* = floor after extraction of the potash seam; *4* = bottom of the 6 m deep crater; *5* = shale bed 4.8 m thick; *6* = fragments of ejected salt.

presence of limited amounts of brines in salt deposits cannot be excluded. Such fluids migrate into the stress-relieved zones above and below mine openings, adding vertical forces to the lateral forces which cause buckling.

Such geological conditions pose severe problems in developing room-and-pillar panels, as the pillars undergo stress-relief creep and allow bed separation at relatively large distances above and below the developing panels. These relationships are dealt with in more detail in Chapter 5 and in Volume 2.

4.5.3 Failure due to buckling

The horizontal component of radial creep into horizontal single openings causes stress-relieved salt beds to respond elastically by buckling.

Buckling is restricted to salt beds inside the "active" surface of the salt rocks around openings, see Fig. 4-3.

The phenomenon of buckling failure is particularly conspicuous in roadways and entries for long-term service when floor beds are buckling; the bed separation voids may open up to considerable widths, the maximum widths normally occurring at the centreline of the opening. These phenomena are described in numerous publications and textbooks; however, buckling cannot occur in laboratory experiments in which the lateral deformation of roof and floor strata is prevented by steel rings to represent what some investigators call the "natural" restraint assumed for in-situ conditions.

Some previous statements are re-emphasized: the "natural" restraint in situ decreases, shortly after the excavation of an opening, by stress-relief creep; buckling rather than conversion creep may occur in response to lateral creep deformations above and below the walls of an opening. It depends entirely on local geological conditions if and when buckling rather than shear failure along the "active" cross-section of an opening occurs; it has been shown that certain minimum widths are required to eliminate shear failure. However, making openings too wide invites buckling failure of separating salt beds; with respect to roof failure, the increasing deadweight of separated beds with increasing opening width also requires consideration.

In many cases, there is a combination of shear failure and buckling failure: shear cracks may develop to intersect prominent bedding planes; then buckling may become dominant, or the elastic behavior of separated salt beds may result in popping-out along shear cracks, i.e., along the walls of roadways.

In view of recent attempts to explain the inconsistency between laboratory creep and actual in-situ creep by bending of separating salt beds above and below openings (Dreyer, 1974, figs. 58 and 59; Serata, 1973, 1974), the following points are stressed: evidently, buckling of horizontal layers is caused by horizontal forces rather than by vertical pressure; the beginning of bed separation in roof and/or floor of openings in conventional mines is conveniently detected in closure graphs such as shown in Fig. 4-4 (thousands of similar graphs are not available for publication): when bed separation commences between a pair of reference points, the corresponding cumulative closure curve deviates from the straight course attained after rapid initial stress-relief creep, indicating the amount of lengthening attributable to the development of one or more bed separation voids.

The allowable amount of roof bed separation depends entirely on local geological conditions, and on the chosen opening widths; as Serata (1972, p. 100) put it: "In the past, numerous finite-element solutions for underground openings have been worked out by various researchers, including the author. Unfortunately, all the available solutions for predicting roof failure are not any better than the judgement of an experienced miner who is familiar with the geological variations of the underground."

However, it is not necessary to rely upon judgement or questionable predictions based on assumptions: once the amount of allowable bed separation has been established for given conditions, a simple warning system can be installed in all mine openings where roof failure would be hazardous: measuring pins anchored at suitable depths into the roof indicate exactly the amount of bed separation between pairs of such pins; if and when the allowable amount is exceeded, an electrical circuit would be closed to light an alarm light, or to set off an acoustic alarm.

Such precautions are recommendable particularly in cases where the effects of reloading creep due to nearby extraction mining cannot be predicted with the degree of certainty that is required for travel ways which are frequented by men. More details on effects of reloading and related roof failures due to buckling are given in Chapter 5.

In some publications (McKinlay, 1972; Serata and Schultz, 1972; Serata, 1972, 1973, 1974; Mackintosh, 1975), the practice of preventing buckling failure by taking the buckling roof strata down, has been given considerable attention, see also Dreyer (1974a). It must be borne in mind that such measures increase the opening height considerably, frequently by 100 per cent, or even more. This certainly prevents impending roof failure due to buckling; however, it also invites increased lateral creep deformation as the wall area is enlargened correspondingly; the resulting long-term deformations must be considered dangerous under certain conditions dealt with in more detail in Chapter 5 and in Volume 2.

The fact emphasized by Serata and Schultz (1972), and Serata (1973, 1974), that the original entry systems in Canadian potash mines had to be abandoned because of roof failure, is mainly attributable to the practice of taking the immediate roof down. By doing so, the free surface of the salt rock behind the walls of travel ways is increased considerably, and this inevitably results in increased horizontal creep deformations above and below the walls.

There is a simple method to prevent buckling of floor strata: to slot the separating beds along the walls, using an undercutter, and to keep the slots open, which may mean that the slots have to be recut when closed by continuing horizontal deformation. It should be borne in mind that such slots increase the effective height of the walls; therefore, no more beds than required should be slotted. Drilling short holes side by side, with their stress-relief creep zones overlapping, has similar effects.

Roof failure due to buckling can be eliminated in this way. The beds which tend to separate must be fully supported by artificial support systems. In cases where artificial support is to be provided by rock bolts anchored in salt rocks, certain types of rock bolts with expanding shells must not be used for long-term purposes, as outlined in detail in Volume 2.

4.5.4 Shear failure

The decisive difference between actual excavation processes in situ, and most laboratory tests from which hypotheses are derived, is re-emphasized: in situ, the stress difference which causes creep deformation is generated by lowering the existing stresses; in the laboratory, the existing stresses, usually resembling the atmospheric pressure, are increased to cause the stress differences required to produce creep deformations.

In situ, rapid stress-relief creep diminishes the stress differences inside the "active" cross-section of an opening, see Fig. 4-3. As the result, corresponding rock sections behave elastically, shearing off the surrounding salt rocks which continue to exhibit plastic behavior. Such spalling (sloughing) resembles that produced in the laboratory by loading; the stress fields are identical: in the laboratory, the interior of opening and pillar models are forced to creep, while the outside tends to behave elastically.

Referring to section 4.2, the significance of the low tensional and shearing strengths of salt rocks (section 3.5) is emphasized again. Superficial slabbing (spalling) in situ is indicative of creep deformations outside the "active" cross-section of an opening; it must not be considered indicative of stress concentrations at or near the surface, as postulated in elasticity theories.

4.6 DISCHARGE OF FLUIDS FROM ISOLATED RESERVOIRS

4.6.1 General remarks

Fluids occluded in and between salt crystals cannot escape from salt deposits unless fracturing occurs, as outlined in the previous section. In contrast, fluids in interconnected cavities discharge when such a reservoir is connected to mine openings or boreholes. Systems of interconnected cavities in salt deposits are not frequently encountered for reasons outlined in sub-section 2.2.3. Even in cases where extensive fissure zones in salt deposits were created by tectonic events, fluids frequently were squeezed out of such zones due to the plasticity of salt rocks.

In cases where fluids were trapped in fissure zones in salt rocks, the pressure in such isolated reservoirs equals the lithostatic stress, which is nearly equal in all directions (near-hydrostatic) because of the low limits of elasticity of salt rocks. Most researchers assume hydrostatic stress fields in deeply buried salt deposits. Further details, and the utilization of these principles in hydraulic stress determinations, are dealt with in sub-section 3.8.2.

It goes without saying that these relationships hold true only for salt rocks, not for other evaporites, such as anhydrite or dolomite, with elasticity limits up to, or even exceeding, values of about 1000 kp/cm^2. Isolated reservoirs in such evaporites may be made up of fissures and/or interconnected pores. In cases where such reservoirs exist in blocks which are surrounded by salt — such geological conditions are frequently encountered in deformed deposits, see sections 2.4 and 3.6 — there will be the tendency for salt to flow into open fissures until the fluid pressure equals the lithostatic pressure. If such highly pressurized reservoirs are lifted closer to the surface, e.g., in rising salt domes, the opposite tendency is to be considered: highly pressurized fluids tend to migrate into the salt, out of isolated reservoirs, and may be occluded in recrystallizing salt.

It follows from these relationships that the pressure of fluids in isolated reservoirs in salt depotis either equals the lithostatic pressure or differs only slightly. This is contrasting to the conditions in reservoirs which are connected to groundwater aquifers; it is re-emphasized that such reservoirs can exist only in anhydrite or dolomite formations, but not in salt rocks below certain depths as fluids are squeezed out of the salt because of its low limits of elastic behavior.

The discharge of fluids from reservoirs which are connected to groundwater aquifers is dealt with in more detail in Volume 2; such dangerous groundwater inflows into salt mines are normally characterized by steadily increasing inflow rates as the passage ways grow larger due to salt solutioning. In contrast, inflows from isolated reservoirs are characterized by inflow rates which decrease according to the gas law shown in Fig. 4-15. The relationships are as follows: unless an isolated reservoir is tapped at the very top, where gas caps may be present, the water content of the reservoir is squeezed out through the opening that had been created in a mine. Provided the width of the outflow opening remains unchanged, the outflow rate reflects the effective gas pressure which decreases according to Fig. 4-15. These relationships have been observed in a great many cases; there are a few cases where measurements were taken. Some measurement results are presented in the following to elucidate the relationships. The data is taken from a research report (Baar, 1964a) unless indicated differently.

4.6.2 Discharge of fluids through boreholes

The usual way in which isolated reservoirs in salt deposits were tapped in mining operations prior to the introduction of cutting machines was by boreholes, either exploratory boreholes or boreholes for blasting in advancing mine openings. For obvious reasons, exact measurements of outflow rates cannot be made in cases where a blast hole hits a fissure system that contains fluids under high pressures: the fluid pressure ejects the drilling rod; in numerous cases reported in the technical literature, drill operators were pierced by drilling rods blown out of boreholes which hit highly pressurized fluids in fissure systems (Baar, 1961a, 1962).

In exploratory boreholes equipped with appropriate instrumentation, the decay of outflow rates in time, corresponding to the decay of gas pressures in fissure systems, can be measured precisely. Fig. 4-17 may serve as an example. The measurements were taken when an exploratory borehole hit a fissure system in anhydrite at the depth of 578 m during shaft sinking in a Zechstein salt diapir in Germany. Apparently, this particular fissure system had only a relatively small volume; the outflow rates decreased within a matter of days to negligible rates.

Fig. 4-18 shows another example which is of particular interest (data from Storck, 1952). After successful sealing of two mine shafts against flooded mine workings — see Fig. 5-10 — attempts were made to reduce the outflow from fissured anhydrite above the shaft seals. On the basis of the assumption that fissure systems in the anhydrite might be connected to the flooded mine workings, pressure grouting to seal the fissure systems was undertaken over many months. It proved impossible to reduce the discharge of brines

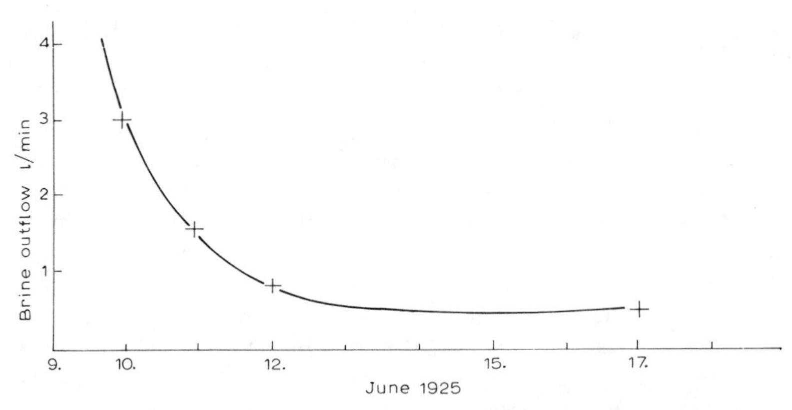

Fig. 4-17. Development of brine outflow rates according to Fig. 4-15: at the depth of 578 m, a borehole hit a fissure system in a potash mine shaft (Baar, 1964a).

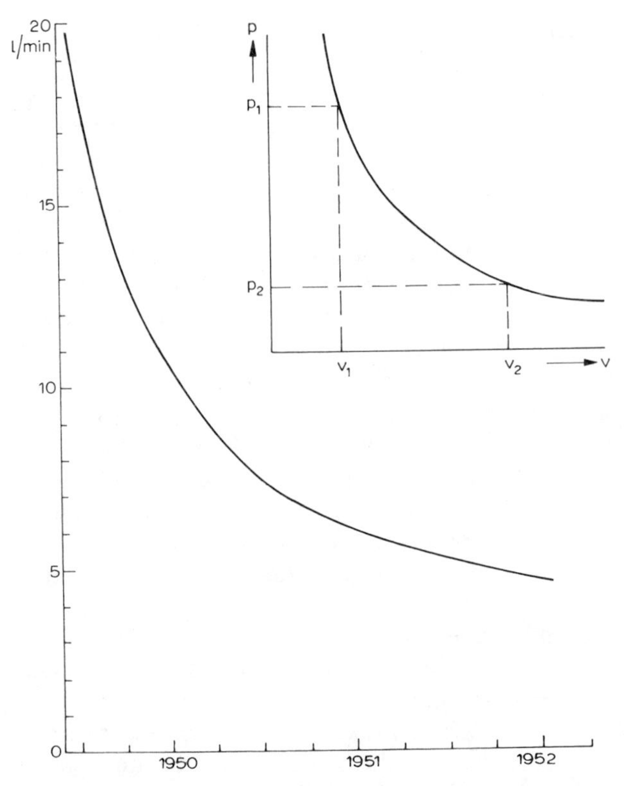

Fig. 4-18. Decrease of outflow rates after pressure grouting; see text for details. (Outflow rates from Storck, 1952.)

from the fissure systems. To the contrary: as the grouting with high pressures resulted in corresponding re-compression of gas and/or air pockets in the fissure systems, the outflow rates usually increased after extensive grouting operations. In addition, existing fracture systems were propagated, and connected to others which had emptied already, but started flowing again. When these senseless operations were finally terminated, the inflow rates decreased regularly as shown in Fig. 4-18.

Similar experiences had been made previously at various occasions. Unfortunately, the apparent relationships between gas pressure and outflow rates were not recognized, but mis-interpreted in a number of publications. It must be emphasized that the application of excess grouting pressures in salt mines is extremely dangerous under geological conditions where the propagation of fissure systems would connect such isolated systems to groundwater reservoirs. Such hydrofracing has resulted in numerous salt mine floodings. There are indications that the most recent flooding disaster, that caused the loss of the Ronnenberg potash mine near Hannover, Germany, in July 1975, was initiated by application of excess grouting pressures in attempts to seal minor inflows from previously isolated reservoirs. More details are presented in Volume 2.

4.6.3 Discharge of fluids through existing fracture zones

When an isolated fracture system in a salt deposit is opened for the first time by an advancing mine opening, the outflow rates frequently increase for some time until a maximum is reached; then the outflow rates decrease regularly, as shown in Fig. 4-17 and 4-18. Such developments are caused by deformations near the mine openings which widen the existing fractures to the effect that the resistance against outflow decreases. As soon as the passage ways remain at constant widths, the outflow rate begins to decrease, reflecting the effective gas pressure as shown in Fig. 4-19.

The measurements shown in Fig. 4-19 were taken in a single roadway at the 794-m level of a potash mine. The brine inflow started when the advancing roadway was driven through a dipping anhydrite bed about 40 m thick (the so-called Main Anhydrite of the German Zechstein; this anhydrite bed was underlain by about 500 m of salt, and overlain by about 100 m of salt, in the mine referred to). The inflow rate increased for about 6 months until the maximum inflow rate of approximately 30 l/min was reached, as shown in the figure. By that time, a borehole had been drilled to intercept the fracture, and to measure the brine pressure that built up when the outflow was shut off. Apparently due to corrosion of the pressure gauge, the latest three pressure values plotted in Fig. 4-19 cannot be regarded correct as the indicated value did not drop below approximately 20 kp/cm^2 while the outflow rate continued to decrease to very low rates over the following years.

It is emphasized that both the measured pressures and the measured out-

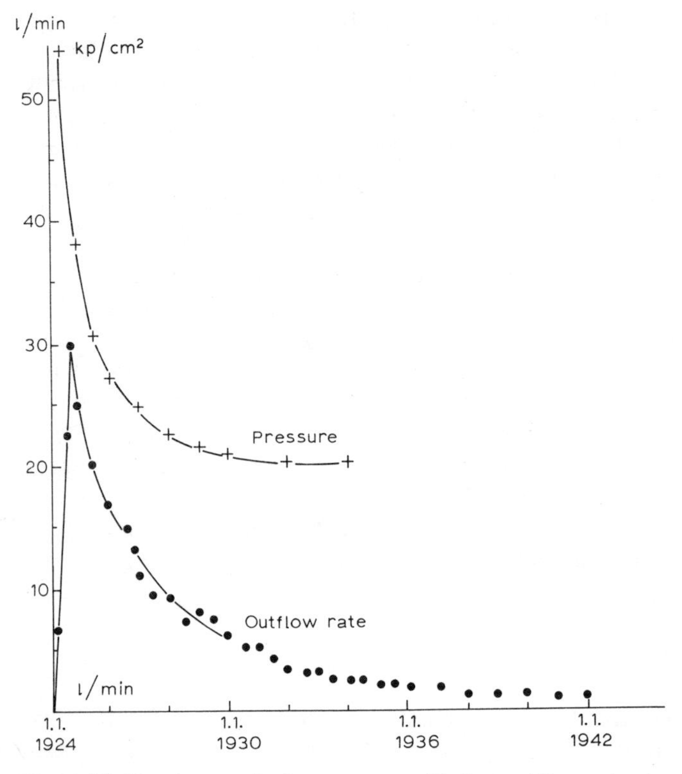

Fig. 4-19. Development of pressures and brine outflow rates in a potash mine; see text for details (Baar, 1964a, fig. 2).

flow rates over the first 4—5 years of measurements reflect remarkably well the gas law shown in Fig. 4-15. The outflow rates continued to decrease to insignificant values over approximately 10 subsequent years.

The total amount of brines discharged at this location was in excess of 50,000 m^3; during each of the first two years, approximately 10,000 m^3 of brines per year had to be pumped out of the mine. These volumes are indicative of the extent of the fracture system in which the brines had been stored for many millions of years. The outflowing brines were rich in $MgCl_2$ and $CaCl_2$, but no solid magnesium-chlorides have been encountered in that particular mine. This led Borchert and Muir (1964, pp. 172—174) to the conclusion that "the asymptotic decrease in the rate of flow with time, clearly shows that the solutions are 'harmless' metamorphic liquors and not 'dangerous' meteoric waters".

This conclusion regarding danger associated with brine discharge from isolated reservoirs in evaporite deposits is certainly correct as far as the flooding potential is concerned. It must not be overlooked, however, that the fluids, which are stored in such reservoirs, are frequently not "harmless" brines, but extremely dangerous gases which are occupying unknown parts of the total volume of such reservoirs. In cases where relatively large volumes of gases are involved, occupying the upper portions of the reservoirs, the gas pressures may not have been reduced sufficiently when the brine levels reach the levels at which the reservoirs had been tapped by mine openings. In such cases, the brine outflows suddenly change to gas discharge, the amounts of gases released per time unit depending largely on the degree of previous reservoir pressure reduction according to Fig. 4-15.

Numerous accidents reported in German publications were caused by such unexpected changes from brine discharge to the discharge of large amounts of gases. To show the danger involved in such events, it may be assumed that the gas pressure in the case shown in Fig. 4-19 was indeed as high as indicated, i.e., 20 kp/cm^2 after the discharge of 50,000 m^3 of brine. It may be further assumed that the gas/brine interface reached the level at which the reservoir had been tapped. This means that 50,000 m^3 of gas would expand to 20 times this volume against the atmospheric pressure in the mine. In other words: 1 million m^3 of gas would be released into the mine workings within a relatively short period of time. Assume the gas consists mainly of methane (CH_4) as is frequently the case in evaporite deposits: to make it "harmless", its concentration in air would have to be reduced to less than 5%. Numerous gas explosions in German potash mines were related to such unexpected gas discharges from isolated reservoirs. Therefore, "harmless" inflows of saturated brines in salt mines must be observed carefully in cases where the inflow rates show the development seen in Figs. 4-17 to 4-19.

Perhaps the most spectacular release of huge amounts of gases after the initial outflow of a few hundred m^3 of brine occurred in 1914 in another German potash mine. As some writers (e.g., Gimm and Meyer, 1962, p. 404; Gimm, 1968, pp. 472—478) continue publishing misleading descriptions and interpretations of this event, the following established facts may be emphasized:

According to published statistics issued by the respective authority (Gropp, 1918), the outburst of "huge amounts" of gases occurred in an inclined exploratory drift at the depth of little over 400 m. The drift was supposed to follow the Stassfurt potash bed, but it came too close to the overlaying shale and Main Anhydrite when, apparently, the inclination of the strata became steeper. Some leakage of brines occurred prior to the gas outburst which, fortunately, occurred at a time when nobody was underground.

An investigation after the mine had become accessable revealed the following: the ventilation system of the inclined drift was destroyed. The last 120 m were filled with brine, through which further release of gases occurred in the following way: every 30 seconds, a large gas bubble was forced out of the reservoir and appeared at the surface of the brine. Over the following months, the time intervals became longer; finally, the release of gas ceased.

At this time, the gas pressure in the reservoir apparently had been reduced to balance exactly the weight of the brine at the end of the drift. Assuming that the reservoir was opened near the face, the hydrostatic head provided by the brine was approximately 1 kp/cm^2; therefore, the reservoir pressure can be assumed safely as 2 kp/cm^2. The brine level in the drift remained virtually unchanged; it was sealed by a dam in 1916 to preclude any possible future release of gas, considering that the volume of the reservoir was not known, and taking into account that the gas volume would increase to twice the reservoir volume when a direct connection to the mine workings would be established.

These facts clearly show that this particular reservoir as well as similar ones, which are frequently encountered in that mining district, are not connected to groundwater aquifers as theorized by some geologists. Such reservoirs are isolated reservoirs; if they are tapped in the mines, the outflow rates develop according to the gas law shown in Fig. 4-15 as soon as the width of the passage ways into the mines has stabilized. In cases where passage ways are created by rock-mechanical consequences of mining methods which are not suitable under such geological conditions, the initial development of outflow rates may differ from Fig. 4-15 for reasons outlined in the following sub-section.

4.6.4 Discharge of fluids due to large-scale deformations

The former potash mine referred to in Fig. 4-19 was used for military purposes since about 1937. Due to the excavation of storage rooms underneath a mined-out area of the 605-m level, approximately 200,000 m^2 of the mined-out room-and-pillar area collapsed, resulting in gas and brine discharge from apparently independent reservoirs in the overlying anhydrite formation. The development of the discharge rates at three different locations *D*, *E*, and *F* is shown in Fig. 4-20. "The solutions trickling into the mine workings near Uslar" (Borchert and Muir, 1964, p. 172) had reached total volumes of 40,000 m^3 at location *D*, and 35,000 m^3 at location *F*, when the measurements were terminated early in 1945. At location *D*, a methane explosion killed two mine inspectors in 1939. The maximum distance between the three locations was approximately 200 m.

Fig. 4-20 shows independent developments of the outflow rates at all three locations, indicating three different fracture systems which were not interconnected. The measurements at locations *D* and *F* show initial rapid increases of the outflow rates to maximum values of nearly 70 and 60 l/min, respectively. Apparently, the passage ways were enlargening during these initial periods due to continuing deformations. After reaching the quoted maximum rates, the inflow rates decreased as is to be expected from Fig. 4-15.

The development at location *E* was probably similar; the inflow was not detected prior to the first measurement. In addition, the curve *E* shows an

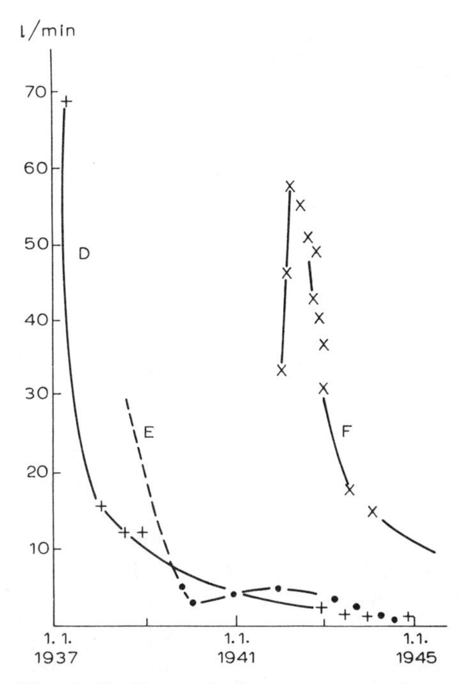

Fig. 4-20. Brine discharge rates at three different locations in a potash mine; outflows were caused by large-scale deformations, see text for details (Baar, 1964a, fig. 3).

increase in discharge rates from 1940 through 1941, indicating continuing deformations which either made the passage ways wider or connected the reservoir to another one which was still under higher pressure.

The measurements at location *F* show considerable inflow rates of 15—20 l/min for the last measurements, indicating corresponding gas pressures in the reservoir. At the end of September, 1945, the mine workings were destroyed by an explosion which may have been a gas explosion. It appears conceivable that the brine level in the reservoir at location *F* had been lowered sufficiently to allow the discharge of approximately 35,000 m^3 of gases under unknown pressure in the reservoir. Such discharge, similar to the one that pushed "huge amounts" of gas into another mine in 1914, would have filled large parts of the mine with explosive gas/air mixtures. It may be mentioned that the ventilation system was not operating at the time of the explosion.

Fig. 4-21 shows the development of brine inflow rates in two other potash mines of the South Harz mining district, East Germany. The brine inflows were caused by large-scale deformations of the floor strata of mined-out room-and-pillar panels with limited extent; the pillars had been destroyed by leaching of carnallite. Details are presented in Volume 2. The Figure is shown here because it clearly demonstrates the development of discharge

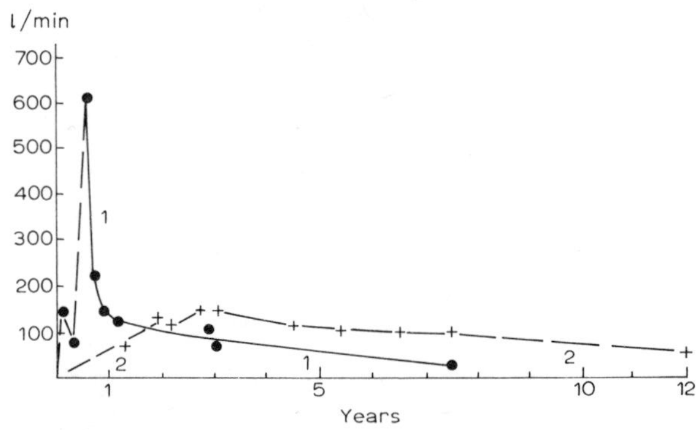

Fig. 4-21. Brine discharge rates in two potash mines (discharge from floor formations); see text for details (Baar, 1964a, fig. 7, supplemented by data from Gimm, 1968).

rates from isolated reservoirs the contents of which are under gas pressure. Initially, the discharge rates increased as the fractures connecting the reservoirs to the mine openings became larger due to continuing deformations. In the case represented by curve *1*, gases escaped together with the brines; as a consequence, the gas pressure in the reservoir decreased faster than would be expected from Fig. 4-15. In the case of curve *2*, the gas volume apparently was relatively large compared to the brine volume, and the small outflow rates had little effect on the gas pressure that forced the brine out of the reservoir.

It is certainly not in the best interest of science, and its practical application to mining engineering problems in salt deposits, if the publication of available knowledge is suppressed, and substituted by questionable publications in which available data is tampered with for the purpose of keeping obsolete hypotheses alive. These remarks refer specifically to the problems dealt with in this section: unexpected discharge of gases in salt and potash mines has repeatedly caused mass accidents. The basic relationships are outlined in publications and published discussions (Baar, 1952b, 1954d) which were triggered by mass accidents in German potash mines in 1951. A detailed research report prepared by this writer is referred to by Fulda (1963), and a paper based on this report was accepted for publication in *Bergakademie* in 1957. However, theoreticians who, at that time, held different opinions suppressed the publication of this writer's paper; its main content can be found in a publication by Gimm and Meyer (1962), in which some fundamental facts are distorted for the apparent reason of fitting them into the hypotheses promoted by these writers.

Apparently, some theoreticians, who manipulate data for the sole purpose of keeping erroneous, extremely dangerous hypotheses alive, are not fully aware of the possibility that their misleading publications might be considered responsible when the application of these hypotheses, which proved erroneous, causes more mass accidents.

Chapter 5

CREEP DEFORMATIONS IN PARTIALLY EXTRACTED SALT DEPOSITS

5.1 GENERAL REMARKS

In salt and potash mining at great depths, it is necessary to distinguish between single openings made for the development of the mine, and excavations made for the sole purpose of mineral extraction. The latter excavations can be abandoned after extraction mining, while main entry systems with certain minimum dimensions must be maintained in safe conditions for time periods of various lengths, regardless of the mining method used for mineral extraction.

It is re-emphasized that stress-relief creep in response to any excavation removes overburden support in addition to that removed by the excavation itself, as shown in Chapter 4. In cases where stress-relief creep zones overlap in pillars designed for full support of the deadweight of the overburden according to depths and extraction ratios, the anticipated support is not provided unless the overburden formations react as quickly by subsidence as the salt rocks respond to excavation by stress-relief creep. Within hours or few days, extensive stress-relief creep zones develop, see Fig. 4-13.

Related design problems are outlined in sub-section 4.3.1: reloading of stress-relieved pillars may commence immediately at a rate dictated by the mechanical properties of the overburden formations, or it may never occur in cases where the mined-out area remains relatively small, the overburden load being transferred to abutment zones; it follows from these relationships that the geological conditions at any particular location are extremely important. Furthermore, if the mine development is not predictable for reasons which frequently apply, the pillar reloading to be expected in time also cannot be predicted.

For these reasons, theoretical calculations are of little, if any, value; they can be based only on assumptions. However, it is possible to derive empirical relationships from observations and measurement data obtained for specific cases; when applying such empirical data to other cases, it is imperative to examine carefully if the prerequisites exist for any comparison. This means in the first place that the geological conditions must be known sufficiently, and must be comparable if any of this Chapter's case histories is used to derive design parameters.

The following parameters is given particular attention for their effects on the support provided by pillars left for overburden support:

(1) room and pillar dimensions; (2) spans of mined-out areas required to initiate overburden subsidence; (3) widths of abutment zones to which overburden load is transferred; (4) vertical distances between mining level and potentially dangerous roof and/or floor formations; (5) development of surface subsidence where applicable, provided data is available.

At shallow depths, salt mining poses no specific problems except those caused by the solubility of salt rocks, as shown in section 2.6.

5.2 STRESS-RELIEF CREEP AND PILLAR RELOADING IN ROOM-AND-PILLAR POTASH MINES

5.2.1 IMC-K 1 potash mine, Esterhazy, Saskatchewan, Canada

Some premature conclusions, drawn from short-term convergence measurements at the IMC-K 1 mine in 1962—1963, are given much publicity in recent publications. Some writers are continuing to refer to such premature publications, attempting to prove hypotheses resembling those applied in the original design at IMC. For details, the reader is referred to sub-section 4.3.2.

In view of the controversial publicity given prematurely to some selected IMC data, it appears indicated to review the long-term measurement results in some detail.

The original mine design at IMC-K 1 was a standard room-and-pillar design (Zahary, 1965, fig. 2; Serata and Schultz, 1972, fig. 3); according to Zahary, "major development consists of a main entry from which block entries are cut at right angles. The entries are a system of usually four or five headings with large pillars on either side. Panels are developed from the block entries by driving three headings up the centre of the panel and mining by retreat. Panel dimensions are set by the capacity of the equipment and are now 2400 ft. (720 m) long and 1400 ft. (420 m) wide."

According to Coolbaugh (1967, p. 73), "mining patterns are designed so as to minimize any possible effects of return mining upon the existing openings, and so that mining can proceed in a retreating direction as much as possible".

Fig. 5-1 shows the first mined-out block, and the locations of convergence measuring sites. Mining the first two panels was completed in 1963. During the same time, the panel entry system for the third panel was cut, and retreat mining began near site 14. This means that the area around site 15 resembled a remnant pillar at the time of mining; it was surrounded by mined-out areas on all sides, the corresponding abutment zones overlapping at site 15 at the time of mining. At the opposite side of the block entry system, the designed mining pattern had to be abandoned because of geological anomalies. Further development of block 1 had to cease as the temporary

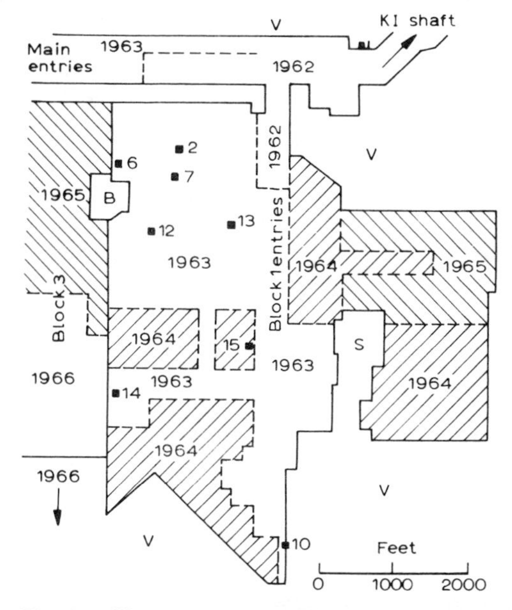

Fig. 5-1. Mining sequence, IMC-K 1 potash mine, blocks 1 and 3 (Baar, 1972a, 1975).
The numbered black squares indicate the location of closure measuring stations. The numbers correspond to the numbers of the creep curves shown in Fig. 5-2.
V = virgin ground, including the shaft safety pillar; S = salt horse (potash bed replaced by barren salt rocks); B = borehole safety pillar.

boundary of the mining lease had been reached.

Fig. 5-2 shows typical cumulative vertical convergence curves measured at the respective sites with identical numbers in Fig. 5-1. The curves were selected from a large number of available similar data to demonstrate typical trends in creep development as caused by mining and pillar reloading after initial stress-relief creep.

The following short summary of the causes of changes in creep rates at each particular site is that given by Baar (1972a, pp. 53—55), with corresponding changes to match the numbers of figures presented in this text. The measurements for each curve began at various times after mining; hence, various amounts of initial stress-relief creep are missing in the curves of Fig. 5-2.

Curve No. 1. This curve shows the continuation of the vertical convergence at the first test site in the shaft safety pillar, located about 50 ft. (15 m) off the main entry system; the predicted creep closure rate is shown in Fig. 3-3, with correction according to the actually measured convergence. Figs. 4-1, 4-2, and 4-4 refer to the same measuring site. Curve No. 1 shows a constant convergence creep rate measured from 1964 through 1968, with

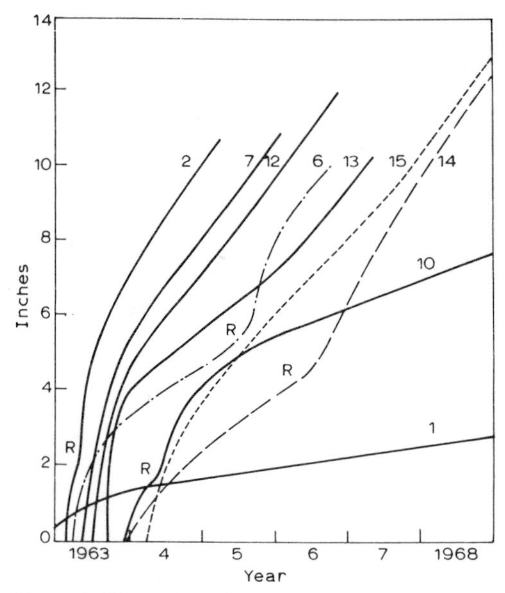

Fig. 5-2. Vertical room convergences, IMC-K 1 mine, block 1 (Baar, 1972a, 1975).
Cumulative convergences measured at the locations identified with the same numbers in Fig. 5-1. *R* indicates the beginning of pillar reloading when overburden subsidence began at the respective locations. Notice that all curves develop into straight lines, indicating constant creep rates, after the establishment of constant loading conditions at any particular location.

no detectable influence from mining in 1965 at distances exceeding 2000 ft. (600 m).

Curve No. 10. The site is located in carnallitic potash, see sub-section 4.2.2. The development of the creep zone around this site is shown in Fig. 4-6. Extraction mining in 1964 caused the increase in the creep closure rate at the indicated point in time. Since 1965, the creep rate is constant, as indicated in Fig. 5-2: the measurement data are represented by a straight line with a constant angle to the time axis.

Curve No. 2. The site is located near the centre of the first mining panel (site 121-A referred to by Zahary, 1965, p. 8). Stress-relief creep changes to reloading creep at time *R*, when retreat mining was in progress in the first panel, and in the adjacent panel. In 1964, the creep closure rate became fairly constant. However, the measurements had to be discontinued as the site was destroyed by roof shearing; this was also the case at other sites for which no data is available after 1965 and 1966, respectively, as shown in Fig. 5-2.

Curve No. 6. The site was located in the first room at the temporary panel boundary. Until 1965, the site was protected by an umbrella provided by block 3; hence, curve No. 6 indicates stress-relief creep to time *R*. Mining block 3 in 1965 caused increased loads which resulted in increased creep. In 1966, the creep rate had approached the one at site No. 2, the curves running parallel. Roof shearing destroyed the measuring system.

Curve No. 7. The site was located in the panel retreat mining area. The initial stress-relief creep, and the subsequent reloading creep due to the beginning of subsidence, follow each other in such a way that the time of beginning of reloading cannot be identified. In 1964, the creep rate became fairly constant; a slight reaction to the 1965 mining of block 3 is indicated by the slightly steeper slope of the curve.

Curve No. 12. Due to nearly identical conditions, the curve duplicates curve No. 7.

Curve No. 13. At the time of mining, the area around this site resembled a remnant pillar; high initial creep rates were caused by the resulting combination of stress relief and reloading creep. The site was temporarily protected by large pillars in the block entry system; this is the reason why the constant creep rate indicated through 1964 and 1965 is much smaller than the one shown by curves No. 7 and 12. When block 1 was abandoned late in 1965, some additional rooms were excavated in the large pillars of the block entry system; in this way, the protection at site No. 13 was removed, and accelerated creep closure developed into a parallel to the other curves.

Curve No. 14. Until 1966, the site was located under an umbrella provided by block 3; therefore, the creep to time *R* can be classified as stress-relief creep. Mining block 3 eliminated the protection; the creep closure curve developed into another parallel to the other curves which indicate full pillar reloading.

Curve No. 15. The measurements for this curve began several months after mining. Hence, a large amount of initial creep is missing; reaction to the 1966 mining in the block entry system gave the curve its final slope parallel to all other curves at sites under full overburden load.

The following conclusions may be quoted from Baar (1972a, p. 55): "It should be noted that all creep curves in mined-out areas develop into straight lines with a final slope which is very much the same for all sites, no matter how any particular curve had developed previously due to the particular loading history at that site. As soon as constant loads are imposed on the pillars, the creep becomes steady-state creep. The final width of the mined-out area shown in Fig. 5-1 exceeds the width required for non-restricted overburden subsidence. Hence, it may be assumed that the pillar loads represent the full overburden load according to depth and extraction ratio at all sites except No. 1 and No. 10."

Figs. 5-1 and 5-2 demonstrate facts which simply cannot be reconciled with some publications by writers who were aware of the data collected during the early development of the IMC-K 1 mine. The following basic design principles were formulated on the basis of the IMC data and earlier observations (Baar, 1972a, pp. 41—44): "In theory, it might be possible to support the total weight of the overburden by stable pillars left in mined-out potash panels, the pillars being designed so that no horizontal creep occurs in the centres of pillars. However, the large extent of the creep zones into pillars makes design principles aiming at stable pillars illusive, unless the extraction ratios were reduced far below economic limits."

"Therefore, another basic design principle in deep salt and potash mines must be to control the inevitable subsidence at any time by pillars, or by other means, in such a way that no excess deformations occur, neither excess creep into mine openings which must be kept at minimum dimensions, nor excess deformations of roof and floor strata which could possibly result into gas or water discharge into the mine. In order to apply these design principles in deep potash mines, it is imperative to know the exact dimensions of the zones of creep around openings at any time. Even more important is the knowledge of the relationships between creep and overburden load redistribution. Here, the different time requirements for rapid initial stress-relief creep, and for overburden subsidence come into the picture. Surface subsidence measurements carried out over many decades above potash mines in Germany have shown that many similarities exist between overburden reactions to room-and-pillar mining of potash beds, and total extraction of other sedimentary beds such as coal seams."

"The larger a mined-out area becomes, the more overburden formations become involved in subsidence, beginning in the immediate roof strata and proceeding to the surface. As a rule of thumb, it can be said that the span of a mined-out area must exceed the depth below surface in order to allow unrestricted subsidence above the centre of the mined-out area. Before this critical span is reached, the overburden weight is not fully supported by pillars in a mined-out potash bed. The remainder of the overburden weight must be supported in an abutment zone around the mined-out area. The width of such abutment zones, and the amount of additional load are important parameters in potash mine design because of the additional stress-relief creep which takes place immediately after excavating an opening in an abutment zone. A mined-out area in which the pillars along an abutment zone are not yet fully loaded, may be termed an 'umbrella zone'. The 'umbrella' may be a temporary one, or it may remain effective over long periods of time."

"Unfortunately, all these vital parameters — the span of a mined-out area at which overburden subsidence begins, the span needed for unrestricted subsidence, the time required for full subsidence, the width of abutment zones, the abutment loads at any time, the width of umbrella zones, their life time, their effectiveness — have to be determined by measurements in the mines, and at surface. It is impossible to determine the behavior of the various overburden formations by testing samples. This is mainly because of the frequent existence of cracks and fissure zones in most formations. The values of some of the above parameters as determined by measurements in the IMC mines, do not apply to other conditions. However, the measurement results provide a general picture of the interaction between creep into openings in potash and overburden load re-distribution after mining."

Mraz (1973), referring to an internal IMC report by Baar (1970b), and to "many controversial statements and conflicting opinions" is emphasizing "that the results of underground tests and observations should be considered much more relevant in the design of underground excavations, because the simulation of the conditions in potash mines is very difficult, if not impossible, in laboratories". "The graphs of the actual closure from mining areas show the effect of pillar reloading as a very significant acceleration of the closure. The steady-state closure rate after reloading is higher as a result of an increase of the pressure gradient in the pillar."

Fig. 5-3 (Baar, 1972a, fig. 9) shows schematically the development of pillar loads in mined-out areas with carnallitic potash.

Summarizing the data published by Baar (1971a, 1972a, 1974a, 1975) and by Mraz (1973), the following original design parameters at IMC may be presented (it is emphasized that these parameters apply to the original standard room-and-pillar design which is no longer used, but was replaced as a result of the data re-evaluation referred to by Mraz, 1973): room and pillar dimensions: see Table 5-I; span of mined-out area required to initiate subsidence: 120 m; width of abutment zones: about 60 m at the beginning of subsidence, over 480 m around large mined-out areas, not exceeding 600 m.

It goes without saying that reloading of mine pillars is possible only after the pillars have been unloaded to a certain degree by stress-relief creep. Apparently, Serata (1973, 1974) has acknowledged the facts outlined in the foregoing, adapting his hypotheses as shown in sub-section 4.4.2.

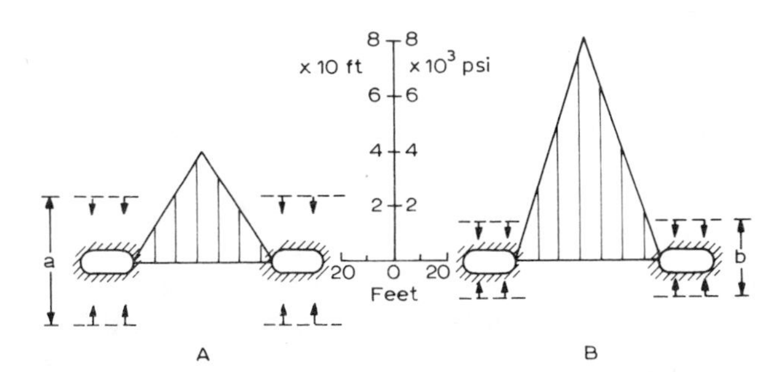

Fig. 5-3. Pillar loads under different conditions, IMC-K 1 mine (Baar, 1972a, 1975).

General conditions as in Fig. 5-1, except for a carnallite content of approximately 10% of the salt rocks involved in creep deformations. Pillar loads shown schematically after room-and-pillar extraction of one third of the potash bed.

a, *b* = vertical dimensions of the creep zones; *A* = one half of the theoretical overburden load is supported in the abutment zone of a mined-out area with limited span; *B* = overburden load fully supported by pillars.

TABLE 5-I

Room and pillar dimensions in extraction mining areas of two Canadian potash mines at depths of approximately 1000 m (conventional room-and-pillar patterns; dimensions in m)

Mine	Year	Rooms		Pillars		Design basis	Reference
		height	width	width	length		
IMC-K1	1963	2.25	6.3	16	42	Serata formula	(Coolbaugh, 1967)
Alwinsal	1972	3.6	8.4	18	44	Dreyer formula	(Duffield, 1972)

In contrast, Dreyer (1972, 1974a) is continuing to refer to premature publications of IMC data, apparently attempting to prove the validity of hypotheses which resemble those applied in the original IMC design. In view of the consequences experienced at another Canadian potash mine — see the following sub-section — extreme caution appears to be indicated.

5.2.2 Alwinsal Potash of Canada mine, Lanigan, Saskatchewan, Canada

Geology and mine development are described in detail in both the English and the German language by Duffield (1972) and by Rininsland (1972). The mine layout resembles the original IMC pattern. "Room and pillar dimensions have been derived using the Dreyer Formula" (Duffield, 1972, p. 224). The room and pillar dimensions in extraction mining areas are compiled in Table 5-I together with the dimensions of the original IMC design. The differences are negligible, although there are significant differences in the geological conditions: the room height is more than 50% greater at Alwinsal than it is at IMC; the opposite relation holds for the thickness of the salt cover above the mine workings: at Alwinsal, it is only 10—24 m; at IMC, it is 30—45 m.

Duffield (p. 223) emphasizes that the overlying Dawson Bay Formation "contains considerable quantities of water in many areas. Although this water is saline, it is usually far from saturated and may migrate down into the workings if roof movement during mining is not carefully controlled".

The potash bed extracted at Alwinsal "is subdivided into five beds by distinct clay partings up to 2 in. (5 cm) thick, the uppermost of which marks the top of the potash and is used as a marker band for grade control in mining".

Local undulations of the potash bed "are generally associated with a washout or collapse, where the normal layer, or the upper portion of the layer (of potash), has been replaced by halite, or occasionally by sylvite. In the more complex collapses, the normal layer has been replaced by a brecciated mixture of halite, sylvite and clay. The extent of these disturbances has varied from a few feet to about 100 ft. (30 m). They have no regularity and are totally unpredictable until a mining face arrives within a few feet of the disturbance" (Duffield, 1972, pp. 223, 224). Apparently, these anomalies resemble those encountered in the neighbouring potash mines—see Baar (1974a) for details.

The overall panel extraction ratio with the dimensions listed in Table 5-I is 38.5 per cent; "the pillar factor of safety is infinity" according to calculations using the Dreyer formula; "as more experience is obtained, it may be possible to mine additional cross rooms to leave 59-ft. (17.7 m) square pillars, which would increase the extraction ratio to 47 per cent with a resultant pillar safety factor of 3" (Duffield, 1972, p. 225).

Such optimistic expectations, based on erroneous assumptions regarding pillar behavior in deep potash mines, proved erroneous as soon as extraction mining resulted in overburden load redistribution and subsidence: as no protecting pillars were left along the main entry systems, "considerable creep rates developed" (Rininsland, 1972, p. 34), indicating considerable abutment loads around developing room-and-pillar mining areas.

The published mine map (Rininsland, 1972; Duffield, 1972) leaves no doubt of the location of abutment load zones and corresponding "considerable creep rates": as all pillars were expected to carry the full overburden load after 38.5 per cent extraction with an infinite safety factor, extraction mining began at the shaft safety pillar, next to underground workshops and ore handling facilities. Much larger pillars in these areas provided the abutments for the extracted areas where stress-relief creep made the smaller pillars ineffective, as shown in Fig. 4-13. These relationships had been shown in detail in a public seminar held prior to extraction mining (Baar, 1971a).

Rininsland (1972, p. 34) is stressing the point that no pillar failure occurred; apparently, this is meant to say that the pillars did not suffer fracture as do model pillars in the laboratory when loaded excessively. There can be no doubt that the panel pillars failed to provide the overburden support they had been expected to provide after re-inforcement by strain hardening, as assumed in the Dreyer formula.

5.2.3 Other Canadian potash mines

All but one of the conventional potash mines in Saskatchewan were originally designed as room-and-pillar mines with standard patterns as described in the two preceding sub-sections. The first changes to the standard design were introduced at the IMC-K 2 mine which began operations in 1967; the reasons are outlined in sub-section 2.6.3: the modern mining equipment used in tabular potash deposits calls for rooms as long as feasible for the use of conveyor belts from the face to the main shaft. Based on the experience at IMC-K 2, particularly after realizing the need for wider rooms to eliminate roof shear failure problems (Baar, 1970b; Mraz, 1973), the neighboring Sylvite mine was designed with long rooms in extraction mining areas—see subsections 4.4.1 and 4.4.2. Attempts to introduce similar changes in other Canadian potash mines resulted in serious difficulties described by Serata and Schultz (1972), and Serata (1973, 1974); it was necessary to develop special stress-relief methods dealt with in detail in Volume 2.

An expensive lesson learned in most Canadian potash mines resembles that learned at Alwinsal: entry systems, and other vital long-term mine openings, must be protected against abutment loads which build up around extraction mining areas, no matter whether standard room-and-pillar patterns or long rooms and pillars are used. As the build-up of abutment loads is caused by stress-relief creep in extracted panels, pending sufficient overburden subsidence to reload the pillars, it is apparent that standard room-and-pillar design is bound to result in relatively higher abutment loads: rectangular short pillars, and square pillars, undergo stress-relief creep in all four horizontal directions; long pillars, however, creep only in two directions; this difference is particularly significant in cases where the pillars dimensions are relatively small compared to the horizontal dimensions of stress-relief creep zones: relatively small pillars do not provide any significant overburden support until corresponding overburden subsidence is taking place. For this reason, the abutment loads around developing standard room-and-pillar panels build up relatively fast, and this requires effective protection of long-term mine openings by large pillars, the widths of which must cover the abutment zones around extracted areas.

In most Canadian potash mines, the original main entry systems were lost, as dramatized by Serata (1972, 1973, 1974). This indicates the magnitude of the errors in the earlier hypotheses on which the design had been based. The main entry system at the Central Canada Potash mine, adjacent to the Alwinsal mine, could be saved just in time in the following way: in 1970, mining of the first two panels at both sides of the standard main entry system was halted at sufficient distance from the entry system, leaving protective pillars to cover the abutment zones of the two panels. In the further mine development, the entry systems were developed according to the principles of stress-relief methods—see Volume 2

5.2.4 Cane Creek potash mine near Moab, Utah, U.S.A.

The development of this deep potash mine — to date, it is the first and only attempt to mine potash at depths of around 1000 m in the U.S.A. — is featured as a "textbook example of cooperation between scientists and practitioners in the practical application of rock-mechanical knowledge" (Dreyer, 1974a, p. 133). In fact, the mine had to be abandoned because of apparent difficulties caused by practical application of erroneous hypotheses. Apparently, Dreyer incorporated a premature publication (1969) with similar statements in his 1974 text, simply ignoring that the mine had to be abandoned several years previously. This demonstrates why "we should proceed with caution with new 'unprecedented' design methods until the field evidence either validates or disproves the method" (Piper, 1974).

The field evidence regarding the Cane Creek mine, published by Wieselmann (1968) and Dreyer (1969), leaves no doubt of the reasons why the mine was lost: the mine maps show typical room-and-pillar patterns, similar to those originally designed in Canadian potash mines; several publications (Adachi et al., 1968; Serata, 1970, fig. 38, which shows a test site for stress-field determination in the Cane Creek mine; Winkel et al., 1972) indicate the application of the Serata Stress Control Method (Serata, 1972), or an earlier version of it.

It appears interesting to note Dreyer's (1969) criticism of the mining patterns at the Cane Creek mine; based on detailed data of room and pillar dimensions and extraction ratios in 23 sections of the complete mine map (l.c., table 1 and fig. 2, p. 442), the mine layout is appraised "most unfavorable and uneconomic as 78 per cent of the potash bed remain unextracted" on account of large pillars left along entry systems and room-and-pillar panels with extraction ratios up to 63 per cent. The so-called barrier pillars and safety pillars are considered "extremely dangerous for the stress concentrations which result in non-uniform stress fields in the adjacent extraction areas, where the pillar loads are reduced".

The latter statement is certainly correct: in some of the extracted panels identified on the mine map, the square pillars were only 11 m wide. Such panels are surrounded by larger pillars to provide stable abutments for stress arches as postulated by Serata in numerous publications. Dreyer (1969, p. 443) cites collapses, and a local rockburst that led to the initiation of a measuring program in 1966. Less than two years later, final conclusions were published, and referred to as "textbook example" of successful application of in-situ research. Extreme caution appears indicated in view of the fate of the Cane Creek mine.

In order to prevent misinterpretations, it is emphasized that the term "barrier pillar" used by Dreyer (1969) does not apply to protective pillars up to several hundred metres wide, which must be left along both sides of entry systems for long-term use in deep potash mines. If no such protective pillars are left, using equal pillar dimensions throughout the mine, the entry systems indeed are "rendered useless" due to excessive closure: "non-elastic media such as potash will never achieve absolute stability around a mine entry. The potash will tend to creep or deform until all entries are closed and the overburden stress is redistributed over what was previously the mine area" (Wieselmann, 1968, p. 2).

As shown in sections 4.2 and 4.3, deformations by creep are not restricted to potash pillars; salt rocks above and below the pillars contribute particularly to failure processes by lateral creep deformation. The published mine map (Dreyer, 1969), and related data, suggest that, following Dreyer's quoted recommendations, the width of protective pillars along the main entries was reduced, rendering the entry systems useless and forcing the conversion of the mine into a solution mining operation. "The expected 'stable conditions' from strain hardening obviously did not materialize" (Baar, 1973, p. 253), as expected in the hypotheses promoted by Serata and by Dreyer.

According to Fig. 3-5 (Wieselmann, 1968), the entry closure rates at the Cane Creek mine became constant after only a few weeks, indicating that strain hardening had no effect on the creep deformations—see sub-section 3.4.2. Effects of reloading of stress-relieved pillars, as shown in Fig. 5-2, are not mentioned in Wieselmann's publication, in all probability due to the fact that measurements had begun less than two years prior to the publication of "final" conclusions. It appears obvious that the design of long-term entry systems, around which extraction mining takes place, cannot be based on short-term measurement data, ignoring the known effects of reloading caused by overburden subsidence.

5.2.5 Werra potash mining district, Germany

Several publications deal with the change-over from mining methods with long rooms and pillars to standard room-and-pillar design with square pillars, and with in-situ measurements to elucidate the rock-mechanical aspects. Measurement data covering any reasonable length of time have not been published, although much publicized measuring programs began over a decade ago.

Neuwirth (1960) published a mine map showing the first experimental room-and-pillar panel developed in the Wintershall-Heringen potash mine; he (l.c., p. 45) believes that extraction ratios of 80 per cent at the 420-m level are feasible with 10-m square pillars in a 3-m potash bed. To support his belief, Neuwirth stresses that, after one and one half year, only negligible deformations of pillars were observed; there was only one measuring site installed in a small test panel. At this site, "little slabbing" to the depth of 1 m into the pillar was observed. According to the mine map (l.c., fig. 10), the span of the mined-out area was approximately 200 m at the time of this observation; the observation was made near the centre of the mined-out area. No conclusions must be drawn from such publications which lack detailed data on mining history prior to, and during, the time period referred to. It appears reasonable to assume that the pillar had not yet been fully loaded as the span of the mined-out area was still too small for unrestricted overburden subsidence.

In the neighboring Hattorf mine, the change-over to standard room-and-pillar mining methods was completed in 1967 (Uhlenbecker, 1971, p. 345).

Results of comprehensive measurements were published by Uhlenbecker (1968, 1971, 1974), including underground convergence measurements, and surface subsidence measurements since 1954, i.e., the subsidence measuring system was set up after a devastating rockburst had occurred in 1953 in the abovementioned mine, see Volume 2 for details.

The first publication of measurement data and evaluations represents the dissertation of the author who, quite understandably, was in agreement with the hypotheses promoted by his academic supervisors (Borchert, Haupt, Dreyer) at that time; in fact, the evaluation of selected measurement data apparently is designed to prove the validity of the strain-hardening concept for mine pillars. This is repeatedly emphasized, and still postulated in the 1971 publication (figs. 3 and 4), which was criticized for this reason by Baar (1972a, pp. 29, 30, fig. 2). In the 1974 publication, the strain-hardening concept is no longer mentioned; this brings the evaluation of the measurement data better in line with the apparent fact that both the underground deformation rates and the surface subsidence rates are virtually constant after constant loads on pillars are established.

In the Hattorf mine, the extraction of the upper potash bed advanced from approximately 600 m depth to approximately 1000 m depth; the room-and-pillar design was adopted at depths of 800 m and more, as shown in the complete mine map published by Uhlenbecker (1971, fig. 2). According to the map, numerous measuring sites were installed in the room-and-pillar mining area; measurement data are not included in any of the publications: the reader is expected to believe in the writer's evaluation of generalized data; it should be borne in mind that the evaluation may be biased by preconceived hypotheses.

Room and pillar dimensions are as follows (Uhlenbecker, 1968, table 1): room widths 12—14 m, height 3.5 m, pillars 16 m square, resulting in extraction ratios of 67—70 per cent, comparing to 57—60 per cent in mined-out areas with long pillars. Neither the mine map nor the texts indicate any protective pillars along entry systems in the mined-out room-and-pillar extraction areas. This is indicative of the belief in adequate overburden support by 16-m square pillars under average loads of 800 kp/cm^2 (Uhlenbecker, 1974, p. 308).

Fig. 5-4, shown repeatedly by Uhlenbecker (1968, 1971, 1974), exhibits selected convergence curves; the history of mine development around the measuring sites is not given in sufficient detail to explain the differences between these cumulative closure curves. These curves were measured in mined-out areas with long pillars; they are shown here to emphasize the apparent similarities with some of the curves shown in Fig. 5-2. These similarities are: curves MP 36 and MP 52 resemble curve No. 2 of Fig. 5-2. They develop into straight lines, i.e., the closure rates decrease regularly to constant small rates. The humps in the closure rate curves, Fig. 5-5, shown at the time of two years, represent an error which was not eliminated, although

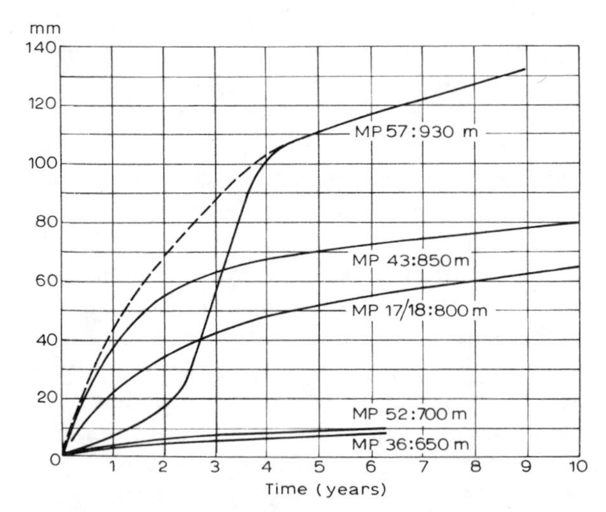

Fig. 5-4. Vertical room convergences, Hattorf potash mine. (After Uhlenbecker, 1974.)
Convergences measured near pillars at indicated depths. Various amounts of initial convergences are missing as the measurements began at various times after mining.

similar errors in previous publications had to be corrected (Baar, 1972a, fig. 2). Curves MP 17/18 and MP 43 resemble curves No. 7 and 12 of Fig. 5-2, indicating combinations of stress-relief creep, and reloading creep shortly after mining; after about five years, constant convergence rates indicate full pillar loads according to depths and extraction ratios. Curve MP 57 resembles those curves in Fig. 5-2 on which the beginning of reloading is indicated by the letter *R* (curves *2*, *6*, *10*, *14*).

Fig. 5-5 shows the convergence rates calculated from the cumulative closure curves of Fig. 5-4, with one addition: MP 71, measured in the developing room-and-pillar area. This curve is of particular interest for the following reasons:

The measuring system was installed at the time when the pillar was still representing a portion of the advancing face of the room-and-pillar panel, i.e., after three sides of the future pillar had been excavated. According to Fig. 5-6 (after Uhlenbecker, 1968, fig. 32), measurements commenced early in 1966. In February 1967, the convergences between the indicated points in roof and floor had reached amounts of 78, 77, and 73 mm, respectively. The maximum initial convergence rate shown in Fig. 5-5 — 18 mm/100 days — apparently is calculated from these measured values for little better than one year; this is incorrect as the convergence rates for the first 100 days are considerably higher, as evidenced by Fig. 5-6: approximately one half of the total convergence after 400 days occurred during the first 100 days; during this initial measuring period, the convergence rate was at least twice as high as shown in Fig. 5-5. As the time since mining is not indicated in the publica-

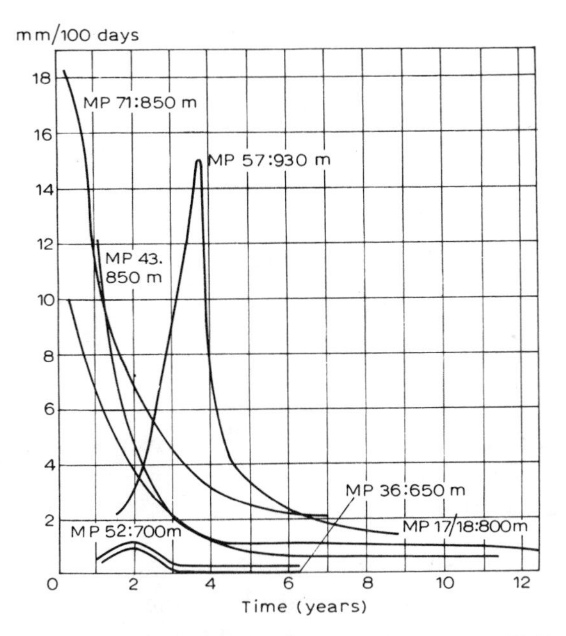

Fig. 5-5. Development of convergence rates, Hattorf potash mine. (After Uhlenbecker, 1974, fig. 5.)

Convergence rates as shown for the cumulative convergence curves of Fig. 5-4. See text for corrections of apparent errors.

tions referred to, the amount of initial stress-relief convergence prior to the beginning of measurements, is not known; it may be of the order shown in Fig. 5-4, probably even higher as blasting is known for increasing the initial stress-relief creep rates (Baar, 1970a, p. 282).

Fig. 5-6 shows another interesting feature which is not shown in Fig. 5-5: in April 1966, the convergence rates had levelled off considerably; this holds particularly for the roof subsidence for the first 100 days. After this period, the roof deformation accelerated temporarily, exactly as shown in Fig. 5-2 by the curves marked with the letter *R*. There is little doubt of the same cause of temporary acceleration of creep after the first 100 days in Fig. 5-6: apparently, the advancing face had reached the span required to initiate effective overburden subsidence to reload the pillar. Full pillar reloading, i.e., pillar loads according the depth and extraction ratio, were achieved after approximately seven years since mining, as shown by curve MP 71 in Fig. 5-5: the constant convergence rate of 2 mm/100 days was finally reached; it should be noticed that this rate is approximately 3 times the convergence rate of MP 43 at the same depth, but in a mined-out area with long pillars.

Fig. 5-7 (Uhlenbecker, 1974, fig. 3) supposedly shows the subsidence at the depth of 900 m within 15 years; it cannot be reconciled with similar

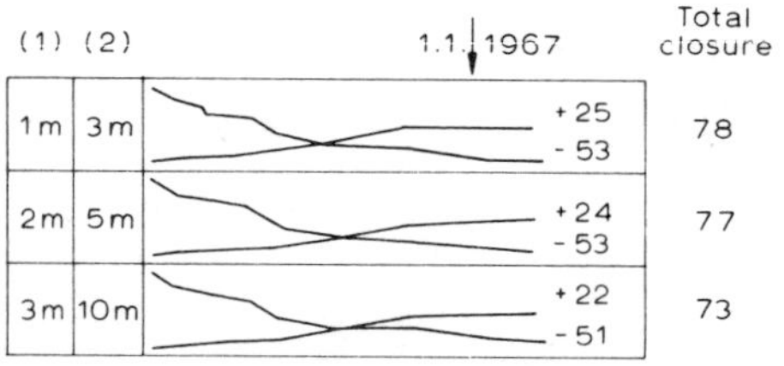

(1) (2)
1.1.1967
Total closure
1m 3m
+25
-53
78
2m 5m
+24
-53
77
3m 10m
+22
-51
73

Fig. 5-6. Convergences in a room-and-pillar panel, Hattorf potash mine.

Cumulative closures measured between points at the indicated depths into floor (*1*) and roof (*2*), starting in January, 1966. + = absolute floor uplift in mm; — = absolute roof subsidence in mm. Measurements at point MP 71 of Fig. 5-5, after Uhlenbecker (1968). See text for corrections.

maps published by the same writer in 1968 and 1971, which show the mine development since 1968, and the related development of the underground subsidence in room-and-pillar mining areas. The following four major corrections are required:

(1) The 60-mm subsidence trough near the NW corner ot the mine map is shown with nearly identical shape and contours on the 1968 mine map (l.c., fig. 22), and on the 1971 mine map (l.c., fig. 11); according to the 1968 map, the 60-mm figure applies to the time period from 1954 to 1966;

-20
-40
-60
Shaft
Shaft
-60
-60
-80
-60
-40
-20
1 km

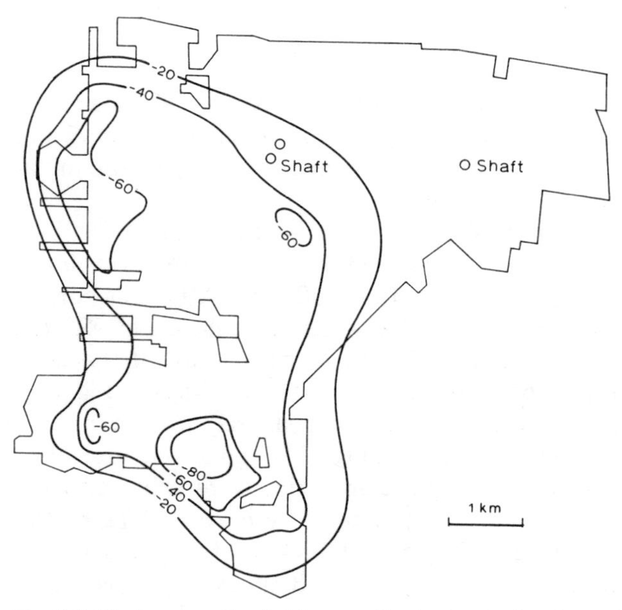

Fig. 5-7. Underground subsidences, Hattorf potash mine. (After Uhlenbecker, 1974.)

Outline of mined-out areas, and subsidences according to Uhlenbecker. See text for comments and corrections.

according to the 1971 map, it was measured from 1954 to 1969; and according to the 1974 map, there would have been no change since 1966. Apparently, extraction mining ceased in that area in 1966; if indeed no further subsidence occurred since 1968, the pillars in that area, including a room-and-pillar panel, would qualify as stable pillars with ultimate strengths to support the full overburden load without further creep deformation, as postulated in the strain-hardening concept. However, the 30-year convergences shown in other figures (Uhlenbecker, 1968, fig. 35; 1971, fig. 17) demonstrate creep closure continuing after 30 years.

(2) The 60-mm subsidence trough near the SW corner of the mine map developed since 1968, in response to room-and-pillar mining of an area over 1 km^2. The maximum subsidence coincides with the area of a 600 m wide remnant pillar which was temporarily left unextracted between the existing mined-out area with long pillars and large remnant pillars to the north, and the room-and-pillar development to the south which advanced to the west—see Uhlenbecker (1971, figs. 8 and 11). The 1971 mine map demonstrates this development which created three overlapping abutment zones in the temporary remnant pillar. The resulting stress-relief creep probably developed quickly into reloading creep, comparable to the development of the curves No. 13 and 15 of Fig. 5-2, i.e., the creep rates must have been considerably high, as confirmed by the subsidence trough shown in Fig. 5-7.

The exact time of mining the remnant pillar is not known; according to the above quoted mine maps, the remnant pillar was created in 1971; if Fig. 5-7 is correct, the 60-mm subsidence trough must have developed within less than three years; the creep rates must have been considerably higher than those shown in Fig. 5-6, curve MP 71. It should be borne in mind that the roof subsidence shown in Fig. 5-7 is only one part of the convergence shown in Fig. 5-6; floor upheaval contributed the other part the magnitude of which is not known.

It is emphasized that such a mining sequence, leaving a remnant pillar 600 m wide in a mined-out area, is typical of those which caused severe rockbursts in other mines of the Werra district; this is shown in more detail in Volume 2. There is reason to assume that a comparable rockburst did not occur in this particular case because the remnant was mined prior to the build-up of tensional stresses in overburden formations sufficient to cause tensile failure. In addition, geological fissure zones in the district generally exhibit approximately a N—S strike; therefore, the development of overlapping tension zones above the remnant pillar was not aggravated by geological features as apparently the room-and-pillar mining at both sides of the remnant pillar advanced in the westerly direction.

(3) The 80-mm subsidence trough in the southern part of Fig. 5-7 covers nearly exactly the room-and-pillar panel around MP 71 of Fig. 5-6. According to Uhlenbecker (1968, p. 51), the measuring site was located in the centre of the panel with dimensions of 1300×900 m. The 900-m span

apparently applies to the N—S direction; only negligible further extraction mining in the southerly direction is indicated on the 1971 mine map. It is not indicated in the publications whether the extraction to the south was terminated for geological reasons, or whether the termination was forced by excessive deformations in the entry systems which were not protected against full overburden loads; if mining ceased for the latter reason, this would merely confirm the experiences in North American deep potash mines described in sub-sections 5.2.1—5.2.4.

Fig. 5-6 clearly shows the following fact: in 1967, only one year after mining, the underground subsidence had already reached over 50 mm. For the second year, Uhlenbecker (1971, fig. 12) shows the average subsidence of about 7 mm/100 days, and this adds 25 mm to the measured subsidence. This means that the 80-mm subsidence trough shown in Fig. 5-7 existed already in 1969 and deepened considerably if Fig. 5-5 is correct. Consequently, Fig. 5-7 must be considered incorrect in this regard; it appears to represent a theoretical construction to prove the correct design of pillars assumed to acquire considerable support capacity by strain hardening. Considering the existence of bedding planes in the immediate roof strata — it is emphasized in various other publications that roof bolting immediately after the excavation of any opening is required to prevent roof failure — it appears highly unlikely that the room-and-pillar panel referred to remained accessible after the extraction ceased in 1969.

(4) After the termination of mining in the panel referred to in the foregoing, further room-and-pillar development to the south was restricted to an area along the eastern lease boundary—see Fig. 5-7. The new extraction area is separated from the older workings by a remnant pillar of considerable size; this configuration represents another rockburst-prone mining pattern, see Volume 2 for details. Apparently, the extension of the 60-mm subsidence contour into the new extraction area, along its western boundary, prevented any dangerous delay in overburden subsidence.

The course of the 20-mm underground subsidence contour line in the extension of room-and-pillar mining to the south and to the west, cannot be reconciled with curve MP 71 in Fig. 5-5 and in Fig. 5-6. Both room-and-pillar areas cover spans exceeding 1 km at depths around 800 m; with such spans reached, the protected zone along advancing faces is only 100—150 m wide, as emphasized by Uhlenbecker (1971, p. 351). At greater distances, the pillars supposedly carry the full overburden load; this means that stress-relief creep and reloading creep together result in considerable short-term convergences as shown in Fig. 5-4; it should be recalled that these convergences are measured convergences, and that the convergences caused by initial rapid stress-relief creep are missing.

The 20-mm subsidence contours in the two new room-and-pillar panels referred to, are shown at distances much greater than 100—150 m from the advancing faces. This is inconsistent with the statements quoted above. And finally, it is emphasized again that the areas were mined in 1971 and later; therefore, the reference to 15-years subsidences in Fig. 5-7 is definitely incorrect.

The need for careful examination of some publications, and for extreme caution in adopting the hypotheses promoted therein, appears beyond doubt. Fortunately, some writers publish data which makes it possible to show their hypotheses erroneous, as emphasized (Baar, 1970a, 1972a) regarding publications by Höfer (1958a, 1964) on which Uhlenbecker is basing his ideas.

As pointed out by Obert (1964, p. 544, with reference to a personal communication from Uhlenbecker): "axial pillar deformation measurements taken in the Hattorf mine over an 8-year period also show that in areas undisturbed by mining the creep rate is virtually constant". Evidently, Uhlenbecker's publications are biased by attempts to prove the validity of parameters derived from laboratory testing over many years; his in-situ measurement results prove exactly the contrary.

5.2.6 Standard room-and-pillar mines at shallow depths

Modern equipment for standard room-and-pillar mining requires sufficient working faces within suitable haulage distances from the stationary haulage system. Equipment developed for coal mining has been employed for decades in the potash mines of the Carlsbad district, New Mexico, U.S.A., at relatively shallow depths of around 350 m; the thickness of the potash beds rarely exceeds 2 m—see Borchert and Muir (1964) or Linn and Adams (1966) for details regarding the geology. Rock-mechanical difficulties of any significance apparently are not encountered; as a matter of fact, secondary pillar mining allows nearly total extraction with controlled subsidence, as shown in more detail in Volume 2.

Standard room-and-pillar design is also the common design in rock salt mines; see, for example, Smith (1966), or Peyfuss and Jacoby (1966) for maps of extraction panels in relatively shallow rock salt mines in the U.S.A., or Hedley (1967) for maps of Canadian rock salt mines. Room heights in rock salt mines are usually much greater than in potash mines, so creep deformations are much less restrained. For this reason, the effects of rapid initial stress-relief creep into newly excavated openings in salt mines are more serious in comparison to potash mines at similar depths.

Hedley (1972) reported on pillar undercutting experiments in a Canadian salt mine which confirm this point. The tests were designed "to transfer the stress concentration from the pillar edge over the pillar itself" (l.c., p. 4, and fig. 5) in order to eliminate roof falls due to "expansion" of salt beds to depths of about 4 m into the roof. Roof deformation curves measured over 290 days show virtually constant deformation rates, indicating that the measurements began after the phase of initial stress-relief creep at decreasing rates. The measurement results indicate: "1. an almost linear relationship between deformation and time; 2. undercutting the pillars produced no major change in the rate of roof deformation; 3. the movements recorded on the 12-ft. (3.7-m) and 16-ft. (5-m) wires in boreholes 1, 3, and 4, were almost identical". Hedley arrives at the conclusion that "the pillar sides were probably destressed prior to undercutting", i.e., by stress-relief creep. The assumption of stress concentrations at the pillar edge is based on elasticity theories which are not applicable to salt rocks, as shown in Chapter 3. In the case of elastic behavior, stress peaks at the pillar edges are transferred deeper into the pillars when near-surface failure occurs.

The salt pillars which were undercut in Hedley's experiments did not show any signs of failure; the conclusion that stress-relief creep destressed the pillar sides, appears inescapable. The depth into the pillars to which the original stresses were reduced by stress-relief creep, are not known.

The effects of stress-relief creep on roof beds were shown in subsection 4.2.3; the fact that roof expansion occurred at constant rates before and after undercutting the pillar sides, clearly indicates that pillar expansion by creep caused the roof failures reported in Hedley's publication. The mechanism apparently resembles a combination of conversion creep and buckling in response to lateral deformation above the pillars.

The roof fall problem improved "when the roof span was reduced from 60 ft. (18 m) to 45 ft. (14 m)" (l.c., p. 6). Apparently, under the given geological conditions, buckling failure problems are reduced in this way; to reduce the shear failure problems, the writer believes "that the openings can be mined to a trapezoidal section that is 20 ft. (6 m) wider at the floor than at the roof" (l.c., p. 6). This opinion is not elaborated on; it is emphasized that such proposals cannot be reconciled with the results of in-situ measurements presented previously in this chapter: to eliminate shearing, room sections should resemble the "active" section of openings in salt rocks—see Fig. 4-3. As a matter of fact, room sections resembling "active" sections proved advantageous in potash mines in the carnallitic Zechstein 2 potash bed—see Pawlick (1943) or Baar (1959a) for details.

Hedley's (1967) "appraisal of convergence measurements in salt mines" is based on limited data from five room-and-pillar salt mines. This appraisal is a typical example of application of elasticity theories to "salt pillars which appear to be stable immediately after mining, (but which) often deteriorate with time and may ultimately fail" (l.c., p. 117). "The rate of convergence is the best parameter for comparing the relative stability of the pillars at different mines or sections of mines. The higher the convergence rate the less time it takes for the pillars to reach instability. The value of deformation at which the deformation rate starts to accelerate is chosen as the point between stability and instability. The time taken by a pillar to reach an unstable condition can be calculated from an experimental value of the maximum permissible vertical deformation and the rate of convergence measured in situ."

Disregarding the possibility of stress-relief creep, Hedley (1967, p. 118) defines the "factors affecting convergence measurements" as follows:

"In room-and-pillar mines the convergence of the roof and floor is made up of two components: the vertical deformation of the pillars because of the weight of the overlying strata; and the local sag or heave of the immediate strata. The component which predominates depends on the relative widths of the pillars and rooms and the geological structure of the overlying and underlying strata. When the strata in the immediate roof and floor are weak, the convergence is principally the result of gaps between layers of strata and does not reflect either the vertical deformation of the pillars or the movement of the mass remote from the excavation. When the pillars are very large compared to the room sizes, the vertical deformation of the pillars is small and the convergence is primarily the local sag of the roof."

In a brief discussion of the factors "which have been found to influence convergence measurements in room-and-pillar salt mines", Hedley (1967, pp. 118 ff.) states: "The reaction of salt to load is time-dependent: after the initial application of load the rate of deformation is high but decreases with time; this is followed by a period when the rate of deformation is approximately constant; and finally the rate of deformation accelerates, resulting in failure."

Evidently, such statements reflect hypotheses based on the so-called "ideal" creep curve obtained in the laboratory—see section 3.4. The effects of stress-relief creep, particularly the additional removal of overburden support by stress-relief creep, as also found by Hedley (1972) in pillar undercutting tests, are ignored.

Fig. 5-8 (Hedley, 1967, fig. 1) shows an example of the "effect of time on convergence measurements at the Meadowbank Salt Mine in Cheshire, England"; the locations of the convergence stations are shown on the mine plan. The stations were installed two to three years after mining; since 1960, the convergence "proceeded at a constant rate".

Hedley (1967, p. 119) confirms that "the loads supported by the pillars are unlikely to be because of the total weight of the overburden, when a room-and-pillar area is only of limited extent. As the extracted area increases in size the load supported by the central pillars increases, until the overburden weight is supported. The pillars in the peripheral zone do not attain full loading conditions, part of the load being transferred to the solid abutments. As a result, the convergence, which is closely related to the loading conditions of the pillars, will vary across an extracted area."

Fig. 5-9 (Hedley, 1967, fig. 2) shows the convergence profile across the centre panel at the Meadowbank mine. "The area was mined previous to 1939 and the rate of convergence at each station is now approximately constant. The convergence is a maximum at the centre of the panel and decreases almost symmetrically towards the solid boundary."

Figs. 5-8 and 5-9 fully confirm observations and measurements in deeper

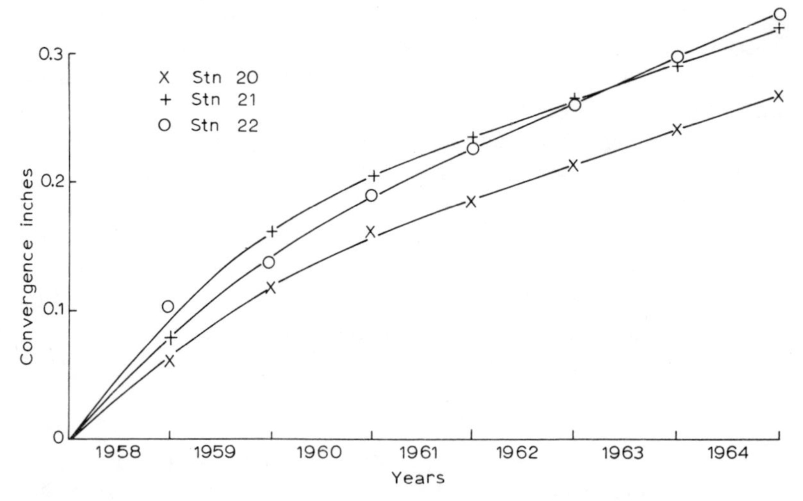

Fig. 5-8. Convergences in a room-and-pillar panel, Meadowbank rock salt mine. (After Hedley, 1967, figs. 1, 5.)

West panel, depth below surface approximately 145 m; rooms 30 m wide, 6 m high; pillars 30 m square; extraction ratio 75%; stations located near the edge of the panel.

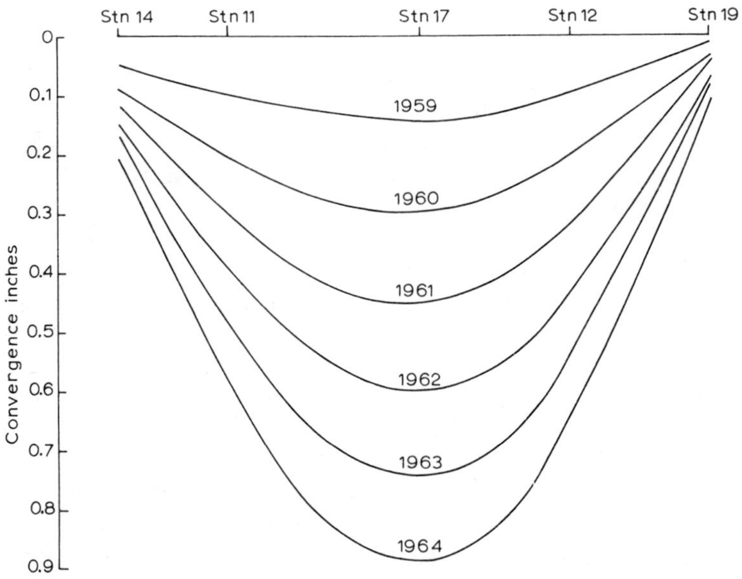

Fig. 5-9. Convergence profile across a room-and-pillar panel. (After Hedley, 1967, fig. 2.)

Meadowbank rock salt mine, centre panel with irregular rooms and pillars, 80—90% extraction; room heights 4.5—9 m; distance between stations 14 and 19 about 300 m.

mines; there is no difference in principles of convergence development in room-and-pillar extraction areas, and in extraction areas with long pillars. Apparently, the initial convergence at decreasing rates cannot be related to pillar loading if the overburden weight in relatively small extraction areas is supported in abutment zones; stress-relief creep is characterized by decreasing creep rates which develop into constant rates under constant loads.

Fig. 5-9 leaves no doubt of the future of the mine, considering the general mine layout: the centre panel is surrounded by other panels with relatively high extraction ratios; the panels are separated by larger pillars which, according to Fig. 5-9, prevent underground subsidence. As a result, tension zones develop above the larger pillars, overlapping at numerous locations. Sooner or later, groundwater will be drawn into the mine. It is difficult to predict whether the mine will be lost due to excess tension above the barrier pillars, or due to tension which develops in the subsiding overburden formations above the centre of the panel.

In contrast to the data shown in Figs. 5-8 and 5-9, Potts et al. (1972) apparently are expecting the underground convergence rates to decrease according to the so-called ideal creep curve, the pillar cores being strengthened to the effect that the creep deformations become negligible in long-term mine planning. Such expectations are based on laboratory experiments which do not reflect the in-situ loading conditions of mine pillars; the effects of stress-relief creep are disregarded.

Disregard of the loss of supporting capacity of salt pillars due to stress-relief creep has resulted in numerous floodings of room-and-pillar mines at relatively shallow depths comparable to those at the Meadowbank mine. Such incidents occurred particularly during the early development of potash mining in Germany during the second half of the past century. Measurement data on surface subsidence caused by room-and-pillar mining is available for time periods exceeding half a century. These measurements are dealt with in detail in Volume 2.

Salt pillars left in shallow mines with relatively high extraction ratios must not be considered stable support elements as designed in civil engineering structures, or in hardrock mines where the rock material exhibits elastic behavior until failure occurs. Salt pillars should be considered effective means to control overburden subsidence, and to reduce subsidence rates according to the requirements at any individual location.

5.3 STRESS-RELIEF CREEP AND PILLAR RELOADING—LONG-PILLAR MINES

5.3.1 General remarks

In the design of mines in rocks which behave elastically until failure occurs, the difference between square and long pillars, with only two pillar sides exposed, is insignificant, although long pillars may offer some advantage in cases where deterioration due to fracture processes at the exposed pillar sides progresses into the pillars: long pillars would lose their supporting capacity at only two sides, the assumed stress peaks near the pillar walls keeping the interior pillars in an elastic state. Some theories claim more efficient support by square pillars as such pillars would provide relatively larger elastic cores.

Such considerations are irrelevant in salt deposits as the effects of rapid stress-relief creep require primary attention: at pillar corners, stress-relief creep occurs in two horizontal directions, with reduced creep rates in each direction; roof and floor failure problems caused by creep deformations above and below the pillars are reduced correspondingly by this "intersection effect" which "extends to distances which are determined by the horizontal extent of creep zones around openings. Since these creep zones also intersect, they result in less horizontal creep near pillar corners where two creep zones overlap" (Baar, 1972a, p. 57).

It goes without saying that these relationships also apply to reloading creep. It must be borne in mind that the effective opening width at intersections is much larger than the widths of the individual openings; in cases where bedding planes allow bed separation in response to lateral creep and limit the maximum opening widths, intersections may produce negative effects as shown in more detail in Volume 2.

5.3.2 Measurements and observations in potash mines, South Harz district, Germany

At the beginning of this century, it was discovered that the famous Stassfurt potash bed exhibits relatively little geological disturbances in areas to the south and to the east of the Harz mountains in central Germany. This discovery led to the development of numerous potash mines at depths between approximately 300 m and more than 1100 m; in large areas of the various mines, the layering is essentially flat or dipping at small angles, offering considerable advantages over the conditions encountered in the Stassfurt anticline and in the salt domes of northern Germany, Poland and Denmark.

The partial conversion of the carnallitic Stassfurt potash bed to sylvinite resulted in high-grade potash ore—see sub-sections 2.2.3 and 2.4.2. However, the extraction of these valuable deposits has been plagued by rock-mechanical problems which apparently have not yet been resolved entirely, in spite of extensive efforts over many decades which resulted in numerous controversial publications. Most writers continue to base their hypotheses on theoretical assumptions, disregarding the effects of stress-relief creep following the extraction of the potash bed. Such disregard has resulted in numerous disasters: rockbursts, mass accidents by gas explosions, mine floodings—see Baar (1966a, pp. 18—20) for the disturbing record of the 1950's. Extensive underground measurement programs have been carried out; the results have been published, and discussed on occasion of international Rock Mechanics conferences over the past two decades. It appears that the results of comprehensive underground measurements to clarify the rock mechanics of disastrous events in the South Harz potash mining district remained largely unnoticed by theoreticians whose hypotheses cannot be reconciled with established facts. For this reason, some of the measurement data is dealt with in detail in the following.

The most comprehensive underground measuring programs were carried out in the Königshall-Hindenburg potash mine near Göttingen, West Germany. The following description of measurement conditions and techniques is that published by Baar (1966a, pp. 18—33); the numbering of figures and tables was changed to match this text.

Fig. 5-10 (Baar 1966a, fig. 2) represents a schematic section through the mine workings; the Zechstein evaporite sequence (salt and anhydrite, see Fig. 2-16 and 2-20 for thicknesses) is overlain by nonevaporitic formations, mainly sandstones, which exhibit near-elastic behavior unless they had suffered deformations in the geological past. The floor strata of the salt sequence are known as reservoirs filled with brines, oil or gases, depending on local geological conditions.

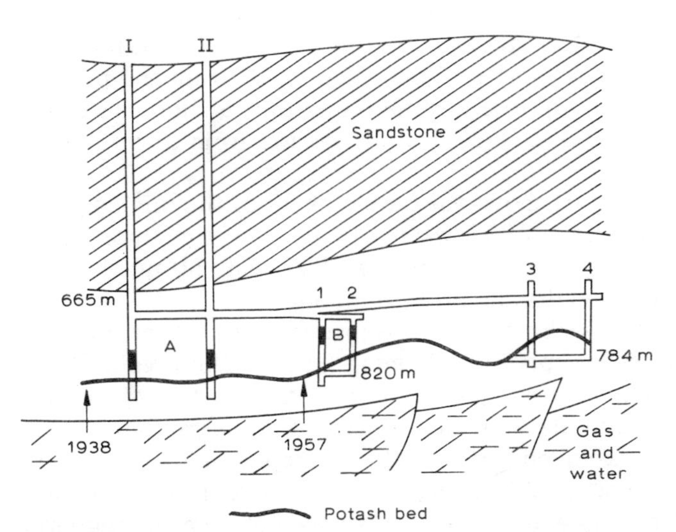

Fig. 5-10. Section through a potash mine, South Harz district, Germany. Königshall-Hindenburg mine near Göttingen — see Figs. 2-19 and 2-20 for general geological conditions (Baar, 1966a).

Main shafts *I* and *II* sealed against mine workings flooded in 1938. Intermediate shafts *1* and *2* sealed against new workings flooded in 1957. Zechstein evaporites overlain by sandstone formation. Stassfurt potash bed with main mining levels as indicated.

In 1938, the mine suffered a water inflow from the floor strata and had to be abandoned. After approximately ten years, the two main shafts were sealed against the flooded mine workings as shown in Fig. 5-10. The 665-m level was extended across the flooded workings, and access to the potash bed was re-established through the two intermediate shafts 1 and 2. This unusual situation made it possible to compare the subsidence at the 665-m level and the deformations caused by extraction mining at the 820-m level.

In 1957, the measured deformations resulted in another water inflow from the floor strata which could not be controlled due to various mistakes, see Volume 2 for details. However, sealing the two intermediate shafts succeeded again, and mining was resumed outside the flooded areas—see Fig. 5-10.

The normal thickness of the salt sequence below the potash bed varies from 30 to 75 m in the South Harz mining district. Unless affected by deformations caused by extraction mining, the floor salt sequence is absolutely impervious and provides for a safe sealing of gases and brines in their reservoirs below the mine workings.

Due to the geological history of sylvite formation, the areas of sylvinitic potash are often relatively small, and surrounded by areas in which no workable ore is found. The sylvinite bed may exceed 10 m in thickness; it was mined in rooms up to 250 m long, pillars of the same length being left between the rooms.

From the geological situation, it is obvious that the remaining pillars have a very important role to fulfill in an extraction mining panel: it is not only indispensable that the pillars support the immediate roof strata and prevent local collapses, but far more important is the prevention of floor deformations which would permit the discharge of water or gases from the reservoirs below the salt sequence.

Fig. 5-11 (Baar, 1966a, fig. 7) shows the development of the mining panel at the 820-m level in which the 1957 water inflow from the floor strata occurred. The salt thickness underneath the panel exceeds 50 m; this had been established by exploration drilling prior to mining.

Fig. 5-11 also shows the development of a subsidence trough at the 665-m level according to measurements by Wilkening (1958). The precise levels of eighteen measuring points across the developing panel were measured at regular time intervals; the traverse was established before extraction mining commenced, and it was extended sufficiently to ground not affected by mining. It is emphasized that no subsidence was detectable until the mined-out panel had reached a span exceeding the distance between mining level and traverse level. Apparently, the immediate roof formations — consisting of shale (30—40 m) and anhydrite (40—60 m) — were behaving like elastic plates until the span required to initiate subsidence was reached. The development of the subsidence trough is discussed in more detail in Volume 2 as it shows the detrimental effects of hot backfill.

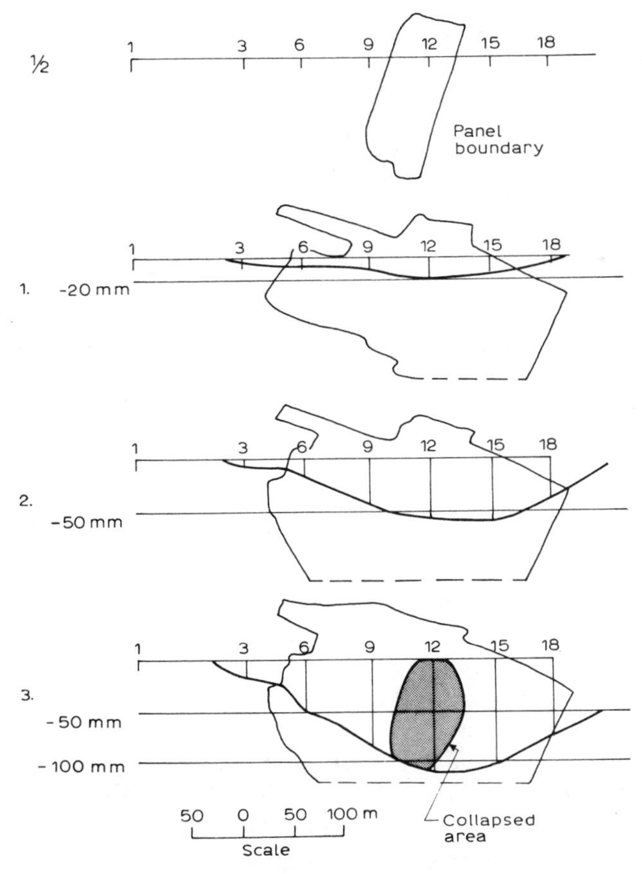

Fig. 5-11. Development of a potash extraction panel, and related subsidences. Panel extracted near the intermediate shafts 1 and 2 of Fig. 5-10. Subsidences measured at the 665-m level, ½, 1, 2 and 3 years after panel mining had begun (Baar, 1966a).

Long rooms and pillars; pillar width 6 m, room width 12 m, final room heights 6—8 m; see Fig. 4-12 for excavation sequence.

Here, the significance of pillar deformations and room convergences measured in the developing extraction panel is discussed. The measurement results are published in detail (Kampf-Emden, 1956); the controversies regarding the evaluation of the measurement data have not yet been settled after 20 years, as emphasized in Chapter 4. The basic question is: do pillar transverse expansions immediately after the excavation of rooms indicate pillar loading, or do such deformations indicate stress-relief creep?

Room and pillar dimensions in the panel shown in Fig. 5-11 were as follows: room width 12 m, room heights 6—8 m, pillar widths 6 m. Extensive spalling during the first year reduced the pillar width at middle height considerably; on some of the first pillars, widths of only 2—3 m after spalling were found in boreholes through the pillars.

The first four pillars in the centre of the panel collapsed after two years; the overlying shale caved in, the following anhydrite bed apparently continued to behave like a plate since the collapse had no effects on the development of the subsidence trough—see Fig. 5-11. A borehole drilled from the 665-m level into the collapsed area confirmed that no bed separation voids had formed above the dome-like cavity which remained accessible during attempts to stop the water inflow from the floor formations.

Measured transverse expansions of the first pillars in the centre of the panel are compiled in Table 5-II. The measurements began one month after mining the first room had started. Measuring time zero is the moment in which the pillar side opposite to the measuring side was increased in height from 2.5 m to the final pillar height of 6—8 m—see Fig. 4-12 for similar mining sequences in another panel. It should be borne in mind that the instrumentation was installed at various times after mining; therefore, considerable amounts of stress-relief deformations immediately after the excavations at any measuring site are missing in Table 5-II.

Some typical pillar expansion curves measured by Kampf-Emden (1956) and discussed by Baar (1966a, p. 24; 1970a, p. 283) are shown in Fig. 5-12. The pillar heights at the measuring sides were 6 m; at the points in time indicated by *B*, the height of the opposite pillar side was increased from 2.5 m to 6 m. Such an increase in pillar height does not change the theoretical pillar load calculated from the deadweight of the overburden and the extraction ratio; consequently, the considerable lateral creep deformations shown in Fig. 5-12 cannot be related to pillar loading, as emphasized by Baar (1966a, p. 24) with reference to in-situ measurement results published by Kampf-Emden (1956), Höfer (1958), and Zahary (1965).

TABLE 5-II

Lateral pillar deformations (in mm) measured on the first pillars in the developing panel shown in Fig. 5-11 (cumulative creep deformations measured since time zero which is the moment in which the pillar side opposite to the measuring side was increased from 2.5 m to the final pillar height as shown in Fig. 4-12)

Time since zero (months)	Pillars [1]				
	1	2	3	4	5,
1	34.1	24.0	24.2	29.5	36.8
2	57.3	43.4	45.4	64.0	66.6
4	97.2	72.6	79.8	106.3	105.6
8	177.4	112.2	117.0	160.1	146.8
10	201.9	123.9	130.9	180.9	157.6

[1] 1—3 = 3-m pillar sections; 4, 5 = 5-m pillar sections

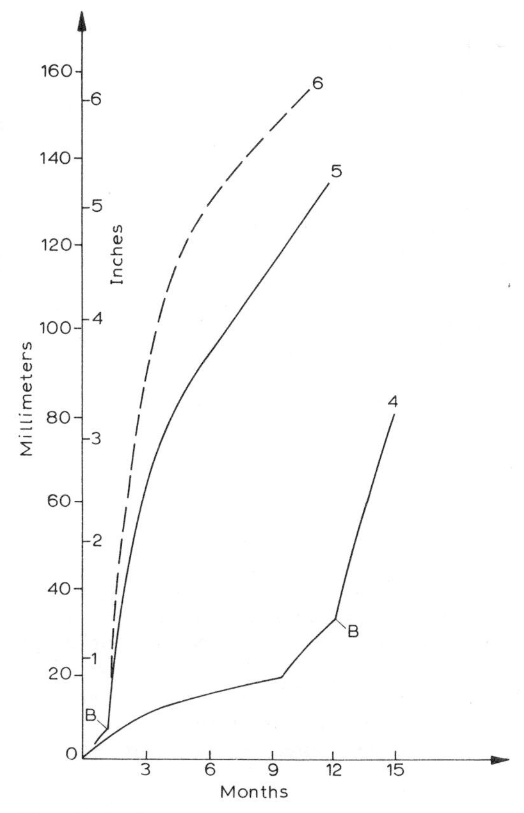

Fig. 5-12. Lateral pillar deformations measured in developing extraction panel (Baar, 1970a).

Pillars between the first rooms of the panel shown in Fig. 5-11. At points *B*, one pillar side was increased in height from 2.5 m to 6 m. Measurements began at various times after mining, i.e., various amounts of initial stress relief creep are missing.

With reference to the discussions by Baar (1959b) and Wilkening (1959), it is re-emphasized that stress-relief creep extends far deeper into the solid salt rocks around newly created underground openings than most theoreticians believe; the stresses in pillars with the dimensions given above are reduced by stress-relief creep as indicated in Fig. 4-12.

Evidently, pillars which are nearly completely stress-relieved and reduced in width by extensive spalling at both sides, cannot fulfill their most important role under the conditions outlined above: such pillars cannot prevent deformations of the floor strata of a developing panel during the time period of negligible roof subsidence as shown in Fig. 5-11. Both Kampf-Emden (1956) and Wilkening (1958) measured considerable room convergences; the convergence rates became constant as soon as nearby mining ceased, indicating extensive creep deformations of the floor salt. Fig. 5-13A shows schema-

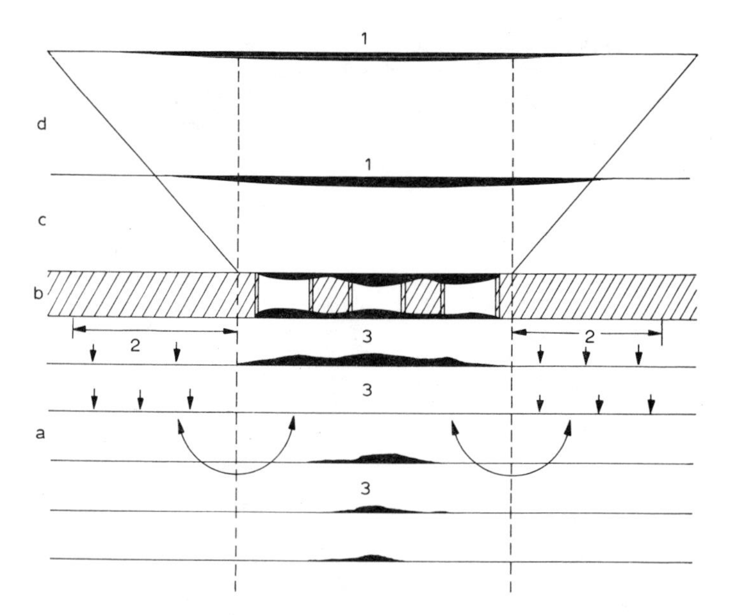

Fig. 5-13A. Schematic summary of measured deformations around a developing potash extraction panel. (Modified after Gimm and Meyer, 1962, who obviously reproduced fig. 24 of Wilkening, 1958.)

Based on measurement data shown in Figs. 5-10 to 5-12. a. rock salt with bedding planes; b. Stassfurt potash bed; c. shale (saliferous clay); d. anhydrite; *1* = roof subsidence; *2* = abutment load; *3* = floor uplift; arrows indicate directions of creep deformations.

tically Wilkening's (1958, fig. 24) summary of his measurement data. The following points are stressed with Baar (1971b, pp. 51—52):

"1. The considerable widening (re. Table 5-II) and shortening of pillars before any subsidence was detectable above the mined-out rooms demonstrates the effects of rapid stress-relief creep around newly mined rooms.

2. In contrast, overburden subsidence requires much more time, in partic-

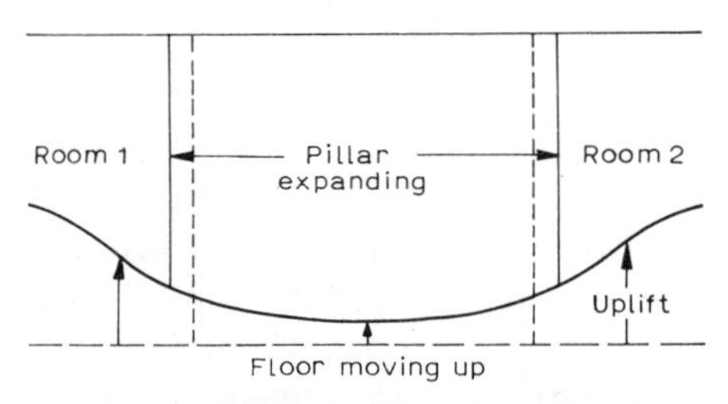

Fig. 5-13B. Floor uplift under conditions of Fig. 5-13A. (After Wilkening, 1958, fig. 22.) Lateral creep deformations as shown in Fig. 5-12 allow floor uplift which is ignored in the original of Fig. 5-13A (Gimm and Meyer, 1962).

ular in developing isolated panels where a certain span has to be reached before the overlying strata begin to subside, and to reload stress-relieved pillars.

3. During this initial stage in panel development in room and pillar mining of salt and potash, considerable additional loads (abutment loads) build up in the virgin ground around a developing panel, regardless of the cause of the delay in subsidence, and how the overburden load not supported by pillars is transferred to the abutments (temporary stress arches in salt rocks; stable stress arches in competent, but fissured rocks; plate-like behavior of stiff, competent formations).

4. The creep of salt rocks out of such abutment load zones causes floor upheaval as demonstrated in various figures published by Baar (1953a,b; 1954a,b; 1959b). Such deformations are known to be dangerous under the given geological conditions because they may damage the natural rock salt seal between the mines and the gas and water reservoirs below the rock salt."

Apparently, Fig. 5-13A finally convinced some theoreticians of the danger involved in the design of barrier pillars as abutments for stable stress arches thought to prevent overburden subsidence in that mining district; after several disastrous gas explosions in 1951, some theoreticians postulated barrier pillars be left after the extraction of the potash bed in a number of long rooms and pillars with standard dimensions. The published controversies climaxed at the International Rock Mechanics Conference 1958 (Baar, Wilkening vs. Höfer, Buchheim, Spackeler, see the Proceedings). The principal objection to barrier pillars for minimizing overburden subsidence in that district is based on the following reasoning (Baar, 1954a,b; 1959a,b): apparently, narrow pillars allow too much floor deformation in developing panels above which the roof formations remain in place due to elastic, plate-like behavior; when large barrier pillars are left unmined after the mined-out span of a developing panel has not yet reached the span required to initiate overburden subsidence, the overburden load is almost completely supported in abutment zones, i.e., in the barrier pillars. Any new panel behind a barrier pillar is again in the dangerous initial stage of delayed overburden subsidence, while the narrow pillars undergo extensive stress-relief creep that allows excessive floor deformations. Gimm (1959, pp. 39—41), referring to Fig. 5-13A, joined this point of view, but also emphasized the need for prevention of constant pillar deformation rates to avoid impending pillar failures as postulated by Höfer (1958) with reference to the so-called ideal creep curve.

To avoid constant pillar deformation rates, the rooms should be backfilled prior to increases in creep rates due to overburden subsidence, or the pillars with equal dimensions throughout mined-out areas should be designed with sufficient widths to ensure continuously decreasing creep rates, leading to ultimate strengths by strain-hardening. It appears that such views are still favored by Gimm (1968, pp. 218—230) and coworkers as they continue to emphasize the validity of creep laws derived from ideal creep curves; the need for establishing "a new equilibrium by ensuring continuously decreasing creep rates in mine pillars" (l.c., p. 226) is particularly stressed.

It must be re-emphasized in the strongest way possible that such hypotheses are erroneous; creep at constant rates is in no way hazardous when caused by subsiding overburden, see Figs. 5-1 and 5-2, even in case the excavated rooms are completely closed by creep deformations within 15 years or so. However, stress-relief creep immediately after extraction mining is extremely dangerous if overburden subsidence is prevented by limiting the span of mined-out areas by barrier pillars to which the overburden load is transferred, resulting in floor deformations as shown in Fig. 5-13A. The danger from water and/or gas reservoirs at relatively small distances into the floor of mined-out panels with limited spans is particularly aggravated when hot refinery waste is used for backfill, as is still recommended by Gimm (1968): in the case referred to above, and in numerous previous cases of gas and/or water inflows from the floor, the temperature increase apparently caused rapid increases in creep rates, and corresponding losses in restraint by pillars against floor deformations; this is shown in detail in Volume 2 for implications with respect to the disposal of radioactive waste.

5.3.3 Sylvite mine, Rocanville, Saskatchewan, Canada

Several misleading publications (Serata, 1972; Serata and Schultz, 1972) dealing with the design principles allegedly developed at the Sylvite mine are corrected in section 4.4. It is re-emphasized that the hypotheses applied by the above writers resemble those promoted by Höfer (1958) until they were finally proved extremely dangerous by several catastrophes (Baar, 1966a, 1970a, 1972a). It is also emphasized again that the need for widening the rooms to prevent roof failures in the eastern portion of the Prairie Evaporites was recognized at the IMC mines (Baar, 1970b, 1972a; Mraz, 1973) where wider rooms were introduced in spite of contrary recommendations and exaggerated warnings by Serata. The misleading publication by Serata and Schultz (1972) made it obligatory for this writer to prepare a detailed confidential paper under a contract which called for review of new publications on potash mining "to safeguard against unadvisable procedures", whatever that may mean. This confidential paper was made available to Serata who referred to it on occasion of a special discussion session held at the Fourth International Symposium on Salt, Houston, 1973, the outcome of which is reported by Piper (1974).

As the confidential content also leaked out of governmental offices, and in view of continuing attempts by Serata (1973, 1974) to promote the obsolete hypotheses first published by Höfer (1958), this writer decided to make his report "Dangerous mining methods in some of Saskatchewan's potash mines must be stopped before it is too late" (Vol. 1, 44 pp., 71 references;

Vol. 2, 51 pp. with figures, 45 pp. with unpublished references; March 1973) available upon request to everyone interested in the technical aspects of potash mining in eastern Saskatchewan.

According to publications by several recognized geologists, the geological conditions in mining the lowermost potash bed in Saskatchewan resemble those encountered in the South Harz potash mining district of Germany dealt with in the preceding sub-section; for details, the reader is referred to Baar (1974a). One particular problem is related to so-called pinnacle reefs which grew locally from the salt basis to the level of the first potash bed which is mined at Sylvite. Such reefs may contain gas, oil, or water, representing a serious safety hazard when connected to mine openings due to application of inadequate mine design principles.

The design principles advocated by Serata (1972, 1974) and by Serata and Schultz (1972) proved dangerous in numerous cases as shown in the preceding sub-section. Although the actual mine design differs greatly from what is claimed in the quoted publications — see sub-section 4.4.2 — the need for the prevention of any stress-relief creep deformations above local reefs is not eliminated as such reefs may have been built up to the level of the potash bed. The authorities responsible for mine safety maintain that adequate precautions are being enforced; details are not available.

The danger involved in applying the present design to extraction of the potash bed above reefs, in evident from data dealt with in preceding sections; this evidence is summarized as follows:

(1) The 50-ft. "yield" pillars, previously considered competent pillars (Serata, 1968), undergo rapid stress-relief creep as shown in Figs. 4-6 and 4-13. It is shown in the related text that the support provided by such stress-relieved pillars is minimal until reloading by subsiding roof formations occurs, see Figs. 5-1 and 5-2.

(2) The room-pillar-room units with 50-ft. "yield" pillars have the span of 184 ft. (55.2 m); these units are separated from each other by 180-ft. (54-m) barrier pillars, apparently in accordance with the design principles promoted by Höfer (1958) which resemble those used in "The Serata Stress Control Method" (Serata, 1972). It has been established at the neighboring IMC mines that overburden subsidence begins when a room-and-pillar panel with 52-ft. pillars has reached the span of approximately 400 ft. (120 m); as the geological conditions at Sylvite are equivalent, it follows that the 184-ft. (55.2-m) spans of the room-pillar-room units do not initiate overburden subsidence and related pillar reloading. Consequently, the overburden load is nearly fully supported by the barrier pillars; this means that the load on the barrier pillars amounts to nearly twice the deadweight of the overburden, i.e., approximately 6000 psi (420 kp/cm^2).

(3) According to Fig. 5-13A, there can be no doubt of considerable horizontal creep into the stress-relieved zones above and below the room-pillar-room units which behave like a single wide opening. As the floor salt

sequence at Sylvite is well stratified by clay and anhydrite bands and beds — see Baar (1974a, fig. 15) — buckling in response to the inevitable horizontal creep away from the highly loaded barrier pillars must be expected, exactly as shown in Fig. 5-13A.

(4) In the case that highly pressurized gases or brines in reservoirs below the mine gain access to bed separation voids in the floor strata of room-pillar-room units, the buckling effect is aggravated by fluid pressure. If and when buckling occurs above reefs filled with fluids, the consequences will resemble those experienced many times under comparable geological conditions, see sub-section 5.3.2. Further details are presented in Volume 2 with emphasis on roof control problems.

(5) The gases contained in reefs of the Prairie Evaporite basin are rich in H_2S, just like the gases encountered in the potash mining district referred to in sub-section 5.3.2. As H_2S is extremely dangerous, even in very small concentrations, particular caution is indicated in the Sylvite mine, as emphasized by Baar (1973, pp. 16 and 38) with reference to accidents in various potash mines, including the one referred to in Fig. 5-13A.

In conclusion, it is felt that those responsible for the safe design of potash mining operations under potentially dangerous geological conditions should be aware of the misleading contents of various publications dealing with the design of the Sylvite mine; the design principles promoted in the publications referred to have repeatedly proved extremely dangerous under comparable geological conditions.

5.3.4 Werra potash mining district, Germany

Mine design with long pillars instead of square or short pillars was adopted during the early 1930's as slusher haulage from the face to the main haulage system was introduced in most potash mining districts; economic considerations limited the lengths of rooms and pillars in extraction panels to approximately 250 m. Examples of mine designs with long pillars can be found in numerous pulications, e.g., Gimm (1968), and in publications referred to in sub-section 5.2.5, in which the recent change-over to standard room-and-pillar design is dealt with. Most of the data presented by Uhlenbecker (1968, 1971, 1974) refer to extraction mining with long rooms and pillars at depths between 600 m and 800 m—see sub-section 5.2.5.

A devastating rockburst on July 8, 1958, destroying 3 km^2 of the upper level and 2 km^2 of the lower level of the Merkers potash mine, initiated serious reconsiderations of the mine design principles applied in the East German potash mines of the Werra district. The following data are taken from related publications since 1960, including another rockburst which occurred in the same mine in 1961; data regarding the most recent rockburst in the adjacent potash mine on June 23, 1975, is not yet available, except for the notification that an area of 0.54 km^2 collapsed, resulting in the heaviest earthquake recorded in Germany since 1911.

The causes of such rockbursts are discussed in detail in Volume 2. Here, some of the recent changes to the design of long-pillar panels, and related in-situ measurements, are presented. Data regarding convergence measurements prior to the 1958 rockburst are extremely scanty as apparently only a few measurements were made.

The room and pillar parameters generally used in the Werra district prior to the 1958 rockburst are compiled in Table 5-III (from Gimm, 1968, p. 362). Apparently, the parameters in sylvinite were the same at all depths from 270 m to more than 800 m. In carnallite at depths exceeding about 500 m, the room width was slightly reduced and the pillar width was increased by 2 m, probably due to considerable heights up to 10 m.

Gimm and Pforr (1961, pp. 74 ff.), referring to unpublished sources, give the results of convergence measurements at eleven measuring sites located on the mined-out upper level of the Merkers mine. The measurements, commencing in 1933, were regularly taken about twice every year, with few years missing around 1945, and continued after the 1958 rockburst. Regrettably, data on the mine development during the measuring time are not included in the publication; this makes any proper evaluation of the measured convergences impossible. It is emphasized that the measured cumulative convergence curves (Gimm and Pforr, 1961, figs. 45, 46) show the characteristics of the curves measured at IMC—see Fig. 5-2; some of the curves demonstrate constant convergence rates over time periods up to 20 years. Most of the curves show temporary increases which apparently are indicative of nearby extraction mining that resulted in higher loads on the pillars near the respective measuring sites. The authors (l.c., p. 74) assume creep deformations according to the "ideal" creep curve, emphasizing their agreement with Höfer (1958) who, like many other theoreticians, believes

TABLE 5-III

Dimensions of long rooms and pillars, Werra district, East Germany (after Gimm, 1968, p. 362); design dimensions (in m) used prior to the 1958 rockburst in the Merkers mine

Mine (location)	Depth	Room width	Room height	Pillar width	Ore type
Dorndorf	270—330	20	2.5	10	sylvinite (Hartsalz)
Merkers	430—480	20	1.8—2.5	10	sylvinite
		20	6.0	12	sylvinite
Menzengraben	525	20	3.0	10	sylvinite
Unterbreizbach	715	20	over 6.0	10	sylvinite
Merkers	480	16	6.0—10.0	10	carnallite
Menzengraben	525	16	3.0—7.0	12	carnallite
Unterbreizbach	715	15	over 6.0	12	carnallite

that pillar failure is inevitable when the stage of secondary creep is reached; when the third stage of accelerated creep is reached, pillar failure is believed to be impending—see sections 3.3 and 3.4. If such hypotheses were correct, various convergence curves shown by Gimm and Pforr (1961, fig. 45) would be indicative of impending pillar failures which should have occurred around 1938 at the locations of their measuring points 5 and 8, shortly after 1950 at points 1, 5, and 9a. At the location of the latter point, pillar failure would again have been impending after 1955. It appears obvious that such hypotheses are untenable.

Gimm and Pforr (1961, p. 74) refer particularly to the measured convergences at site 9, located near the centre of the rockburst area on the upper mining level at the depth of 627 m. Fig. 5-14 (Gimm, 1968, fig. 4/96, and p. 249) shows these convergences and the development after the 1958 rockburst; it is emphasized that the latter publication is obviously erroneous in locating the measuring site in the centre of an area where carnallite pillars were completely destroyed. According to the location map shown by Gimm and Pforr (1961, fig. 47), sites 9 and 9a are located in an area where relatively large remnant pillars were left, apparently because of barren zones in the potash bed; it is known from numerous publications that such barren areas are surrounded by non-carnallitic potash. This is also shown by Gimm and Pforr (1961, fig. 35).

As the mining history around sites 9 and 9a is given in detail in the location map referred to, the causes of changes in convergence rates in Fig. 5-14 can be determined rather precisely. Striking similarities with Fig. 5-2 are outlined as follows:

(1) The area around site 9 was developed in 1930/31 by a double-entry system; during the same time period, a few rooms were mined to the south from the entry system, at distances over 100 m from the future measuring site 9, located approximately 40 m north of the entry system. These excavations may have caused slight increases in the original stresses around the future site 9 (abutment loads).

(2) The room in which site 9 was installed in 1934, had been mined in 1932. Two years of initial stress-relief creep are missing in Fig. 5-14. The room remained a single opening until 1935, at which time the adjacent parallel rooms were mined at the eastern side. The western side remained unmined, providing a large abutment. Increasing abutment loads may have caused the relatively high convergences during this period of time; the excavation by blasting may have been another contributing factor.

(3) After 1937, when mining around the site ceased completely, the convergence rate became virtually constant and remained so until 1950, when mining began to the west of the site, removing the abutment. The temporary increase in convergence rates resembles similar increases in Fig. 5-2 identified by R; such increases are caused by pillar reloading, as shown in sub-section 5.2.1. Apparently, the loads on the pillars around site 9 increased only

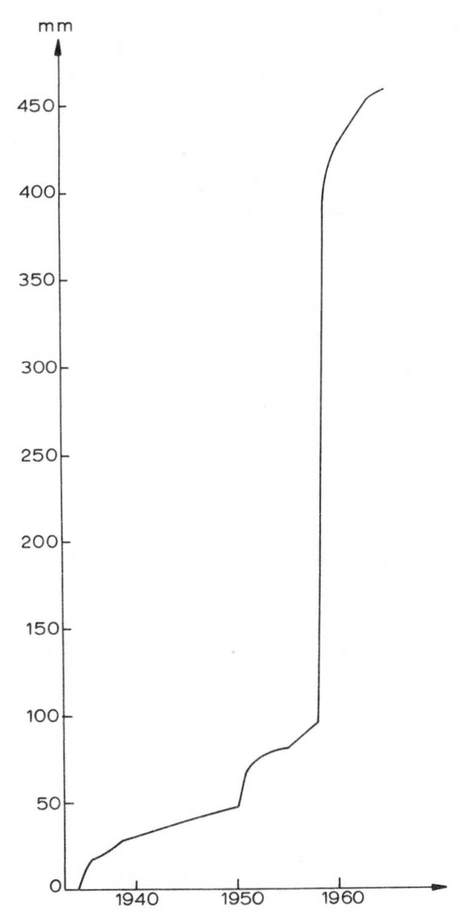

Fig. 5-14. Room convergences measured in a rockburst area. (After Gimm, 1968, fig. 4/96.)

Measurements near the centre of a 3-km^2 area of the upper level of the Merkers potash mine, Werra district, Germany, destroyed by a rockburst in 1958.

slightly; curve *10* in Fig. 5-2 shows a course which resembles that from 1948—1955 in Fig. 5-14. If any comparison is made with ideal creep curves, the creep from 1950 to approximately 1953 would have to be classified as primary creep as it was apparently caused by loading of stress-relieved pillars.

(4) The cause of the accelered creep since 1955 until the rockburst occured in 1958, cannot be identified from the available data; it is assumed that further pillar loading due to subsidence of overburden formations became effective, leading to the rockburst by tensile failure of overburden formations; these relationships are shown in Volume 2.

(5) The rockburst apparently resulted in full pillar loads according to the depth and the extraction ratio. Fig. 5-14 shows the convergence rates again decreasing to much lower values; it appears that constant convergence rates according to the new loading conditions were reached by 1964. This is also indicated by the development of the measured surface subsidence rates (Gimm, 1968, fig. 4/95) to constant rates at the end of 1964. The sudden loading of the pillars around site 9 due to the rockburst may have resulted in some elastic reaction (buckling) of roof and/or floor beds. Therefore, the 250 mm of convergence immediately after the rockburst are not indicative of the amount of load put on the pillars due to the rockburst. However, the convergence creep after the rockburst again resembles the primary creep of an ideal creep curve produced in the laboratory by loading; the secondary creep at constant rates is clearly indicated by the last measurements shown in Fig. 5-14.

Obviously, the spalling off the pillars emphasized by various writers is not indicative of pillar failure; it rather indicates plastic behavior of the interior pillars as shown in sub-sections 3.3.2 and 4.5.4. Sudden loading of pillars with shear fractures already developed, or in the process of propagation, is expected to result in abrupt completion of the slabbing process, leaving the rooms with a near-ideal cross-section—see Fig. 4-3.

With reference to pillar transverse expansion measurements reported by Höfer (1958, p. 85 and table 16), Gimm and Pforr (1961, p. 77) emphasize that the measured lateral pillar creep at the Merkers mine was comparatively small. It should be borne in mind that Höfer's data refer to pillars 2.2—2.4 m high, the creep rates being calculated from time periods up to 10 years, without adequate account for initial stress-relief creep and possible influences by mining. It is not known for a fact, but it appears most likely, considering the irregular mining patterns which resulted in the creep development shown in Fig. 5-14, that the overburden weight was supported mainly by remnant pillars rather than by regular pillars, until the rockburst in 1958 resulted in overburden load redistribution. The calculation of creep rates without proper attention to the length of time since mining any particular opening, is misleading; this is obvious from Fig. 5-2; it is stressed particularly by Baar (1970a).

Caution is also necessary with regard to lateral pillar deformation rates calculated from few long-term measurements in the Unterbreizbach mine, in which the 1975 rockburst occurred. Höfer (1958, pp. 85—88, tables 17—19) and Gimm and Pforr (1961, pp. 77—79, table 6) apparently refer to the same unpublished measurement data, e.g., for the site listed as No. 1 in their respective tables. Höfer gives the depth below surface as 690 m, the other writers are listing 803 m; there is no way of telling who is in error. Neither is there any way of finding out which of the differing creep rates listed in the two publications should be given preference. Höfer (l.c., pp. 87, 88) emphasizes the decrease in creep rates with time; however, the creep rates listed for various periods of time vary considerably in his table 19: for the first 416 days of measurements, a creep rate of 0.24 mm/100 days is calculated; apparently, the measurements commenced after the rapid initial stress-relief creep had died away. For the following 272 days, the listed creep rate is 2.94 mm/100 days. For the following 78 days, it is 6.41 mm/100 days, decreasing to 1.36 for the following 588 days. There is one more increase followed by a decrease, and for the very last 33 days of measuring time, the calculated creep rate is zero, as expected in theory. The creep rate calculated for the total measuring time of 1698 days is calculated as 1.7 mm/100 days, the depth is given as 700 m.

Such varying creep rates evidently reflect various stages of pillar reloading after rapid initial stress-relief creep, as shown in Figs. 5-2 and 5-14; the listed final creep rate of zero for pillar loads which must be the highest ones, after repeated additions of load, is obviously the result of an error in measurement or calculation.

Gimm and Pforr (l.c., p. 78, site No. 2) refer to the very same measuring site; the listed room and pillar parameters, and the calculated creep rate of 1.7 mm/100 days for the total measuring period are identical; however, the depth is listed as 776 m, the length of the measuring period is listed as 3216 days. Again, there is no possibility of telling who is right and who is wrong; the need for extreme caution against such data evaluation appears evident.

On the basis of such and similar questionable interpretations of in-situ measurement data, and on the basis of erroneous assumptions regarding the elasticity of salt rock formations, the authors referred to above arrived at a panel design which promptly resulted in another rockburst in the same mine (Merkers) in 1961. The design principles are of particular interest as they are still promoted in some publications.

The basic error is the assumption of elastic, plate-like behavior of rock salt formations in the case of limited extents of mined-out areas. According to Spackeler et al. (1960, p. 595), the rigidity of a rock salt formation 110 m thick is assumed 15 times the rigidity of an anhydrite bed 30 m thick; the thicknesses apply to the roof formations of the mines in the Werra district, and in districts where the Stassfurt potash bed is mined, respectively. Fig. 5-11 shows measured subsidences for the latter conditions.

According to Gimm and Pforr (1961, pp. 169—171, fig. 99), the following design parameters were adopted in the Merkers mine after the 1958 rockburst: panel width 163 m, the panels separated by barrier pillars 76 m wide; room width 16 m, the rooms separated by 5-m pillars.

The only difference from previous designs postulated for mines overlain by thick, competent anhydrite (Höfer, 1958), is the assumption of plate-like behavior of rock salt formations instead of the assumption of stress arches, which are still postulated by Serata (1972, 1974).

The 1961 rockburst proved the above design parameters dangerous before they were published. Gimm (1968, p. 257) cites the rockburst as an example of relatively rapid subsidence of rock salt formations above very small mined-out areas: "even the span of only 37 m (two rooms with a 5-m separating pillar) resulted in measurable subsidence" at the level approximately 50 m above the excavations. Such subsidences are apparently due to stress-relief deformations, confirming the principles outlined in sub-section 4.4.2—see Figs. 3-6 and 4-13. The significance of stress-relief creep is stressed in various publications by this writer since 1959. These results of in-situ measurements are ignored by Gimm (1968) and others who continue to believe in laboratory parameters, even after application to mine design had proved such parameters dangerous.

After the apparent failure of the above quoted design, new designs were based on the model pillar formulas developed by Dreyer (Gimm, 1968, pp. 365, 366). For sylvinite with the compressive strength of 338 kp/cm^2, the new design called for room widths of 20 m, and pillar widths increasing from 7,5 m at the depth of 400 m to 11.4 m at the depth of 800 m; these parameters are applied in 4-m-high rooms.

For carnallitic salt with the compressive strength of 150 kp/cm^2 and room heights of 6 m, the new design resembled that applied at the Canadian Sylvite mine (see sub-section 5.3.3): room-pillar-room units 42 m wide (two rooms 16 m each, 10-m pillar) were separated by larger pillars for overburden support, the pillar widths increasing from 18 m at the depth of 400 m to 34 m at the depth of 800 m. Measurement results to show the effects of the latter design are not published; application of the above principles in shallow sylvite mines led to what is called the pillar-splitting method (Gimm, 1968, p. 372): in mined-out areas with 10-m pillars, the pillar widths were reduced to about 5 m; this resulted in considerable subsidences dealt with in more detail in Volume 2.

5.3.5 Horizontal creep deformations, Borth rock salt mine, Germany

Comprehensive data on horizontal creep deformations into mine roadways (entries) in rock salt at depths around 800 m is published by Erasmus (1965). The horizontal closure was measured on pins anchored at 1 m depths into the walls over periods up to 10 years. The published data was re-evaluated (Baar, 1971a); in the following, a shortened version of the re-evaluation is presented.

Prior to production mining, two entry systems were developed: an exploration entry system at the 740-m level, and a mining entry system at deeper levels, according to the slope of the salt bed to be mined. The dimensions of the entries are not given in the original publication; apparently, the entry height was approximately 3 m (Dreyer, 1974a, p. 140), while the entry widths may have varied from 6 to 8.5 m. The horizontal closure measurements referred to in the following were carried out at numerous locations in the entry systems.

The designed room dimensions in mining panels are as follows: room height 18 m, width 20 m; pillar width 30 m. The entries are not protected against influences from extraction mining—see Fig. 5-15 which shows a schematic cross-section with entries at the 740-m and the 776-m levels.

Erasmus (1965) shows a series of mine maps covering the time period from January 1, 1950 — at this time, extraction of the first three rooms was in progress — to January 1, 1958 — at this time, an area up to 800 m wide and nearly 900 m long had been extracted, except for a borehole safety pillar approximately 250 m square. Apparently, this large pillar prevented the full loading of the pillars in extracted areas around it; as a result, the final measured horizontal convergence rates vary considerably throughout the extracted area, reflecting the local loading conditions. The measured cumulative closure curves—see Fig. 5-16, resemble those shown in Fig. 5-2, with one important difference which follows from the different final loading conditions: apparently, the span of mined-out areas required to allow full pillar reloading according to depths and extraction ratios, was never reached, particularly around a large borehole safety pillar. In these areas, the final convergence rates developed into constant rates considerably smaller than those in larger extracted areas, see Fig. 5-16 for some curves selected to show the similarities to Fig. 5-2.

Even in the larger area to the north of the borehole safety pillar, the final span of the extracted area is only approximately 500 m; the span was limited to this extent by barrier pillars. In view of the depth below surface, full pillar loads cannot be expected even in central parts of this extracted area; the transfer of considerable loads to abutments is clearly indicated by the results of convergence measurements in the abutment zone which remained approximately 200 m wide. "These abutment loads were not removed by overburden subsidence in the central parts of the mined-out area. This means that an umbrella remained effective along the edge of the mined-out area" (Baar, 1971a, p. 107).

At the end of the measuring period, the width of umbrella zones along the edge of the 500 m^2 mined-out area identified above was between 100 and 150 m; final constant creep rates smaller than those measured in the central parts clearly indicate less overburden load along the edge of the extracted area.

The following conclusions based on the re-evaluation of the published data may be quoted (Baar, 1971a, p. 113): "Erasmus's measurements clearly demonstrate how sensitive creep is in indicating stress changes and overburden load re-distribution in salt and potash mines. They also demonstrate that long-term measurements are required for reliable conclusions. Temporary increases and decreases in creep rates, as demonstrated by the numerous creep curves, definitely exclude the possibility of predicting future creep from measurements over short periods of time, unless the creep rate has become constant, and no further changes in pillar loads occur."

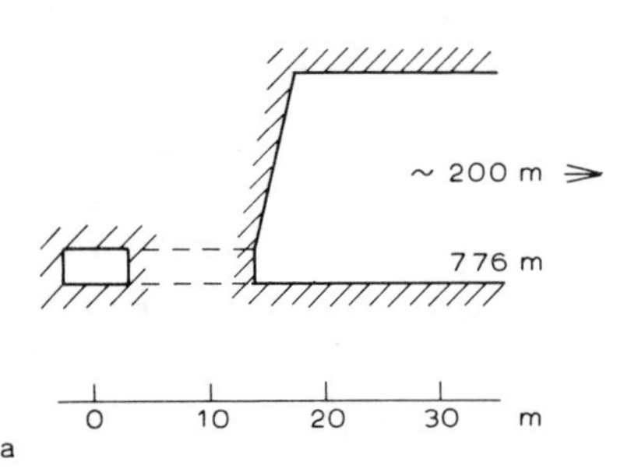

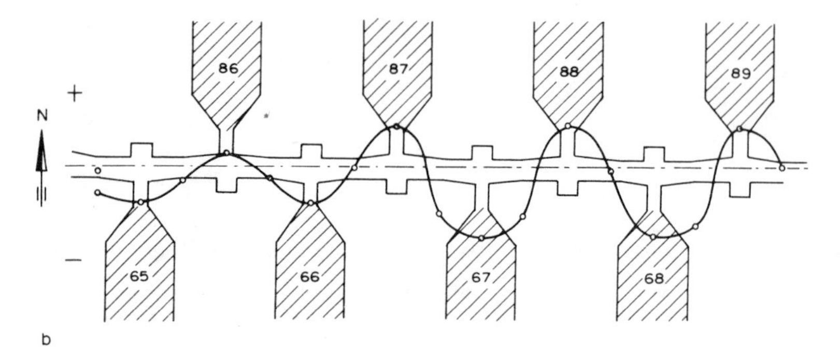

Fig. 5-15. Horizontal displacements of a roadway in the Borth rock salt mine, Germany. (Baar, 1971a, modified after Erasmus, 1965.)

a. Schematic section through entries and room. Exploration entry at the 740-m level, haulage roadway at the 776-m level. Design dimensions of rooms: length approximately 200 m, width 20 m, height 18 m, pillars 30 m wide. b. Measured horizontal displacements of the axis of roadway 8 East. Reactions to the excavation of rooms according to section a: + = displacement to the north; — = displacement to the south. See text for details.

It is emphasized that the measuring sites in the entry systems were installed at varying times after mining the entry systems; for this reason, the initial stress-relief creep at decreasing rates is missing in the curves published by Erasmus. At most stations, the measurement data are represented by fairly straight lines at the beginning of the measurements, indicating virtually constant closure rates after the initial stress-relief creep, and prior to extraction mining. This means: the initial straight sections of Erasmus's curves resemble the straight section of curve *1* in Fig. 5-2, which was not affected by extraction mining. Depending on the reloading history at any particular location, Erasmus's curves develop into straight lines with steeper slopes, the courses of most curves resembling those shown in Fig. 5-2.

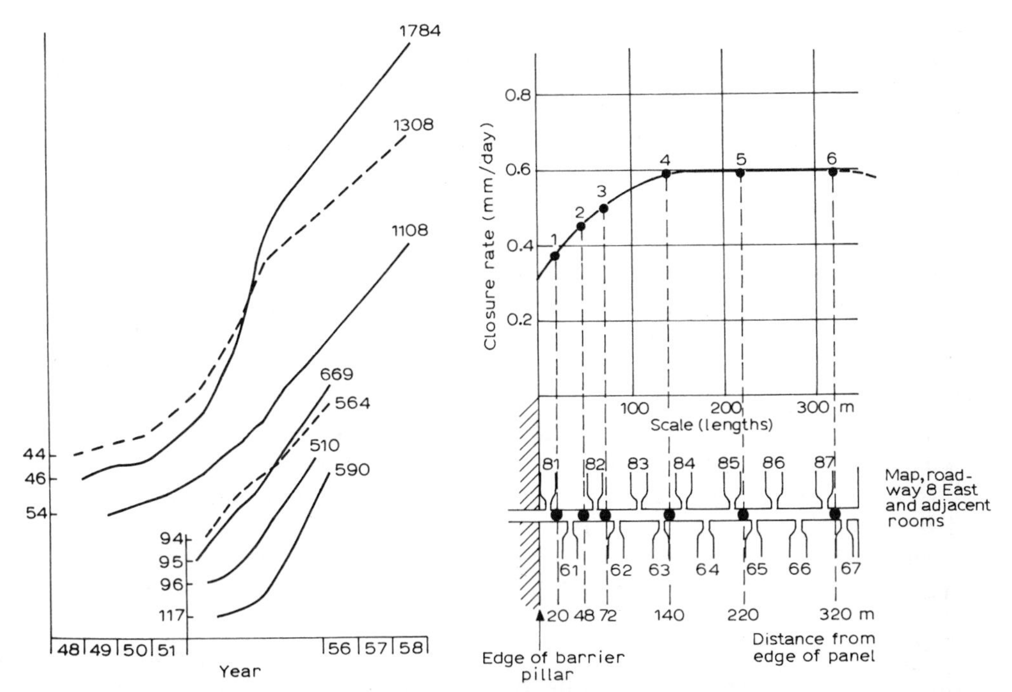

Fig. 5-16. Horizontal creep convergences of roadways in the Borth rock salt mine, Germany, roadways 8 East and 6 East. (Modified after Erasmus, 1965.)

Location of measuring stations: 94 (= *1* in Fig. 5-17) between the first two rooms at the western edge of the panel: 81 (north) and 61 (south) of roadway 8 East. 95 (= *3* in Fig. 5-17) between rooms 82 (north) and 62 (south). 96 (= *5* in Fig. 5-17) between rooms 85 (north) and 65 (south), see Fig. 5-15b. 117 (near *6* in Fig. 5-17) east of room 67 (south) of Fig. 5-15b. Roadway 6 East: 44 between the first two rooms at the western edge of the panel with the span of approximately 300 m (borehole safety pillar left at the eastern edge). 46 at the distance of approximately 150 m from both the western and the eastern edge (borehole pillar) of the panel. 54 east of the room next to the borehole safety pillar at the eastern edge of the panel.

Notice: for each curve, the starting point indicates zero deformation according to the measurements; different amounts of initial stress-relief creep are missing. The latest measured values (mm) are given at the end of each curve.

Fig. 5-17. Horizontal closure rates of a roadway in the Borth rock salt mine, roadway 8 East. (Modified after Erasmus, 1965, and Dreyer, 1974.)

See Fig. 5-16 for identification of stations, and Fig. 5-15 for locations. The measurements commenced after the termination of extraction mining, indicating various degrees of reloading after stress-relief creep in response to room extraction according to Fig. 5-15.

When comparing Erasmus's measured curves, it should be noticed that their final slopes over several years, i.e., the creep rates after termination of extraction mining, differ greatly, contrasting in this regard to the curves shown in Fig. 5-2 (with the exception of curves *1* and *10*). This difference is due to the varying pillar loads at the Borth mine: barrier pillars, the borehole pillar, and large abutment zones in virgin ground were supporting considerable additional loads transferred from extracted areas. Erasmus (1965, pp. 157—159) emphasizes that his results cannot be reconciled with the stress-arch concept of load transfer; his conclusion is that no further increase in pillar loads will occur in the centre of mined-out areas which have reached twice the span of umbrella zones, i.e., approximately 300 m. Considering the depth of 785 m of the entry in which the basic measurements were taken for less than one year (see Fig. 5-17), and particularly considering the fact that the total span of the extracted area was limited to about 500 m, the validity of this conclusion appears questionable.

It is based on earlier assumptions by Dreyer (1955) and Höfer (1958) who postulate "ultimate strengths" of mine pillars due to strain hardening; Erasmus (1965, p. 157) theorizes that strain hardening during the stage of primary creep of an ideal creep curve caused the relatively small creep rates measured in 1952 and 1953 at two locations near the centre of the subsequent extraction mining (curves 96 and 117 in Fig. 5-16); temporary increases in creep deformations during the period of extraction mining are mainly attributed to effects of blasting. Erasmus expects the cumulative creep curves to approach an asymptotic end value; the stage of secondary creep at constant rates would never be reached. The reader is reminded that, at the time of Erasmus's publication, most European theoreticians considered constant creep rates indicative of impending pillar failure—see section 3.4 for details. Gimm (1968, p. 223, figs. 4/66 and 4/67) refers to curve 46 of Fig. 5-16 as "proof of the validity of the ideal creep curve for salt pillars, as previously proved by Höfer (1958)".

It is emphasized here that the curves 94 and 95 of Fig. 5-16 show virtually constant creep rates from 1952 through 1955, when the measurements were terminated, in contrast to most other series which continued until 1958 or 1960, respectively. The exact measurement data is presented in Erasmus's figures; the total measured horizontal convergences at the stations shown in Fig. 5-16 are listed at the end of the respective curves.

For the sites 94 and 95, the convergence rates per day calculate as 0.25 and 0.3 mm from 1952 to 1956; at site 117, the convergence rate per day increased to nearly 1 mm in 1954, although there was no extraction mining activity near this site until after January 1, 1955, as shown on the respective mine maps. At that time, extraction mining was approaching from the west; it had resulted in similar accelerated creep at site 96 in 1953, although this site also could not have been affected by nearby extraction which began in 1954 (Erasmus, 1965, fig. 8).

Both Erasmus (1965, figs. 12 and 13) and Dreyer (1974a, figs. 68 and 69) emphasize the build-up of abutment loads around developing extraction mining areas in the Borth mine. It goes without saying that such abutment load zones develop all around developing extraction areas; when extraction mining proceeds in one direction, every new room is excavated in the abutment zone which travels ahead of the advancing extraction front. This is the reason why the creep accelerated at site 96 in 1953, and at site 117 in 1954, while the creep rates remained virtually constant at sites 94 and 95; these two sites are located in the umbrella zone provided by virgin ground to the west.

The loading conditions at site 117 became aggravated after January 1, 1955, when the extraction mining front approached from the west, and another abutment zone developed due to extraction mining at the distance of little over 100 m to the east. As the result, two abutment zones were temporarily overlapping around site 117. The consequences are shown in Fig. 5-15 (Erasmus, 1965, fig. 6, supplemented by an instructive photograph): site 117 was located between the rooms 88 and 68; the measurement results plotted in Fig. 5-15 demonstrate the combined effects of horizontal stress relief creep into the large excavated rooms, and of subsequent reloading creep due to continued extraction. The whole roadway moved in the direction of the nearest large room, resulting in a wave-like horizontal deformation of the base line indicated in Fig. 5-15.

The absolute measured amounts of horizontal movements of the indicated base line are of little significance as the exact date of the first survey is not given. Fig. 5-15 shows results of a survey carried out on November 1, 1958. In January 1959, the base line east of room 67 was "re-aligned"; the next survey was made on November 1, 1959, showing absolute movements of 62 mm towards room 88, and of 44 mm towards room 68 during this period of time. The maximum movements prior to November 1, 1958, are given as 201 mm towards room 67, and 112 mm towards room 88. Apparently, the measured movements are relative to the point east of room 89 for which zero movement is listed in both surveys.

It appears obvious that such extensive creep deformations into large openings, including horizontal displacements of the whole 8-m roadway, cannot be reconciled with theoretical concepts of stress peaks and stress envelopes near such large openings; if such stress concentrations exist, the displacements must occur in the opposite directions, away from the respective openings. On the other hand, considering the width of the displaced roadway, it appears conceivable that the displacements of the roadway walls next to any extracted room are greater than the displacements of the opposite roadway walls; this would result in some stretching of the sections of the roadway, i.e., in widening of the roadway, unless creep into the roadway compensates such widening, or exceeds it to the effect that horizontal creep closure of the roadway continues. Regrettably, measurements to elucidate these relationships are not available. Figs. 5-16 and 5-17 show that possible stretching of the roadway section did not prevent small horizontal creep closures.

These interpretations contradict those given by Erasmus (1965), and by

Dreyer (1974a); for this reason, a more detailed discussion of the measurement results is indicated.

With reference to convergence rates listed above, it is stressed that site 5 of Fig. 5-17 is identical with site 96 of Fig. 5-16; site 6 is located near site 117 where the horizontal convergence rates through 1955, during the time period of extensive extraction, were constant at nearly 1 mm per day. At the time of extraction, two abutment zones were overlapping around sites 6 and 117, and this explains the relatively high convergence rates from 1954 through 1955. After extraction of the temporary remnant pillar around site 117, this area became the centre of surrounding extraction mining, and this leaves no doubt of relatively high loads imposed on the pillars between sites 6 and 7 of Fig. 5-17. For this reason, the highest horizontal closure rates after extraction should have occurred at sites 6 and 7, if the convergences reflect the pillar loads. Fig. 5-17 shows particularly for site 7 that this is not the case.

Erasmus postulates that the pillars between sites 4—6 carried the maximum possible load, as indicated by the measured horizontal convergences. Regarding the displacements shown in Fig. 5-15, it is postulated that the roadway is located in the zone of maximum load on the roadway pillar, the total width of which equals the widths of the pillars between the rooms (30—35 m). The reader is reminded that Erasmus follows Dreyer's (1955) hypotheses, which call for envelopes of increased stresses very close to openings and exclude stress-relieved zones around the roadway.

Erasmus assumes maximum vertical stresses along the axis of the roadway pillar, i.e., above the roadway itself; it is further assumed that the stress concentration above the roadway is pushing the salt into the nearest rooms; the resulting shear deformations are considered detrimental to the structural stability of the whole mined-out area. The roadway deformations are exclusively attributed to overburden load, see Fig. 5-17 in which site 5 is shown in the centre of the area of assumed maximum possible pillar loads.

As the wave-like deformation of the roadway axis is also attributed to the loads on the roadway pillar, the maximum horizontal displacements should also be expected around site 5, which is located between rooms 65 and 85, see Fig. 5-15. This Figure, however, shows the maximum displacements east of room 87, i.e., between sites 6 and 7 of Fig. 5-17; hardly any displacements occurred to the west of room 65, i.e., between sites 4 and 5. These facts show that the horizontal roadway convergences and the horizontal roadway displacements cannot be attributed to a common source.

Dreyer (1974a, pp. 140, 141 and 146, 147, with several figures) offers an entirely different view of the development of the convergences in roadway 8 East, ignoring the measurement results since 1952 shown in Fig. 5-16, ignoring the detailed re-evaluation of Erasmus's data (Baar, 1971a) which had been made available to him, and using rather questionable methods of "proving" constant convergence rates for the first 300 days after excavation.

According to Dreyer's table 13, and to the figure cited above, the horizontal closure in 1960 took place at the following constant rates (in mm per day): 0.39 at station 1, at the distance of 20 m from the edge of the panel; 0.49 at station 3, at the distance of 72 m from the edge of the panel; 0.59 at stations 4 through 6 (with site 4 at the distance of 140 m from the edge of the panel, see Fig. 5-17).

For the latter three stations, the constant convergence rate of 0.9 mm/day is given in the text (l.c., p. 147); this appears to be an error as Fig. 5-17 shows the convergence rates quoted above, and Dreyer's fig. 61 shows perfectly straight lines for the cumulative horizontal convergence at stations 1 through 6 of Fig. 5-17. The denomination "horizontal convergence during the first 300 days after excavation" obviously represents another error, as the writer refers to the measurements initiated by Erasmus (1962) "approximately one year after the extraction of the adjacent rooms had been terminated" (l.c., p. 140). Although the dissertation referred to is not published, there is little doubt that Fig. 5-17 (after Erasmus, 1965) shows the measurements to which Dreyer refers; these measurements commenced on November 27, 1959. According to the mine maps published by Erasmus (1965), the extraction of rooms adjacent to stations 1—3 was terminated in 1955; all extraction mining to the north and to the south of roadway 8 East was terminated in 1957, and this covers the rest of the stations shown in Fig. 5-17. As a matter of fact, the extraction of an area 200 × 400 m east of station 7 is also shown as terminated in 1957.

Consequently, the convergence rates shown in Fig. 5-17 were measured during a time period which commenced at least 2 years and up to 5 years after the extraction of adjacent rooms had been terminated. In 1960, the roadway 8 East was at least 8 years old, as the measurements shown in Fig. 5-16 commenced in 1952.

Dreyer (1974a, pp. 140—142, figs. 61—63) presents Erasmus's data as "proof of constant creep rates" for infinite periods of time after an initial period of decreasing creep rates. This represents a rather drastic change in hypotheses as the same writer postulates exponentially decreasing creep rates in numerous publications over the previous 20 years, the creep convergence approaching an asymptotic end value "beyond which no further closure is possible" on account of postulated strain hardening—see section 3.4.

Dreyer's "proof" of constant creep rates was certainly overdue in view of numerous publications in which the results of in-situ measurements are reported, e.g., Höfer (1958a). However, like Höfer whose methods of proving continuously decreasing creep rates had to be critized (Baar, 1970a), Dreyer is using equally questionable methods in his attempts to prove the contrary: parts of measured in-situ creep curves which do not match the "new" concept of creep at constant rates are simply ignored, as becomes obvious from Fig. 5-16. "There should be no question as to the scientific value of these arbitrary procedures. . . any wished-for curve could be found by selecting and tailoring suitable values" (Baar, 1970a). As mentioned previously: Gimm (1968) selected curve 46 of Fig. 5-16 to "prove" the validity of so-called ideal creep curves for salt pillars, and many theoreticians continue to postulate that creep at constant rates is indicative of impending pillar failure because of tertiary creep. If such postulates were justified, most of Erasmus's published curves would indicate pillar failure.

The following point is particularly emphasized: Dreyer (1974a, pp. 140—161), re-interpreting in-situ measurements by Erasmus (1965) and other writers, attempts to prove creep at constant rates; however, the rest of his text is devoted to alleged proof of the validity of formulas which promise in-situ creep at continuously decreasing rates. The need for extreme caution in applying such formulas to actual mine design appears to be obvious from the comprehensive in-situ measurements at the Borth rock salt mine.

5.4 CREEP DEFORMATIONS IN PANEL-AND-PILLAR MINING

5.4.1 General remarks

Some writers, particularly Wardell (1968), speak of panel-and-pillar mining in cases where relatively large pillars are left between extraction mining panels in which much smaller pillars may or may not be used for control of the roof; the large pillars are thought to provide structural stability for the whole mine, supporting the deadweight of the overburden and allowing only negligible subsidence at the surface. According to Wardell (l.c., p. 275), panel-and-pillar mining has been used extensively in coal mining in Britain, France, and in other countries, in iron-ore mining in Lorraine, and in potash mining in Alsace. It may be added that the barrier pillar system as advocated by German writers, e.g., Höfer (1958a) or Spackeler et al. (1960), for mining flat salt and potash deposits, is based on identical principles; it was introduced in the past century after the first potash mine floodings had been caused by inadequate pillar design. Since that time, the question of whether or not barrier pillars are suitable, particularly at great depths, has been dealt with in numerous publications; the controversial issue has not yet been settled, although the barrier pillar concept, when applied under certain geological conditions, has repeatedly caused disaster. The reader is referred to sub-section 5.3.2 with respect to controversies which resulted in the abandonment of barrier pillars in a potash mining district.

Another application of the panel-and-pillar principle reported in sub-section 5.3.4 caused a severe rockburst, so the design was dropped immediately. Erasmus (1965) called for elimination of barrier pillars in future mine planning on the basis of the measurement results discussed in sub-section 5.3.5, rejecting particularly the concept of stress arches which would provide structural mine stability.

Wardell (1968, p. 271) leaves no doubt of his contrary opinion: "The theoretical and practical studies in rock mechanics which have been made so far are, to some extent, fragmentary and imprecise." His statements (l.c., p. 272) may be quoted: "The author has frequently been responsible for mine design and its implementations, particularly in the mining of stratified deposits of coal, potash and iron-ore. This has led to a detailed consideration of partial extraction systems and to the development of a structural concept of mine design in which movement of the main strata is controlled by appropriate dimensional design of the areas of extraction and the pillars to be left."

Statements such as these cannot be ignored when they appear in the Proceedings of a Rock Mechanics Symposium, sponsored by scientific institutions and committees, and published by a governmental department. Considering the serious consequences attributed by others to the application of Wardell's structural concept of mine design, it appears necessary indeed "to enquire about the evidence" to support the structural concept which "postulates an arch or dome theory in most, if not all, mining conditions. That is to say, any given mine excavation should have critical maximum dimensions beyond which the main rock mass will no longer behave structurally and span the excavation, but will fail in some way or another."

Wardell considers accurate field studies of surface subsidence caused by underground extraction the most important evidence for the validity of his concept, referring to the mass of data accumulated over more than fifty years in European coal mining. "In Britain, the idea of a so-called 'pressure-arch' has, of course, been prevalent in the thinking of coal mining engineers for several decades; it was given approximate quantitative form", which is presented in Wardell's publication. Except for the notice that the stress arch concept is not even mentioned any longer in Salamon's (1974) summary report on recent developments in rock mechanics of underground excavations, controversial aspects of surface subsidence caused by coal mining cannot be dealt with in this text.

Regarding subsidence caused by iron-ore mining in Lorraine, it is emphasized that the authors referred to by Wardell definitely reject the stress-arch concept (Tincelin and Sinou, 1960, figs. 3—6). This is also shown in the presentation by Fine et al. (1964) in which the mathematical theory of elasticity is used to prove "that a certain breadth of working must be reached" before bending of the overburden begins.

To prove the validity of his concepts in potash mining, Wardell (l.c., p. 273, and fig. 2) "shows the observed subsidence over a caved longwall extraction at a potash mine" in Spain. "The maximum observed subsidence was only" 40 mm for an extraction area of 450 × 125 m at a depth of 200 m, the thickness of seam extracted was 2.5 m, according to a plan and a section shown in the figure. At the rear solid edge of the extracted area, the subsidence trough is shown 35 mm deep, approximately 5 mm deeper than above the advancing face; the maximum depth of the trough is shown at the distance of 50 m from the face, contrasting in this regard to the profiles shown in Fig. 5-11; these show a more normal development such as expected because of the time required for full subsidence. As Wardell's figure shows only one subsidence profile measured at a time after mining which is not indicated, any conclusion based on it must be considered premature. However, it should be noticed that the surface subsidence trough extends to about 200 m ahead of the face; Wardell emphasizes that "thick, strong individual beds of rock were present in the overlying strata". These two observations, and the maximum observed subsidence of only 40 mm after extraction of a seam 2.5 m thick, strongly suggest plate-like behavior of the main overburden formation.

For further proof of his ideas, Wardell (l.c., p. 278) refers to panel and pillar workings he designed "in consultation with colleagues from Mines Domaniales de Potasse d'Alsace"; the design is similar to one for the extraction of a coal seam underlying the University of Nottingham, in which case "panels and pillars were chosen largely on the customary basis of minimizing surface subsidence". Wardell, referring to his M.Sc. Thesis (1965) at the same University, gives the impression — by writing that he "also" designed the panel-and-pillar system in the French mine — that he designed the extraction scheme at Nottingham which resulted, within 8 months, in observed convergences of 45—55% of the thicknesses extracted; these percentages apply to the centres of the panels. At the edges of the pillars, up to 25% convergences are indicated (Wardell, 1968, fig. 10). "Nevertheless, there was no material spalling at the pillar edges" (l.c., p. 277). The seam thicknesses were 1.38 m and 3.38 m, respectively, and "some effects must have occurred before the measurements began". The observed surface subsidence is shown as a regular trough extending over the edges of the extracted area; the depths of the surface subsidence trough, and its development in relation to the panel extraction in 1959 and 1960, are not indicated in Wardell's (1968) publication. However, regarding the experimental mine layout at the French potash mine, the observation of subsidence is considered "the primary control, supplemented by stress and deformation measurements in the underground pillars and by laboratory tests on potash samples from the mine. The latter were conducted in collaboration with Potts (1964) and McClain (1964, Ph.D. Thesis Univ. Newcastle-on-Tyne, December)".

There is some obvious disagreement in the publications by McClain (1964) and by Wardell (1968), not only regarding McClain's Ph.D. Thesis the presentation of which the author himself dates one year earlier; as a matter of fact, Wardell's involvement in the design of the French mine referred to in both publications is not even mentioned by McClain (1964). The underground measurements reported and discussed by McClain are considered extremely important (Baar, 1971a); some details are discussed in the following sub-section.

5.4.2 Experimental mining at the Mine Fernand, Alsace, France

Neither McClain (1964) nor Wardell (1968) explain the reasons why the experimental mining, using the panel-and-pillar layout, was performed at depths of 400—500 m. According to Wagner (1955, figs. 1 and 5, see Borchert and Muir, 1964, fig. 17.1), the mine is located approximately 5 km north of the city limits of Mulhouse in an area where the Upper Potash bed is not present. If Wagner's contour map (fig. 5) is correct, the Lower Potash bed is dipping from 200 m below the surface at 1 km distance to the west of the mine shafts to 500 m at 2 km distance to the east, reaching at this distance the axis of a syncline which strikes north—south; farther to the east, the distance between the 500-m and the 400-m contour lines is approximately 600 m, as is also shown on Wardell's mine map (fig. 12). It could be speculated that the rock salt beds, which provide for a safe cover of the mines at greater depths, are missing in the experimental area, and that this was the reason for using the panel-and-pillar layout; according to Wardell (l.c., p. 278), the potash bed is only 2.25 m thick, compared to more than 4 m at greater depths in the other mines of the district.

Another possible reason for experimenting with the panel-and-pillar layout may have been the necessity to reduce the surface subsidence to an acceptable minimum. The geological situation in the experimental area is similar to that known from the eastern part of the Buggingen mine, see Fig. 2-25 and 2-26; a more detailed section, showing the potash bed at depths similar to those at the Mine Fernand, can be found in Borchert and Muir (1964, fig. 17.2). At the Buggingen mine, the 3.5-m potash bed was totally extracted by long-wall mining, see Gimm (1968, fig. 6/34) for details; backfill was brought in immediately after advancing the face to minimize the surface subsidence. However, subsidences of the order of 1.5—2 m could not be prevented, and mining sequences such as used in coal mining had to be followed in cases where sensitive surface structures required protection; in this way, the main European north—south railway was settled without traffic interruption. Details are not available for publication. It may be emphasized, however, that surface subsidences caused by potash mining in Germany have been surveyed since the beginning of this century, as shown in more detail in Volume 2; therefore Wardell's (l.c., p. 278) notion that "there was no observed data on which to judge either the optimum panel or pillar width" for minimal surface subsidence, appears to be indicative of the disadvantage of language barriers, particularly in the field of rock mechanics in potash mining where, for obvious reasons, the vast majority of publications are written in German.

Fig. 5-18 (modified after Wardell, 1968, fig. 13) shows a section of the experimental area at Mine Fernand, and observed surface subsidences. The subsidence observation line represents approximately the centreline of the experimental area; the average panel lengths are approximately 500 m, with

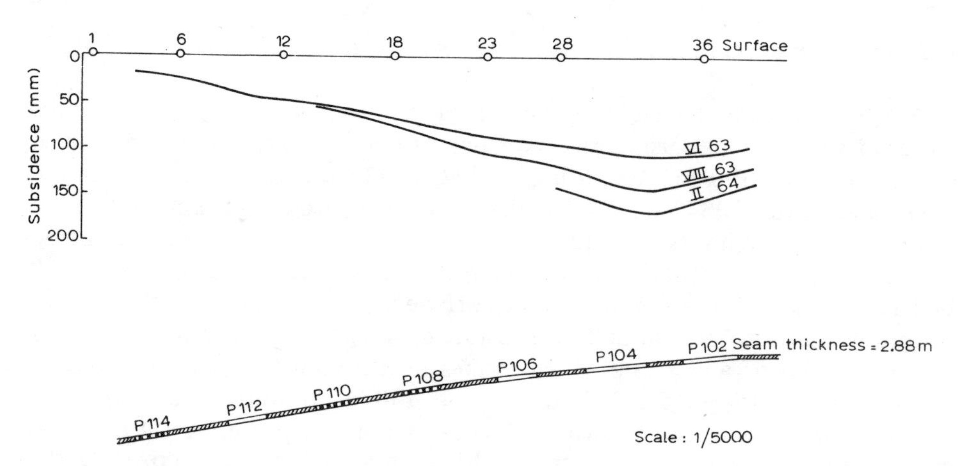

Fig. 5-18. Cross-section through panels, Mine Fernand, Alsace. Surface subsidences measured along the centreline of the extracted block of panels. (After Wardell, 1968.)

the exception of panel 104 (460 m) and panel 102 (280 m) due to previous extraction. The caved area of previous extraction was separated from the experimental area by a 30-m pillar. It appears that the areas to the south and to the east of the experimental area were virgin ground during the time period of the experimental mining. The exact time periods of mining each panel are not indicated on the maps published by Wardell (1968) and McClain (1964). It is possible, however, to reconstruct the times of extraction of the panels 102, 104, 106, from McClain's figures in which the results of extensometer and stress-meter measurements are shown, see Figs. 5-19 and 5-20.

Wardell (l.c., p. 278) gives the following account of the mining sequence in the experimental area, distinguishing three phases:

(1) "Initially, panels 108, 110, 112 and 114 were planned at width of 130 ft. (39 m, 0.08—0.09 × depth) with intervening pillars of 250—300 ft. (75—90 m). The panels were to be worked by room and pillar with the smaller pillars left intact. Panels 110 and 114 were begun first, followed later by panels 108 and 112. When panels 108, 110 and 114 were virtually complete and panel 112 had advanced about 425 ft. (128 m), the observed subsidence was virtually negligible. It was then decided to increase the width of panel 112 to 165 ft. (50 m, 0.10 × depth) and to extract the seam completely in this panel. The effect of this was marginally to increase the surface subsidence."

Fig. 5-18 shows the surface subsidence profile observed in June 1963, with approximately 50 mm subsidence above panel 112, comparing to approximately 25 mm above panel 114 which was extracted prior to panel 112, the last one in the group of four panels surrounded at least on three sides by virgin ground. The time period of the extraction of this group is not indicated; at appears safe to assume that the extraction of this group was

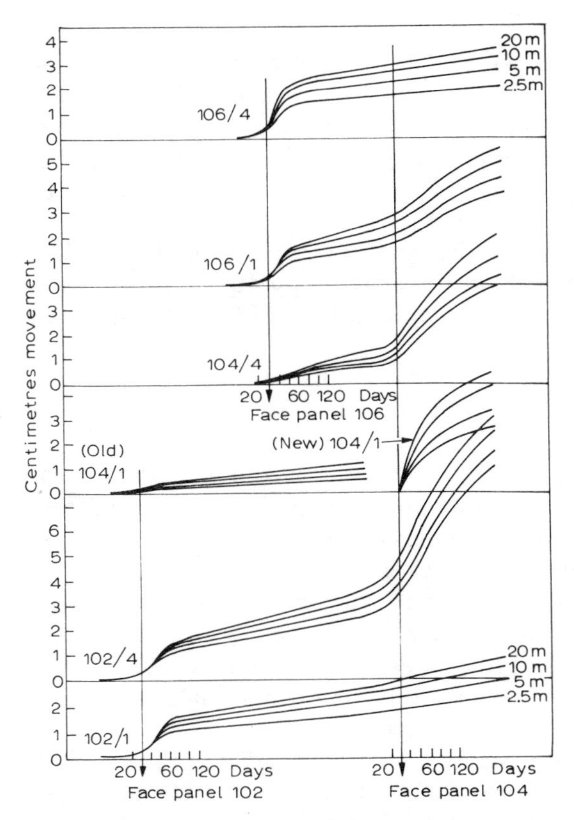

Fig. 5-19. Mean horizontal pillar deformations, Mine Fernand, Alsace. (Data from McClain, 1964.)

Measurements in pillars underneath points 23—36 of the surface subsidence measurements shown in Fig. 5-18. See text for details and discussion.

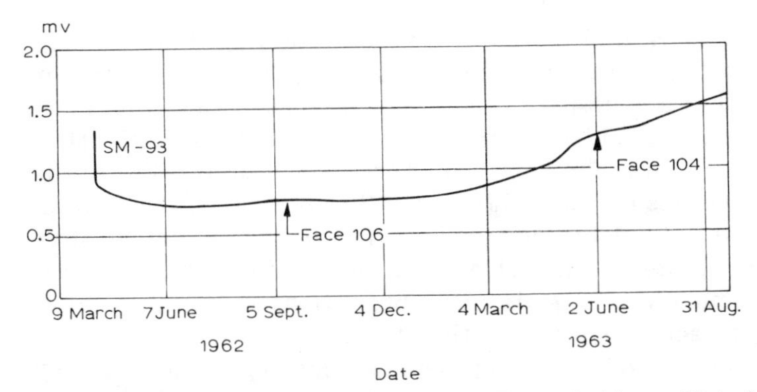

Fig. 5-20. Typical stress-meter results, Mine Fernand, Alsace. (Data from McClain, 1964.)

complete prior to the extraction of the next group which began late in 1961, according to the data presented in Fig. 5-20. However, the exact time periods do not really matter with regard to the subsidence trough which apparently developed during the time period of mining the first group of four panels. The considerable deepening of this early trough due to widening, and complete extraction of panel 112 over most of its length, is still reflected in the 1963 survey shown in Fig. 5-18.

The denomination "virtually negligible" and "marginal" for that amount of measured subsidence above panels with spans of only 40—50 m, and separating pillars 75—90 m wide, appear debatable in view of the constant creep rates observed by McClain—see Fig. 5-19.

(2) In the second phase of mining, "further dimensional changes were therefore made in relation to panels 102, 106 and 104... Fig. 5-18 shows the finally observed surface subsidence."

It is emphasized here that the sequence of listing in the above quotation was also the sequence of mining, contrary to the sequence given by Wardell (l.c., p. 278); this is an important difference which explains some of the characteristics of the cumulative creep curves shown in Fig. 5-19. These characteristics in turn explain the development of the surface subsidence above this group of panels; this is shown together with the analysis of Figs. 5-19 and 5-20. Here, the development of the experimental area according to Wardell (l.c., pp. 278, 279), and the reasons given for further changes to panel and pillar dimensions, may be quoted as follows:

(3) "The maximum panel width was that of 104 at 245 ft. (74 m, 0.18 × depth), the minimum pillar width was 102/104 at 165 ft. (50 m, height/width ratio 1/24). The maximum hypothetical pillar stress was in the order of 3500 psi (245 kp/cm^2). Again, a calculation... was rather difficult because of the complete extraction area to the north-east, and the small pillars left in panels 108, 110, 112 and 114. An approximate calculation for the three adjoining total extraction panels 102, 104 and 106 was made assuming an abutment support of 260 ft. (78 m) around this area of extraction."

"Two conclusions were drawn from these data. The first, that the panel widths could be further increased without destroying the arch formation. The second, and rather less positive from a design point of view, that pillars with height/width ratio of 1/24 were structurally stable under the stress imposed upon them in these circumstances. These conclusions led to further changes in the dimensions of panels 116 and 118 to the west... The surface subsidence is not yet complete but broadly confirms the continuing structural stability of the system. The widening of the panel dimensions will be continued as the mining area extends until observations suggest that critical limits are being approached. Up to this time it seems that it will be possible to achieve an overall rate of 65—70 per cent at this depth and still maintain stability and control."

In his concluding comparison between panel-and-pillar and room-and-pillar systems, Wardell (l.c., p. 282) summarizes his views as follows:

In room-and-pillar mining, "the main problem is to determine the optimum pillar dimensions. This requires firstly an assessment of the maximum stress to which a pillar will be subjected in the mine and secondly an assessment of the capacity of the pillar to sustain stress." With reference to some disasters in room-and-pillar mining, Wardell states that "the problem of design for small pillars must obviously increase with working depth and increasing seam thickness, and the room and pillar system is then only operable either at a relatively low rate of extraction or at a marginal factor of safety".

The panel-and-pillar system "has a much higher intrinsic factor of safety and also offers generally the possibility of higher extraction ratios. Moreover, it is probable that a combination of room and pillar working — within areas which were more generally supported by stabilizing pillars — could be used to considerable advantage both from the point of economic extraction and safe mining conditions."

Finally, "so far as room-and-pillar working is concerned, it can be envisaged that this could be used for extraction panels of planned dimensions separated by appropriately dimensioned barrier pillars. This might well serve to extend the range of application and the safety of small pillar mining."

It should be noticed that the controversial question of load transfer to pillars and abutments — stable arches vs. elastic plates — is no longer mentioned in this summary which concentrates on pillar strengths and pillar loads in mines. Wardell (l.c., p. 274) refers to the laboratory testing by Dreyer (1964) who "deserves special recognition. His work confirms and consolidates the bulk of the results from the compressive testing of prepared samples." Wardell emphasizes particularly Dreyer's finding that "it was impossible to induce failure in a test sample beyond a certain limit of height to width. That is to say, beyond this limit, samples were capable of sustaining very high stress indeed without failure in the accepted sense." "Similar tests have been carried out by the author and his colleagues. . . These results confirm Dreyer's findings. All these tests were made at a uniform rate of loading."

Wardell, utilizing the concept of practically unlimited strength beyond a certain limit of height to width, argues "that a single pillar of a given superficial area would sustain greater stress without failure than a larger number of smaller pillars with the same total area. On the other hand, it would be imprudent to import quantitative values from laboratory tests into the design of mine pillars without a substantial factor of safety. . . The author has used laboratory tests of the form suggested by Dreyer but with a safety factor of at least 4. The hypothetical maximum pillar stress was calculated according to what Coates has defined as the 'tributary area theory'. This is certainly a crude method of assessment but the degree of extrapolation necessary from a controlled laboratory situation to the mine can hardly be recognized in any other way. So far, measurements of stress and deformation in mine pillars underground have not clarified the situation sufficiently to justify a more optimistic approach."

These statements leave no doubt of Wardell's assessment of the results of extensive measurements of pillar deformations and related stresses "conducted in collaboration with Potts (1964) and McClain (1964)" in the experimental panel-and-pillar mining area at the Mine Fernand. These results are discussed in the following on the basis of McClain's (1964) publication; again, in view of the controversial nature of the subject, it appears indicated to resort to quotations.

(p. 491): "The rock-mechanics instrumentation was installed in maintained roadways adjacent to and just outside of the working panels (102, 104, 106). The principal instruments used in the investigation were a borehole extensometer, which measures axial borehole deformations and the 'stress meter' which measures diametral deformations. These instruments were developed by Prof. E.L.J. Potts at the University of Newcastle, England and have been described in detail in the literature (Potts, 1957, 1957a)." The location of all of the instrumentation is shown on a mine map. "In general, the instruments were installed in groups, near the middle portion of each panel and directly beneath a line of surface subsidence measurements. The pillar measurements were made from both sides of the pillars, before, during and after the mining operations. The horizontal extensometer wires were installed at depths of 20, 10, 5 and 2.5 meters from the ribside. Stress meters were installed up to 35 m deep."

"The sequence of mining operations is of material importance in the discussion of any rock mechanics results and even more so in a material which exhibits strong time-dependent properties. All three panels were mined in retreat" from the mined-out, caved area to the north and to the north-east.

The significance of this mining sequence may be pointed out here as it is not mentioned in either of the publications (McClain, 1964; Wardell, 1968): at the time of mining the three panels, the area was representing a remnant pillar surrounded by the caved area to the north and to the northeast, by a three-entry system approximately 20 m wide to the east, by a four-entry system approximately 28 m wide to the south, and by the four previously extracted panels to the west. The pillar left between panel 102 and the three-entry system was approximately 55 m wide, the one left between panels 108 and 106 was 85 m wide, with a single entry approximately along the centreline of the pillar.

The widths of the abutment zones around these pre-existing openings is not known; this width may well have been approximately 200 m, as established by measurement under comparable geological conditions. The reactions to mining at various distances, shown in Figs. 5-19 and 5-20, apparently confirm load transfer over considerable distances.

Resuming McClain's (l.c., p. 491) description, Fig. 5-19 "shows the horizontal deformations of the pillars measured at different depths into the pillar. The pillar sides adjacent to panel 102 responded to the workings in that panel by showing a rapid deformation immediately after the face passed. The deformation rate decreased slowly until a constant strain rate was established about three months after the working of the panel. This constant strain rate of the pillars next to panel 102 remained unchanged throughout the extraction of panel 106. The movements adjacent to panel 106, the second to be worked, were nearly identical to those around panel 102."

It appears indicated, at this point in the quotation from McClain's description, to emphasize the following: the measurements were taken in the development roadways at both sides of each panel; these roadways were maintained during the extraction of the panels, and were separated from the extracted areas by small pillars the width of which is not given in the publications referred to. As the total width of a pair of development entries plus separating pillar is shown on the mine map as approximately 10 m, the design width of the pillar appears to be 2—3 m; the actual width may differ slightly. These small pillars along the panel roadways apparently do not influence the load redistribution caused by panel extraction.

The denomination 102/1 indicates the roadway along the eastern side of panel 102, and 102/4 is the one along the western side; the same scheme applies to the other panels. The time lag between driving the development roadways and panel extraction is not known. It is emphasized that the measurements for curves 102/1, 102/4 and 106/4 began approximately two months before the face of the panels passed the measuring sites. The measured deformations began to increase about ten days before the face passed. This could be interpreted as re-loading creep — see sections 4.1 and 4.2 — or as response to blasting and corresponding increased stress-relief deformation. However, the initial parts of these curves clearly resemble the straight lines which develop after rapid initial stress-relief creep that slows down to constant small rates according to the original loading conditions—see curve *1* in Fig. 5-2 which was not affected by reloading.

It should be noted that the first reloading due to extraction of panel 102 is also reflected in the slight increase in creep rates in roadway 104/1 (old) at the distance of 75 m from panel 102. This shows the width of the abutment zone around panel 102 over 75 m wide; the extraction of panel 106, at the distance of 130 m from 104/1 (old), is not reflected, indicating an abutment zone less than 130 m wide. Panel 106 was only 50 m wide, comparing to 70 m of panel 102. The creep rate in roadway 104/4, at the distance of 75 m from panel 106, was slightly increased by the extraction of panel 106.

With reference to sub-section 5.2.1 for comparison with the width of abutment zones at IMC (standard room-and-pillar mining), it is emphasized that, at the beginning of underground subsidence, the widths of the abutment zones are similar; however, the span of the mined-out area required to initiate measurable increases in creep rates is more than twice the span at the Mine Fernand, reflecting the effects of the 15-m pillars in the original IMC design.

McClain's (1964, p. 491) description of the creep curves shown in Fig. 5-19 continues as follows: "Finally, when panel 104 was extracted, between the two previously caved panels, deformations throughout the entire area were reactivated. The ribside deformations adjacent to panel 104 were much larger than those next to the two outside panels. More significantly, the opposite sides of the two central pillars, that is, the deformations measured at 102/4 and 106/1 showed a drastic increase. The deformations measured in 102/4 taking place as a consequence of the extraction of panel 104 a full pillar width away were many times larger than the movements caused by the working of the adjacent panel 102."

These reactions — see Fig. 5-19 — show conclusively the mechanism of load redistribution, and therefore deserve a more detailed discussion presented in the following. The denominations 104/1 (old) and (new) result from the decision to reduce the design width of pillar 102/104; this made it necessary to drive the new development entry 104/1 (new); it was driven as a single entry in the abutment zone of panel 102 at the distance of 50 m from this panel. Apparently, the extensometer sites were installed shortly after the excavation of 104/1 (new). The course of the respective curves in Fig. 5-19 resembles the course of curves *7* and *12* in Fig. 5-2, indicating a combination of initial stress-relief creep and reloading creep.

It is re-emphasized that the remnant pillar in which panel 104 was extracted, had the width of 200 m; the panel and the pillar 104/106 each were 75 m wide, leaving 50 m for the pillar 102/104, while the pillar 104/106 remained at the original design width. The courses of the curves at the time when the face panel 104 passed the respective extensometer sites, clearly reflect the distances from panel 104, i.e., the amounts of abutment loads transferred to each of the measuring sites. Site 106/4, at the distance of 125 m, shows no reaction; creep continued at constant rates.

Site 102/1, also at the distance of 125 m, shows a slight increase in creep rates which may be related to the location in the abutment zone of the caved area to the north-east.

Site 102/4, at the distance of 50 m, shows the strongest reaction, indicating considerable loading of the 50-m pillar 102/104. This is also indicated by the increase in the stress-meter output, see Fig. 5-20; this stress meter was installed near the centre-line of pillar 102/104 (McClain, 1964, fig. 2). Fig. 5-18 shows the deepest depression measured at the surface in June, 1963, developing above pillar 102/104. Note the considerable subsidence of about 40 mm during the following two months, and the shift of the deepest depression towards panel 104 in response to the excavation of this panel.

Site 104/4, installed in the ribside of panel 104 in the 75-m pillar 104/106, shows considerably less reaction compared to 102/4 at the opposite side of the adjacent 50-m pillar.

Site 106/1, installed at the opposite side of the 75-m pillar, indicates considerably less loading of the 75-m pillar compared to the 50-m pillar; this suggests that the creep increase at site 104/4 may partly reflect stress-relief creep into the newly excavated panel 104, as also indicated by site 104/1 (new).

McClain (1964, p. 194) drew the following conclusions from the horizontal pillar deformations shown in Fig. 5-19:

"(1) There appears to be a zone of very large strains at the edge of the pillars. As the distance into the pillar is increased, the strain very rapidly decreases. Strain at the edge of the pillar of over 2.0% has been reduced to less than 0.1% within the first five meters. It must be concluded therefore that the edges of the pillars have deformed so much that they no longer are supporting their portion of the redistributed cover load but have transferred part of it further into the pillar."

"(2) The extremely small strains measured in the core of the pillars (0.025—0.040%) would further suggest that although they must be supporting a considerable load, they are not undergoing the amount of deformation which might be expected. The only reasonable explanation for this lack of deformation is that the width of the pillars (50—70 m) has provided a horizontal constraint sufficient so that the material cannot yield."

"The mechanism by which these pillars provide support for the overlying strata is therefore seen as a dynamic process whereby the edges of the pillars yield and deform plastically, transferring load toward the center of the pillar while at the same time providing constraint to enable the central cores of the pillars to support that load without developing sizable strains."

These conclusions disregard the effects of rapid initial stress-relief creep that removes overburden support in addition to the support removed by the excavation into which the horizontal creep deformation occurs; such plastic yielding must not be mistaken for the yielding of laboratory samples which deform elastically, and fracture when loaded too fast. However, the effects of stress-relief creep can also be seen as a dynamic process whereby the load previously supported in the creep zone is transferred to the core of the pillar — provided the pillar width suffices to prevent the overlapping of stress-relief creep zones, as shown in Fig. 4-14 for 15-m pillars with heights similar to those at the Mine Fernand. Considering the large opening widths of the panels in the latter case, considerable creep above and below the large inter-panel pillars can be assumed, particularly after roof caving which increases the effective pillar height considerably.

However, the lack of horizontal deformation of the 50—70-m pillars is evident and cannot be explained satisfactorily, in view of the measured surface subsidence, unless horizontal creep above and below these pillars allowed the pillar cores to be pushed into roof and floor; this in turn would have contributed to the horizontal forces acting on the roof and floor strata of the panels.

If the existence of structurally stable stress arches across the extracted panels is postulated with Wardell (1968), it is also necessary to assume that the pillar cores were pushed into the floor salt, resulting in horizontal floor deformations as shown in Fig. 5-13.

McClain (1964, p. 494) emphasizes that "a more interesting, but as yet less conclusive feature" of the measurement results shown in Fig. 5-19 can be seen: "The initial portions of these curves are seen to be essentially the same as classical creep curves: there is a primary phase of decreasing strain rate followed by a secondary phase of constant strain rate. These constant strain rates are shown", in a figure resembling figures shown by Zahary (1965) and Coolbaugh (1967), "as a function of the depth into the pillar". Initial creep at decreasing rates, followed by creep at constant rates, is a perfectly conclusive feature if related to the actual stress relief caused by excavation rather than to inadequate laboratory testing, with loads applied to stress-relieved specimens. McClain's (1964) conclusion in view of the discrepancy between laboratory loading tests and in-situ measurements is remarkable: "In fact, it is felt that these results constitute a serious indictment of the validity of applying laboratory test data to underground behavior in plastic materials."

The typical stress-meter results shown in Fig. 5-20, and their interpretation by McClain (1964, p. 498), also require some discussion as the results are considered to confirm extensive stress-relief creep rather than full pillar loads at any time according to theoretical calculations on the basis of extraction ratios. "The borehole deformation gage or 'stress meter'. . . is installed with a sizable prestress on two diametrically opposed wedges. With no load change occurring on the pillars, this initial prestress is partially lost as the material immediately around the instrument wedges responds to the prestress load by flowing away. This effect is roughly equivalent to a relaxation test and is approximately exponential."

With reference to section 3.8, it is emphasized that a stress-relieved zone exists around the borehole in which the stress meter is installed; the sizable prestress on the wedges results in creep away from the borehole. This initial creep indeed "is approximately exponential"; it must not be mistaken for elastic after-effects in relaxation tests. The exponential decay of the instrument output, be it the pressure indication in hydraulic cells or the millivolt output from the stress meter, is indicative of the equilibrium established by creep around the instrument. Any change in pillar stresses after an equilibrium had been established, is indicated by both the hydraulic cell and the stress meter, the hydraulic cell indicating directly the change to the stress field around the cell, the stress meter responding by an increase in millivolt output as shown in Fig. 5-20.

"The slight increase in indication caused by the extraction of panel 106, 140 m away", indicates how sensitive salt rock is in responding to increased stresses—and how sensitive the stress meter is in indicating such changes. The extensometer results shown in Fig. 5-19 for the same pillar 102/104 do not indicate the change in pillar load—however, the curves may have been over-idealized. It appears difficult to reconcile the slight increase in millivolt output over a period of about three months — two months prior to, and one month after the face 106 passed the stressmeter site — with "probably mostly elastic" response.

"The sharp increase of output as the adjacent panel 104 is extracted" can definitely not be "seen as mostly elastic response to the transferred load"; this particular increase began prior to December 1962, and it accelerated in May 1963. At the time when the face 104 passed, the output had reached a constant rate of acceleration, i.e., the load transfer to the pillar also occurred at a constant rate.

"The levelling-out of the curve" coincides with the decrease in the rate of surface subsidence—see Fig. 5-18. "The behavior of high modulus inclusions in plastic materials" appears not of "complex nature" if not biased by prejudice; "a quantitative analysis of the stress-meter results" involves difficulties similar to those experienced in the evaluation of sonar velocities—see section 3.8; however, the calibration of stress-meter outputs appears possible by using hydraulic cells, as previously suggested regarding measurements of sonar velocities (Baar, 1959b, p. 138).

In conclusion of this detailed discussion, it is felt that the evaluation of in-situ measurements should not be based on questionable pre-assumptions; the consequent use of common sense frequently can resolve anticipated problems. The relatively smooth surface subsidence trough which developed over the panel-and-pillar extraction area, shows that extensive horizontal deformations must have taken place to compensate for the 2-m difference in underground convergences between pillars and panels, respectively. As the caving of the panels increased the effective hight of the pillars considerably, most of the compensating horizontal deformation may have been brought about by creep of the salt rocks above the pillars. These relationships are dealt with in more detail in Volume 2.

5.4.3 Panel-and-pillar mining in other potash mining districts

Panel-and-pillar mining — or long-pillar mining with barrier pillars left after a number of small pillars — has been tried repeatedly in the past. One example is mentioned in sub-section 5.3.4; 163-m wide panels with 5-m pillars were separated by 76-m barrier pillars; the design proved dangerous by resulting in a rockburst. Apparently, the carnallite pillars up to 10 m in height did not provide the expected roof support as the 5-m pillars deformed extensively due to stress-relief creep. Detailed data is not available. This holds also for the room-pillar-room design with "stabilizing" pillars based on hypotheses promoted by Dreyer (Gimm, 1968) and Serata (Piper, 1974)—see sub-section 5.3.4 for the former design, and sub-section 5.3.3 for the latter design. The design dimensions may be listed here for comparison; the first number is the design dimension (in m) in the Werra district, Germany, the second one applies to the Sylvite mine, Canada: room widths 16 and 20, pillar widths 10 and 15, barrier pillar widths 18—34 and 54. Considering the differences in ore composition (carnallitic ore in the former case) and in pillar hights (6 and 2.4), the latter design certainly allows much less subsidence, as is required by the presence of a water-bearing formation only 33 m above the mine workings—see sub-section 4.4.2 for details.

The long-lasting controversies regarding the feasibility of panel-and-pillar design — or the feasibility of barrier pillars — in mining the Stassfurt potash bed are dealt with in detail in Volume 2. The apparent need for better overburden subsidence control was particularly demonstrated by a rockburst in 1940 which claimed 42 lives. The problems are aggravated in mines where the potash bed consists mainly of carnallite in considerable thicknesses, the room and pillar heights frequently exceeding 10 m. Under such conditions, pillars with widths of 15 to 25 m, or more, undergo stress-relief creep which leaves the pillars without noteworthy supporting capacity.

In addition, the immediate roof of the mines consists of shale and anhydrite approximately 50—80 m thick. These strata do not exhibit creep deformations as shown schematically in Fig. 5-21. Excessive bending of these strata due to insufficient support by pillars may result in excess tensile stresses; sudden tensile failure of these strata, or of other overburden formations, causes destructions as shown in Volume 2.

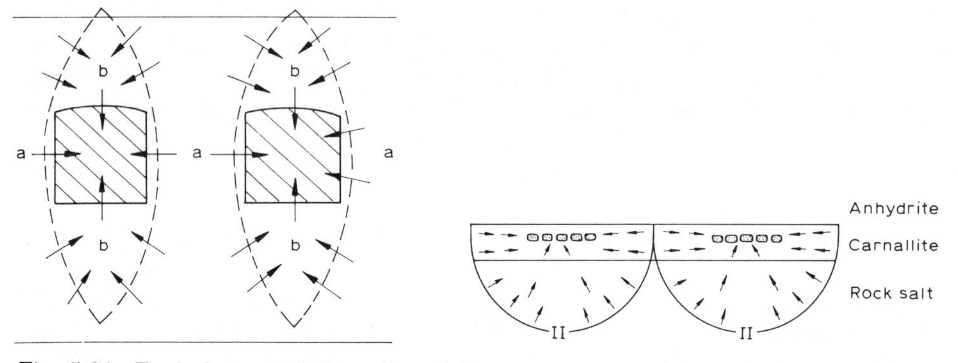

Fig. 5-21. Typical creep deformations into rooms excavated in a thick carnallite bed (Baar, 1953a, fig. 10).

Stassfurt potash bed, Krügershall mine, Teutschenthal/Halle, Germany. Section of rooms 12 m high; carnallite bed approximately 40 m thick. a. pillars; b. stress-relieved zones above and below the rooms.

Fig. 5-22. Creep deformations in panel-and-pillar mining with backfill (Baar, 1959a).

Conditions as in Fig. 5-21. *II* = expected extent of creep zones below panels with long pillars for roof support during extraction.

Under such and similar conditions, the span of mined-out areas must be limited until creep deformations of the abutment zones and the floor strata have resulted in adequate support of the overburden in mined-out panels. The panel-and-pillar design proposed for that specific purpose (Baar, 1959a) is shown schematically in Fig. 5-22: panels 100—120 m wide are mined with long pillars for support of the immediate roof; the mined-out rooms are backfilled as soon as possible to prevent excessive damage to the pillars by spalling. Creep deformation of the salt rocks as indicated by arrows results in uniform load distribution over panels and large pillars, without developing excessive tensile forces in the slowly subsiding overburden. It is emphasized that such designs must not be used in cases where thick salt strata to perform the stress redistribution by plastic deformation are not present, e.g., under geological conditions as shown in Fig. 5-10.

Panel-and-pillar design similar to that shown in Fig. 5-22, but without backfilling, which is not required as the rooms are only about 3 m high and are allowed to cave-in after panel extraction, is being used in some of the potash mines near Saskatoon, Canada, provided that no water is to be expected in the immediate roof formations; these designs were developed because of the need for application of stress-relief methods. For this reason, they are dealt with in detail in Volume 2.

5.5 CREEP DEFORMATIONS CAUSED BY IRREGULAR EXTRACTION PATTERNS

5.5.1 General remarks

As the result of geological conditions during and after the deposition of potash beds, most potash deposits do not resemble blankets with uniform thicknesses and potash contents, "extending for miles in every direction", and allowing mining patterns which would not be interrupted when a mine develops and produces simultaneously (Schultz, 1971, p. 39). This was again experienced in developing potash mines in North America, as outlined in Chapter 2. Some publications — e.g., Shitkov and Permyakov, 1972, discussed by Baar, 1973 — indicate similar experience in the development of the new Russian potash mines near Soligorsk, Belo-Russia. Such irregularities in the German Zechstein potash beds have been described in numerous publications over the past 50 years or so.

There is no point in extending a planned extraction scheme into barren zones in a potash bed, and this results in remnant pillars with various configurations. It goes without saying that these natural barrier pillars have effects which resemble those caused by planned barrier pillars. The problem in cases where a mine develops and produces simultaneously, appears obvious from Fig. 2-10. The consequences of leaving remnant pillars in extracted areas may be serious as shown in detail in Volume 2.

The problems related to the extraction of potash beds are particularly severe in salt domes: the tendency of total extraction of isolated lenses results in huge cavities as shown, for example, by Langer and Hofrichter (1969): the vertical or near-vertical dimensions may exceed 200—225 m, with horizontal dimensions of 80—100 m in one direction, and several metres, depending on the thickness of the potash bed extracted, in the perpendicular horizontal direction. Theoretical calculations of the plastified zones around such cavities (see also Albrecht and Langer, 1974) — if based on laboratory formulas which call for decreasing creep rates on account of strain hardening, and on creep limits which are about 10 times higher than the in-situ creep limits of salt rocks, see section 3.3 — necessarily lead to grave underestimation of the extent of creep deformations caused by such extraction methods. In-situ measurement results to show these relationships are not available—however, the recent disastrous flooding of the Ronnenberg mine near Hannover, Germany, leaves no doubt of extensive creep deformations caused by extracted cavities with the dimensions listed above.

Some details of floodings caused by similar underestimation of the creep phenomena caused by potash extraction in salt domes and similar structures are presented in Volume 2.

5.5.2 Lyons Mine (Project Salt Vault), Kansas, U.S.A.

McClain and Bradshaw (1967, p. 245) summarize their interpretation of "stress redistribution in room and pillar salt mines", based on in-situ measurements in the Lyons Mine, as follows: "During the excavation of four rooms for an experiment demonstrating the disposal of radioactive wastes in salt mines, a number of unique rock deformation measurements were made. Since the new rooms were higher stratigraphically than the surrounding older workings, it was possible to observe deformations as the working face passed directly over instrumented boreholes. The results dramatically illustrate the presence of an abutment pressure in the floor travelling along with the face."

This interpretation of the measurement data is untenable (Baar, 1971a, p. 124). The data is interpreted in similar ways in a number of other publications, e.g. Empson et al. (1966), Lomenick and Bradshaw (1969), Bradshaw and Lomenick (1970), Empson et al. (1970). A detailed discussion appears indicated; the following re-interpretation is the one published by Baar (1971a, pp. 124—137, figs. 55—59). A general description of the geological conditions may be quoted from Lomenick and Bradshaw (1969, p. 18):

"The Lyons Mine of the Carey Salt Company. . . was opened in 1890 and closed in 1948. The salt was mined by the room-and-pillar method, with salt extraction varying from about 60—65%. The floor of the mine lies a little over 1000 ft. (300 m) below the land surface. . . At the Lyons Mine, the Hutchinson Salt Member consists of about 300 ft. (90 m) of nearly flat-lying beds of salt, shale, and anhydrite, with salt comprising about 60% of the sequence. The mine was operated in the lower part of the member. Within the mined unit, 1- to 6-in. (2.5—15 cm) layers of relatively pure sodium chloride are separated by clay and shale laminae that are usually less than 1 mm in thickness."

Fig. 5-23 (McClain and Bradshaw, 1967, fig. 1) "shows the location of the newly mined experimental area in its relationship to the surrounding older workings. The irregular area to the northeast of the experimental area was mined during the period from 1905 to 1935, while the square checkerboard pattern to the south of the experimental area was mined between 1935 and 1948."

The most important aspect — from a rock-mechanical point of view — of the experimental area is not mentioned in any of the listed publications; it becomes apparent from a general plan of the Lyons Mine published by Empson et al. (1966, fig. 1, p. 434): this plan shows no extracted mine workings to the west and to the east of the experimental area, with the exception of the two entries shown in Fig. 5-23 (2nd *south* and a nearly parallel one); these two entries end in a large remnant pillar which was apparently left for the protection of two railroad lines which cross each other above the pillar.

The experimental area shown in Fig. 5-23 was excavated in this remnant pillar. According to Empson et al. (1966, p. 434), "this location for the site of the demonstration was chosen because (1) it was located at the periphery of the mine", so there can be no doubt of the location of the experimental area in the remnant pillar which was protruding from virgin ground, separating the older workings to the north from the younger workings to the south.

According to Bradshaw (1970, p. 170), "the minimum floor-to-ceiling convergence rate is about 0.1 inch (2.5 mm) per year" in the extensive area of the older workings "which was developed about 70 years ago 14 ft. (4.2 m) high pillars, and theoretical average pillar stress of about 2500 psi (175 kp/cm^2); the area mined is large enough that the pillars have probably taken up the theoretical overburden load. At the south end of the Lyons mine, in an area of about 30 years age, but of limited extent (actually, at the end of a drift), the rate is about 0.02 inch (0.5 mm) per year. Those are about the lowest rates we have found."

Such differences in minimum convergence rates clearly indicate that, at the time when the experimental area was excavated, the redistribution of the overburden load over the previously extracted area was still in progress; in other words, there were still umbrella zones with less than average pillar loads along the periphery of the extracted area, and corresponding zones of abutment loads in the virgin ground surrounding the mined-out area. In the remnant pillar chosen for the experimental excavation, three such abutment zones were overlapping, resulting in original stresses higher than those calculated from the depth and the assumed gravities of overburden formations. These conditions must be borne in mind when interpreting the measurement results obtained in the extensive program described as follows (McClain and

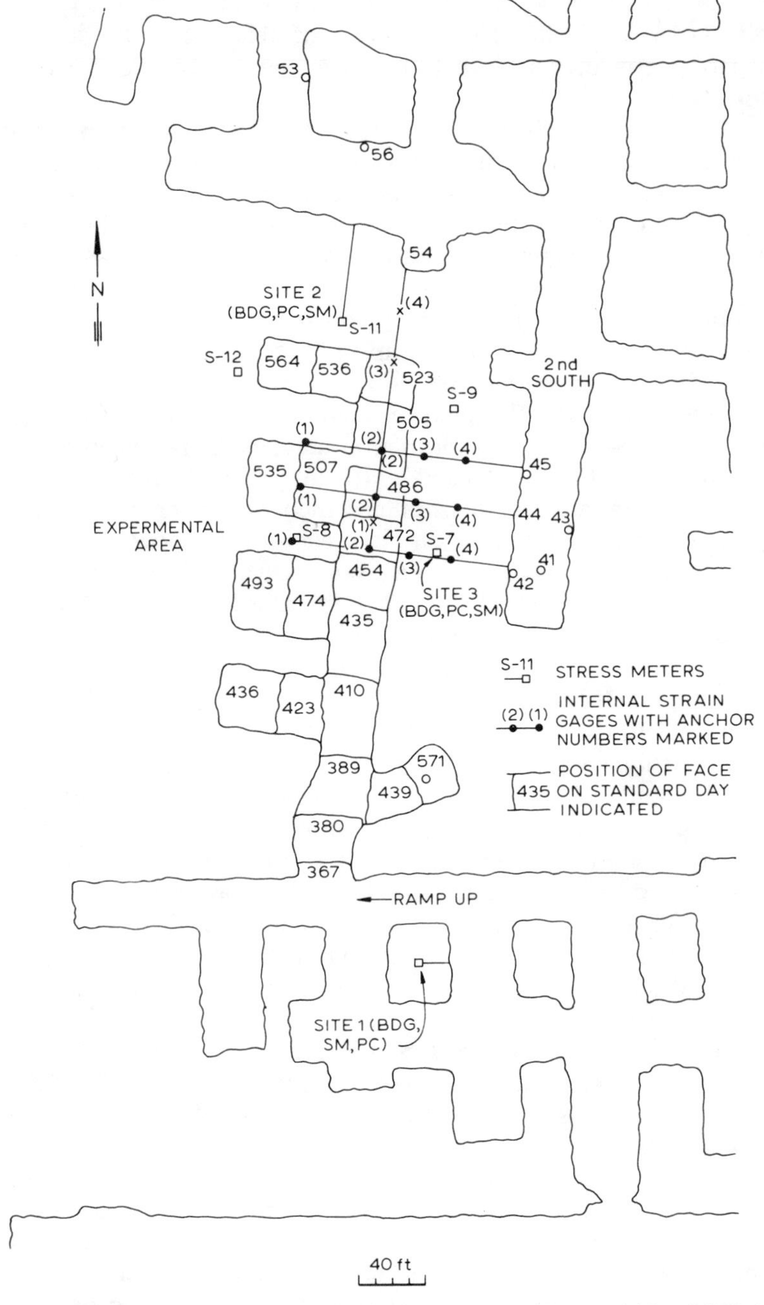

Fig. 5-23. Plan of experimental area, Lyons mine, Kansas. (After McClain and Bradshaw, 1967, fig. 1.)

SM = "stress meter"; *PC* = pressure cell; *BDG* = borehole deformation gauge. See Fig. 5-28 for location relative to older mine openings: the experimental area was excavated between the N-arrow and older workings to the south, north, and east.

Bradshaw, 1967, p. 247):

"The rate of mining (in the experimental area) was relatively slow. . . Fig. 5-23 indicates the position of the face at various times in terms of 'standard days' (standard day 1 = 1 September 1963). The location and identification of the gauges which will be discussed are also shown on Fig. 5-23. In general, three types of rock mechanics instruments were used: (i) borehole deformation gauges; (ii) internal strain indicators; and (iii) convergence gauges."

"Borehole deformation gauges. Three types of instruments to measure diametral borehole deformation were employed. The first was obtained from Leonard E. Obert of U.S. Bureau of Mines (Obert et al., 1962), and is a low modulus gauge (*BDG*). The second type was the high modulus "stress meter" (*SM*), developed by Professor E.L.J. Potts (Potts and Tomlin, 1960). The third type of borehole deformation instrument used was the U.S. Bureau of Mines borehole pressure cell (*PC*), developed by L. Panek (1961)."

"Internal strain indicators. These instruments measure axial borehole deformations by sensing the elongation or shortening of a wire anchored at some distance down the borehole relative to a fixture at the collar of the hole (Potts, 1957)".

"Convergence gauges. Vertical closure of the rooms was measured with convergence gauges, consisting of two pieces of ½-inch (1.25-cm) pipe, anchored vertically and in line several inches into roof and floor."

McClain and Bradshaw (1967, p. 252) give the following interpretation of vertical convergence measurements at "about 60" sites shown on a partial plan of the mined-out area outside the experimental area; the sites "extend to the mine shaft area, more than 1000 ft. (300 m) from the experimental

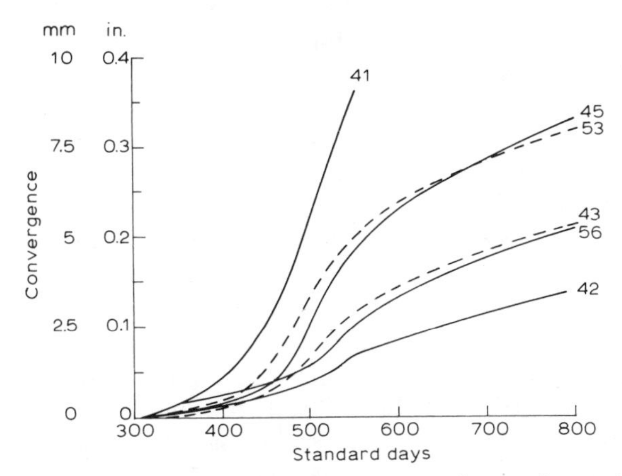

Fig. 5-24. Vertical convergences around experimental area shown in Fig. 5-23 (measuring stations identified with the same numbers as in Fig. 5-23). (After McClain and Bradshaw, 1967, fig. 10.)

area (to the north), and considerably farther to the south" (Empson et al., 1966, p. 440):

"The redistribution of the load transferred from the newly mined openings over a considerable area can be observed from the results of the convergence gauges installed throughout the old workings. Fig. 5-24 shows the measured convergences at several points (Fig. 5-23) where the effect of mining in the experimental area is indicated by the step in the curve. (Another figure) highlights this increased convergence by illustrating it as a strain rate against time. The strain and the strain rate increase as the vertical stress in the vicinity of the gauge increases due to load transference from the mining area."

This interpretation appears to confirm the existence of umbrella zones north, east and south of the remnant pillar in which the experimental area was excavated. With the progress in excavation, the abutment load on the remnant pillar was transferred to pillars in the umbrella zone; the pillars, and the salt above and below them, responded by increased creep into the old openings. The horizontal component of the creep above and below the pillars resulted in the measured vertical convergences, exactly as outlined in Chapter 4. The courses of the curves shown in Fig. 5-24 resemble the courses of corresponding curves in Fig. 5-2. The curve measured at point 41, see Fig. 5-23, appears to indicate bed separation.

The mechanism of load transfer from the newly mined experimental area "over a considerable area" outside the surrounding large pillars cannot be reconciled with high abutment pressures near the advancing faces of the experimental area, as postulated in the quotation at the beginning of this sub-section.

McClain and Bradshaw (1967, p. 252) believe that the results of horizontal extensometer measurements — see Fig. 5-23 for the location of internal strain gages with anchor numbers marked (1) to (4) at each extensometer — "in general, confirm the presence of a wave of compressive stress travelling with the advancing face, and they suggest that the peak stress in the wave is fairly close to the face even when it is advancing slowly and in a highly plastic material. They also confirm that when the face stops, this travelling compression wave continues to move out away from the face to its equilibrium position at some depth into the rock."

These interpretations resemble those discussed in sub-section 5.4.2, concerning similar extensometer measurements in a French potash mine in the Alsace district to which the authors refer as follows: "Large tensional strains develop in the pillar separating the new experimental entry from the older workings and compressional strains of about the same magnitude develop in the floor across the experimental entry. All these strains are increasing but at a constantly decreasing rate. This behavior reflects the usually observed effect of creating a new opening in plastic material (McClain, 1964)."

It is emphasized here that Fig. 5-19 shows constant horizontal strain rates after load redistribution; so do some of the curves shown by McClain and

Bradshaw (1967, figs. 2—9), while others do not, apparently due to the fact that the figures show only the results of measurements over about 300 days after the experimental area had been excavated. The curve for the last 100 days of the measuring period are virtually straight lines, so there is no justification for any conclusion regarding the future course of these curves. Considering the fact that creep deformations were measured in workings up to 70 years old, the development of constant creep rates in the experimental area after termination of the pillar heating experiments — see section 3.7 — should be expected with Bradshaw (1970), as quoted previously.

It appears that the assumption of peak stresses near the advancing face of the experimental area is biased by comparison to advancing long-wall faces in coal mining; the effects of stress-relief creep after the excavation in a remnant pillar in salt are ignored. Apparently, McClain and Bradshaw (1967) assume stress peaks at the locations of strain peaks. This may hold true in laboratory experiments when pillar models are loaded, the lateral expansion in "roof and floor" of such models being prevented by steel rings for the prevention of any deformation in these regions. It is shown in section 4.2 that such testing procedures are misleading as it is the stress difference that causes creep deformations, no matter how the stress difference is created; in salt under high original stresses, the first and most effective stress difference is created by any new excavation, and the highest strains occur initially at new faces as stress-relief creep. These strains exhibit an exponential decay until stress gradients corresponding to the local conditions are established; the subsequent creep under constant loads is characterized by constant creep rates—see sub-section 3.4.2 and the results of in-situ measurements presented so far in this chapter.

There are two lines of conclusive evidence for stress-relief creep that occurred immediately after any step of excavation of the experimental area shown in Fig. 5-23; the evidence is obscured by the way in which the measurement data is presented in the respective figures by McClain and Bradshaw (1967). This has been shown in detail by Baar (1971a) regarding the extensometer data, and by Baar (1972a) regarding the diametral borehole deformation data published by McClain and Bradshaw (1967); here, the following short summary may outline some of the shortcomings in the latter publication.

The data obtained from the extensometer, with wires anchored at the points *1* to *4* in each of the four boreholes passing 3—4.5 m below the floor of the experimental area, Fig. 5-23, are shown in two series of graphs. In the first series (figs. 2—5), the plots represent the cumulative movement of the anchor points relative to the borehole collars. "In this form the measured deformation appears very complex and difficult to interpret, especially with respect to those wires which show a reversal" in the movement of the respective anchor points. Such reversals occurred only at the points (*1*) of the extensometers 42, 44 and 45, after the face of the advancing excavation had passed over the anchor points (*2*) located under the centreline of the experimental corridor, i.e., the entry from which four rooms were excavated to the west.

As these plots show the changes in the total distances between the anchor points and the borehole collar in the 2nd *south* entry, the changes are only minimal, and the curves run parallel, until the advancing face passed over the respective extensometer lines. Up to this time, the anchor points were stationary, and the changes in distances from the collar were caused by movement of the collars due to creep around, and into, entry 2nd *south.* These movements of the collars increased when the face approached — see Fig. 5-25 — and levelled off only after the excavation ceased, i.e., when the loading of the pillars along entry 2nd *south* levelled off.

Assuming that the anchor points (*2*), below the centreline of the experimental corridor, remained stationary in the horizontal distance from the centreline of entry 2nd *south,* curve *2* in Fig. 5-25 reflects the creep deformation into entry 2nd *south.* It is emphasized that its course resembles the course of the vertical convergence curves of Fig. 5-24, indicating reloading creep. This reloading creep near entry 2nd *south,* and in the direction away from anchor point (*2*), may be called positive creep; it resulted in the lengthening of the distance between point (*2*) and borehole collar indicated by curve *2* of Fig. 5-25.

Point (*4*), at the distance of 35 ft. (10.5 m) from the collar, shows more than the total lengthening of the distance between point (*2*) and the collar for the first section next to the collar, i.e., negative movement relative to

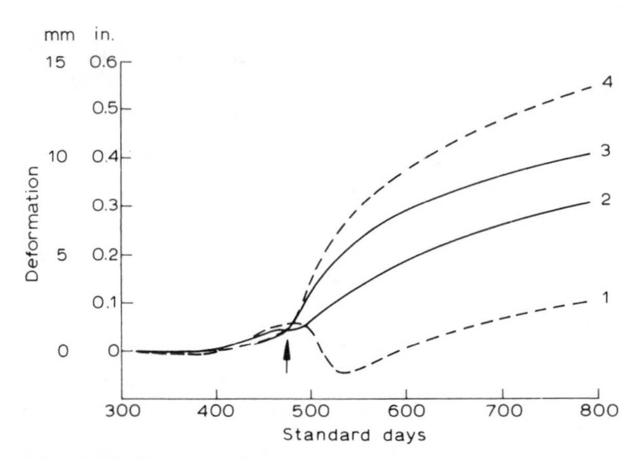

Fig. 5-25. Horizontal strains measured at site 44, Fig. 5-23. (After McClain and Bradshaw, 1967, fig. 3.)

Notice the "negative" displacement of anchor (*1*) in response to excavation between anchor and borehole collar in *2nd south.* Arrow indicates time of excavation.

point (*2*). Such negative movements are caused by stress-relief creep into newly excavated openings, as shown in section 4.3 and Figs. 4-13 to 4-15. Fig. 5-25, curve *4*, clearly shows that the negative movement of point (*4*) due to stress-relief creep into the newly excavated experimental area continued for approximately 100 days after the face had passed. After 100 days, the curves *2* and *4* run parallel, indicating that no further changes in the distance of point (*4*) from point (*2*) occurred. This means that point (*4*), located in the centre of the newly created pillar between entry 2nd *south* and experimental area, became stationary after about 100 days if point (*2*) is assumed stationary throughout the total measuring period.

Point (*3*), located next to the experimental area, continued to exhibit negative movement after the initial stress relief creep, as indicated by curve *3* of Fig. 5-25. This continued negative movement after 100 days since the face passed, indicates reloading creep.

Point (*1*), located 45 ft. (13.5 m) to the west of point (*2*) at the opposite side of the experimental corridor, shows the most dramatic positive movement towards point (*2*) after the face passed (day 486). This movement due to stress relief creep into the newly excavated corridor ceased at about day 535 when a room to the west of point (*1*) was excavated. Fig. 5-25 shows curve *1* after day 535 virtually parallel to curve *2*.

The plots of the two other sets of extensometer data (42 and 45) show similar reactions to the excavation of the experimental area; these reactions are complex, but they are not difficult to interpret provided that the effects of stress relief creep into newly excavated openings are taken into appropriate consideration.

The latter statement applies particularly to the data plotted in Fig. 5-26. This set of data was obtained with the extensometer 54 which was installed below the experimental corridor along its axis, see Fig. 5-23 for the locations of the anchor points which are marked *x*.

McClain and Bradshaw (1967, p. 251) give the following description of the anchor movements: "As the working face approached and passed over the anchor (*1*), it began to move rapidly toward the newly excavated opening. During this period, all of the other anchor points indicated relatively little displacement. The face then approached and passed over anchor (*2*), and it behaved in an identical manner. The movements detected at anchor (*3*) are not a further duplication of the previous two, because the excavation did not pass over and beyond this anchor. Notice, however, that the movement shows at the collar anchor at a delayed date."

Apparently, the statement regarding "the presence of a wave of compressive stress travelling with the advancing face" is mainly based on the interpretation of the creep deformations quoted above. As the data plotted in Fig. 5-26 is given such significance, an apparent error in the above quotation must be corrected in the first place: the excavation of the experimental area did pass over anchor (*3*), as shown in Fig. 5-23 and in Fig. 5-26; conse-

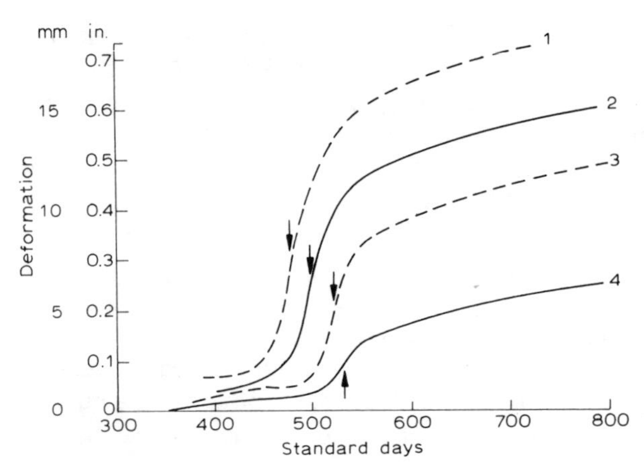

Fig. 5-26. Horizontal strains measured at site 54, Fig. 5-23. (After McClain and Bradshaw, 1967, fig. 5.)

See text for corrections to the original figure and its interpretation.

quently, curve *3* in Fig. 5-26 shows a course which resembles that of curves *1* and *2*, with only one minor, but important difference: curve *3* levels off much more rapidly into a parallel to the two other curves, undoubtedly because, on day 523, the excavation of the corridor stopped just north of anchor point (*3*). It must be borne in mind that all of the four curves include the creep movement of the collar into the old mine opening to the north. There is reason to assume that this creep at the location of collar 54 was much smaller than at other collar locations, due to the configuration of the old opening around collar site 54.

Consequently, the displacement of the "collar anchor", which is anchor (*4*), cannot be related to displacement away from the excavated area; in such a case, a corresponding displacement would have been indicated by all of the anchors. The increase in the distance between collar and anchor (*4*), as shown by curve *4*, is another "duplication" of the displacements of the other three anchors during the respective time periods when the face was approaching. Apparently, anchor (*4*) was affected by stress-relief creep into the newly excavated area, similar to the movement of the anchors (*4*) of the three other extensometers.

Such creep from distances of about 50 ft. (15 m) into the excavation definitely rules out the existence of peak stresses "fairly close to the face" which travel with the advancing face. The measured displacements rather confirm the rapid development of stress-relief creep zones which travel ahead of slowly advancing faces. If and when reloading occurs, the direction of the displacement of anchor points may be reversed, and this results in non-parallel courses of the respective curves of cumulative displacements, i.e., in widening of their distances as also indicated by curves *3* and *4* in Fig. 5-26 after the face had passed over anchor point (*3*).

The interpretation given above appears to explain conclusively the extensometer data published by McClain and Bradshaw (1967); the re-interpretation was made available to the authors after it had been discussed publicly (Baar, 1971a); no objection to the re-interpretation was received, or published.

The second line of evidence of extensive stress-relief creep around the experimental area shown in Fig. 5-23 is provided by the results of measurements of diametral borehole deformation. McClain and Bradshaw (1967, p. 252) consider "the quantitative interpretation of results obtained from "stress meters" installed in salt or other highly plastic materials extremely complex. However, it is valid to make use of the results in a qualitative manner. The readings obtained from four stressmeters installed around the experimental area. . . show the effect of the new excavation as an increase in indication, which can be interpreted as an increase in stress."

With reference to the detailed discussion in sub-section 3.8.1, and to the further discussion of stressmeter data in sub-section 5.4.2, some cautioning against the interpretation of increases in stress-meter outputs as increases in stresses appears indicated: axial borehole deformation caused by stress-relief creep may result in diametral deformation which would increase the stress-meter output, although the stresses decrease.

McClain and Bradshaw (1967, p. 154) emphasize that "one of the objectives of employing three different types of borehole deformation gauges was to make a comparison of their performance in rock salt. In order to facilitate this comparison, the three different types of gauges were installed in a close array at three different locations"—see Fig. 5-23. For the reasons given above, the strainmeter data (*BDG* and *SM*) will not be discussed any further. However, "the results of pressure cell measurements are of particular interest as they indeed allow a quantitative evaluation" (Baar, 1972a, pp. 51—55, fig. 12); a shortened version of this re-evaluation is presented as follows.

At site 3 (see Fig. 5-23 for location), the pressure cell (*PC*) was installed at the distance of 12 ft. (3.6 m) from the wall of entry 2nd *south.* It indicated an equilibrium pressure of about 575 psi (40 kp/cm^2) for about 75 days prior to standard day 400, at which date the excavation of the experimental area was approaching at the distance of about 200 ft. (60 m). This indicates the width of the abutment zone travelling ahead of the advancing face.

From day 400 to day 460, the cell pressure increased to 700 psi (49 kp/cm^2). This is the last value reported; the course of the curve, according to the plotted values, indicates a further increase in pressure. The reported values at the distance of 12 ft. (3.6 m) from an old opening agree fairly with previously reported values measured in creep zones around openings in salt—see sub-section 3.8.2. It may be emphasized that the average pillar stress of about 2500 psi (175 kp/cm^2) was calculated for the mined-out area, see p. 251.

At site 2, the *PC* failed before any results could be obtained. At site 1, the

PC was installed at the depth of 12 ft. (3.6 m) into a 40-ft. (12-m) square pillar. This pillar was affected primarily by the raising of the ceiling for the ramp up to the experimental area.

The original cell pressure of about 750 psi (52.5 kp/cm^2) decreased slowly to about 675 psi (47 kp/cm^2) over a period of about 40 days; this may be due to overpressure given to the cell when it was installed, or it may have been caused by stress-relief creep related to the increase in pillar height.

Pillar loading, apparently related to the excavation of the experimental area which started around day 367, is clearly indicated by the regular increase of the cell pressure to about 1200 psi (84 kp/cm^2) at day 500. At this date, the cell pressure was levelling off, indicating the termination of pillar loading. The irregular mining sequence disallows to relate extraction at specific distances to the regular increase in cell pressure.

According to the principles outlined in sub-section 3.8.2, the decreasing cell pressures at the beginning of the measurements indicates the maximum local stress, i.e., the vertical stress. The subsequently increasing pressures indicate the minimum local stress, i.e., the horizontal stress in the pillar at the distance of 12 ft. (3.6 m) from the edge of the pillar, where the horizontal stress equals the atmospheric pressure.

The stress gradients which calculate from the results of pressure-cell measurements, are similar to those established in stress relieved zones at other salt mines—see section 4.4. Although only a few hydraulic cells were employed, and the measurements were terminated prior to the termination of load redistribution, the results confirm the conclusions drawn from the extensometer data: the loads which were no longer supported in the experimental area due to excavation and due to stress-relief creep into the newly excavated openings, were at least temporarily supported in abutment zones around the experimental area; the depths of stress-relief creep zones into the salt rock around new excavations, and the widths of the related abutment zones resemble those found in other salt deposits with comparable conditions. The assumption of stress peaks near the walls of the excavations is disproved by the results of the in-situ measurements published by McClain and Bradshaw (1967).

In view of the publicity given to the theoretical design principles for the experimental area by numerous writers in numerous publications over the past 15 years — some, e.g., King (1973), Serata (1974), Dreyer (1974a), Nair et al. (1974), to name a few, are continuing to postulate the validity of power laws developed in the laboratory for the prediction of creep convergences in situ — it appears important to compare the above discussed data with the anticipated "correlation of convergence measurements in salt mines with laboratory creep-test data" (Bradshaw et al., 1964). According to these authors (l.c., p. 501), "at the Oak Ridge National Laboratory" (ORNL), salt mine stability was studied "to determine the design parameters for the ultimate disposal of high-level radio-active wastes in salt mines".

"S. Serata in 1959 (reference to an unpublished report; such listings are not included in this text) suggested that creep rates (or closure rates) of mined openings always decrease with time. In 1959, both ORNL and Serata installed several vertical and horizontal convergence measuring stations in the Carey Salt Company's mine at Hutchinson, Kansas. Since that time, ORNL has added additional stations at the Carey mines in both Hutchinson and Lyons, Kansas. These stations were designed to determine the effects of both the age of the opening and pillar stress on creep rates. Data from these stations tended to indicate that Serata's hypothesis of steadily decreasing vertical convergence was correct, at least for the pillar stress levels encountered in the Kansas mines. In March 1963 Serata indicated that he had successfully extrapolated creep data from laboratory pillar models to creep rates in mine openings."

Serata's hypothesis could not be reconciled with Obert's (1964) results of pillar model testing under constant loads; Obert found "that the creep rate became constant after a relatively short transient initial period", and his "data was re-interpreted by Bradshaw et al. (1964). They found that the model pillar creep could be fitted by a power law" (King and Acar, 1970, p. 227). Plots of creep rates which decrease according to a power law, versus time on a log-log diagram result in straight lines with a certain slope depending on the time function—see Fig. 3-3 with the straight line of predicted opening closure rates according to Serata's hypothesis.

Fig. 5-27 (Bradshaw et al., 1964, fig. 4) shows the creep rate decreases under various loads according to their re-interpretation of Obert's data, compared with vertical convergence rates in both the Hutchinson and the Lyons mines. The basic data as of September 1963 are presented in two tables. The locations of the measuring sites are shown on a schematic plan of the Hutchinson mine, and on a partial plan of the Lyons mine; the latter is shown as Fig. 5-28 for use in the following discussion. The same maps are also found in Hedley's (1967) publication.

According to the schematic plan of the Hutchinson mine, the mine shaft was protected by an irregular pillar; the area north of the mine shaft was not extracted, apparently for protection of a railroad line which is shown crossing approximately east-west about 500 ft. (150 m) north of the shaft, running all along the extracted area to the south. According to the data listed in the respective table, the oldest openings were approximately 40 years old at the time when the measurements were taken. The older panels were extracted by the excavation of rooms with the dimensions of 50 × 300 ft. (15 × 90 m), 6 ft. (1.8 m) high; the pillars were 20 ft. (6 m) wide.

Approximately 20 years before the measurements began, the extraction method had been changed to standard room-and-pillar mining with rooms 50 × 50 ft. (15 m) wide and 10 ft. (3 m) high in the south-eastern area of the mine. One panel, 1000 ft. (300 m) wide, was extended far to the south; according to the map, the adjacent area remained unextracted. In this panel,

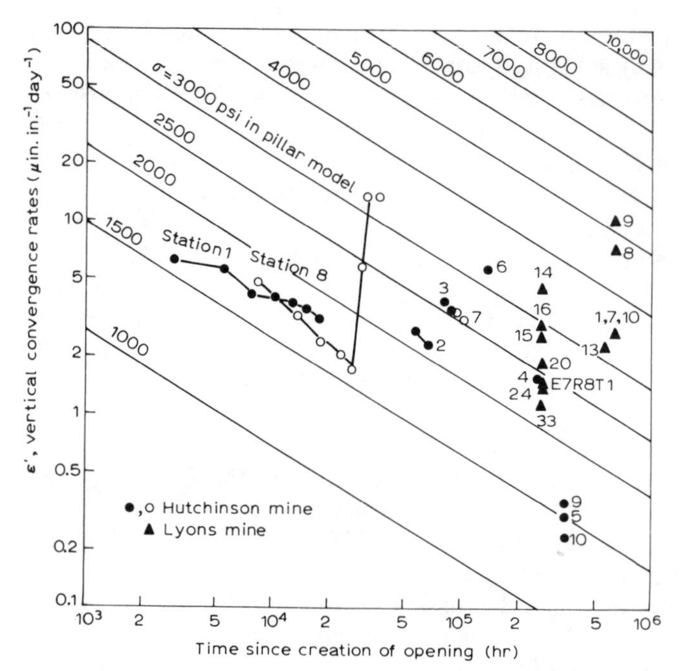

Fig. 5-27. Measured mine convergence rates superimposed on plots of equation derived from model tests. (After Bradshaw et al., 1964, fig. 4; Bradshaw and Lomenick, 1970, fig. 5.)

the pillars were 40 ft. (12 m) square. Stations 6 and 7 (see Fig. 5-27 for plotted convergence rates) were located in this panel near its eastern boundary.

Stations 2 and 3 were located in another panel with about the same span, and unextracted areas to the north and to the south, at the eastern boundary of the mined-out area; the pillars left in this panel were 50 ft. (15 m) square. Stations 1 and 8 were located in the North panel being developed at the distance of 1500 ft. (450 m) from the shaft with the same room dimensions, but 12 ft. (4.2 m) high, and 50 ft. (15 m) square pillars. The panel span was approximately 800 ft. (240 m).

The other stations for which the convergence rates are plotted on Fig. 5-27, were located in the older workings with long pillars. Stations 5, 9 and 10, with the smallest convergence rates, were located next to the shaft pillar; station 4 was the only one located in the central area of the mine, extracted 28 years prior to the beginning of the measurements.

Bradshaw et al. (1964) apparently realized that theoretical calculations of pillar loads under conditions as described above are rather meaningless, stating that "most of the gaging stations in the Hutchinson mine are located near the boundaries of the mined-out area and thus the adjacent pillars

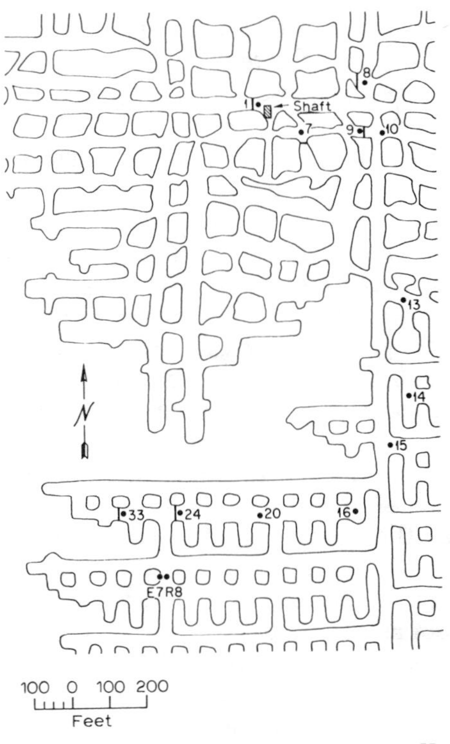

Fig. 5-28. Partial plan of Lyons mine, Kansas, showing locations of convergence measuring stations. (After Bradshaw et al., 1964, fig. 3; Hedley, 1967, fig. 8.)

would not be expected to be supporting the calculated dead weight of the overburden as should be the case if the pillars were in the center of a very large panel of rooms. That this is true (at least for openings up to a few years of age) is shown in Fig. 5-27 by the behavior of Hutchinson station No. 8."

"A short time after the excavation in the area of Station 8, mining was discontinued on the west side (of the isolated north panel, with station 8 near the temporary panel boundary) and shifted to the east side of the north panel (with station 1 near the panel boundary). When the opening at station 8 was about 3 years old, the apparent stress in the pillar was about 1700 psi (see Fig. 5-27). At this time, mining was resumed near station 8 and the western boundary was extended. The rapid rise in vertical convergence rate indicates that the pillar had not been fully loaded after the initial mining; however, it should not be assumed that the load increased to more than 3000 psi (as might be inferred from Fig. 5-27). With the incremental load added some three years after the opening was created, the time scale would have to be shifted to take into account the fact that the creep due to the incremental load essentially begins at a new zero on the time scale. With the points plotted, as shown, as a function of time after the original loading (initial creation of the opening), the convergence rate would have been expected to show a more rapid decrease as the pillar stabilized under its new load. Unfortunately, the gaging station had to be removed before this stage was reached. It is expected that station 1 will show a similar behavior when mining is again resumed on the east side of the north panel."

This interpretation of the development of the convergence rates at stations 1 and 8 is fully quoted because it shows clearly the basic error in attempts to calculate pillar loads from measured creep rates without paying attention to the fact that the initial creep after creation of an opening in salt is caused by stress relief rather than by "original loading (initial creation of the opening)". With reference to Fig. 3-3, and to Fig. 5-1 and 5-2, it is emphasized that the plotted creep rates for station 1, and the decreasing creep rates at station 8 — see Fig. 5-27 — are typical stress-relief creep rates. The loads on the 4.2 m high, 15 m square pillars at this mine were probably relatively more reduced by stress relief creep than the loads on the pillars referred to in the reference figures; the latter were 2.25 m high, 15 m wide, but sufficiently long to prevent stress-relief creep in the direction of the longer pillar axis. This gives every reason to assume that the vertical stresses in the two Hutchinson pillars were smaller than those shown in Fig. 4-14. The temporary increase in convergence rates at station 8 indicates the beginning of pillar reloading — see Fig. 5-2 — which depends on the development of mined-out areas, as shown in detail in sub-section 5.2.1.

Bradshaw et al. (1964, p. 508) acknowledge that "Hutchinson stations 5, 9, and 10, located near a northern boundary as well as the large irregular pillar left around the mine shaft, indicate stresses considerably below the calculated values". Considering the established fact that constant creep rates develop even in virgin ground where any reloading after stress-relief creep is excluded, these creep rates have definitely no relationship to laboratory creep rates determined under questionable conditions of confinement at the roof and the floor of model pillars.

Bradshaw et al. (1964, p. 511) "felt that a better comparison of the effects of age and extraction ratio on creep rates could be made if all stations were similarly located", i.e., near the periphery of mined-out areas, where pillar loads less than the theoretical loads calculated from depth and extraction ratios are expected. The only station located in the centre of the mined-out area was station 4; according to theory, it should have shown the highest creep rate; in fact, however, it showed about the smallest rate. This requires an explanation, so the data was checked, and it is shown in the following that age and theoretical pillar loads cannot be regarded the only factors that determine creep rates: at station 4, it is apparently the room height of only 2.4 m that causes relatively small creep rates of 1 mm/year next to the pillar, and 1.25 mm/year at the centreline of the long room, the age of which is listed as 28 years.

At stations 5 and 9, age 40 years, protected against full loading by the shaft pillar and by virgin ground, the convergence rates are similar: 0.5 mm/year next to the pillar, 0.75 mm/year at the centreline of long rooms. The room height of 4 m is the apparent cause of these relatively high convergence rates under stresses considered "considerably below the calculated values."

The effects of pillar heights on convergence rates are also demonstrated by the results obtained at the Lyons mine, see Fig. 5-28 for the locations of a number of measuring stations. The vertical convergence rates as of September 1963, i.e., prior to the excavation of the experimental area just east of the *N*-arrow, are listed for some of the stations in Table 5-IV. The convergence rates are plotted on Fig. 5-27 for comparison with the theoretical assumptions made by Bradshaw et al. (1964) who comment as follows on the reasons why great differences in the measured convergence rates occur:

"Lyons stations 1 through 13 are located in the area around the shaft (no large shaft pillar was left in this mine) where the mining pattern was very irregular and the estimated extraction ratio is no better than a guess. Therefore the calculated pillar loads for these stations (Table 5-IV) may be too high."

"The vertical convergence rates at stations 8 and 9 are anomalous, and the

TABLE 5-IV

Creep measurements in the Lyons mine, Kansas (after Bradshaw et al., 1964, tables 2 and 3); data as of September 1963; opening widths varying, see Fig. 5-28; due to rounded openings, most stations are in the center of the openings

Station No.	Approximate age of opening (years)	Vertical closure rate (mm/year)	Effective pillar (room) height (m)	Estimated average pillar stress (kp/cm^2)
1	70	5.0	5.1	280
7	70	5.5	5.4	280
8	70	13.0 *	4.6	280
9	70	9.3 *	2.4	280
10	70	5.5	5.1	280
13	60	4.3	5.1	280
14	30	6.3	3.6	208
15	30	3.5	3.6	208
16	30	4.9	4.4	175
20	30	2.0	3.0	175
24	30	2.5	4.6	175
33	30	2.0	4.6	175
E7R8-T1 (center of opening)	30	1.5	2.7	175
E7R8-T2 (next to pillar)	30	1.3	2.7	175

* Sagging roof slabs.

anomaly is definitely known to be due to a parting separation and resultant ceiling sag."

Fig. 5-27 reveals that only the stations 1 through 10 are located in the area around the shaft. As the pillar sizes differ greatly, equal pillar loads must not be expected as the stress gradients around openings with similar heights are similar—see section 4.4; the measured vertical convergence rates at stations 1, 7 and 10 apparently confirm equal stress gradients in the adjacent pillars.

At station 8, located in the center of a room intersection, the effective opening span is considerably larger than at the other stations, and this accounts for the higher convergence rate due to sagging. The sagging, of course, reflects buckling of stress relieved roof beds caused by horizontal pillar and pillar roof creep, as shown in detail in sub-section 4.5.3.

At station 9, the measured horizontal closure rate is about the same as at the other stations around the shaft, approximately 5 mm/year. The opening heights is only about one half of the height at other stations. The narrower opening width, compared to station 8, suggests shear failure rather than buckling at station 9. In both cases, there is definitely no relationship to the creep law represented by the slope of the straight lines shown in Fig. 5-27.

Stations 13, 14 and 15 are located in the umbrella zone provided by the large remnant pillar shown in Fig. 5-28. The reasons for the estimate of full pillar loads at station 13, and reduced pillar loads at the two other stations, are not given. At station 13, the greater opening height (5.1 m compared to 3.6 m) calls for smaller stress gradients, i.e., for reduced pillar loads, contrary to the listings in Table 5-IV.

The assessment of the data listed in Table 5-IV continues as follows (Bradshaw et al., 1964, p. 511): "Stations 16 through 33 are progressively nearer a western boundary (of the mined-out area, see Figure 5-28) and the convergence rates seem to reflect this fact, although it is believed that there may be some ceiling sag contributing to the convergences at 16 and possibly at 20."

The fact that the respective stations are progressively nearer the western boundary of the mined-out area, is undisputable. However, Fig. 5-28 also shows that the stations 16 through 33 are located parallel to the boundary of the large remnant pillar to the north, the openings being separated from this boundary by a row of 12 m square pillars, and a long room the width of which is about 10 m according to both Figs. 5-28 and 5-23. As a matter of fact, station 33 is located in the second long room south of the large remnant pillar; the adjacent pillar is the one in which site 1 of Fig. 5-23 was installed for pillar stress determinations. Apparently, both of the mine plans are not intended to represent exact surveyed plans; the considerable deviation from the north—south direction shown for the north—south entries, including the experimental corridor, in Fig. 5-23 does not affect the rock-mechanical conclusions based on the fact that the stations 16 through 33 are located in the second long room parallel to the edge of a large remnant pillar.

Relatively small convergence rates of 2—2.5 mm/year are in line with the conclusion that the pillar loads were considerably smaller than listed in Table 5-II, with the exception of station 16; this station apparently did not receive the same protection by an umbrella along the remnant pillar because of its location at the southeast corner of the remnant. In addition, some extraction had taken place just north of that corner, see Fig. 5-28.

Finally, station E7R8 is located at the distance of about 75 m from the remnant pillar to the north, and at about the same distance from virgin ground to the southwest, according to the mine plan shown by Empson et al. (1966, fig. 1). This location, and the room height of only 2.7 m, explain the convergence rate of only 1.3 mm next to the pillar, and 1.5 mm at the center of the opening.

It appears rather difficult to agree with the conclusions drawn by Bradshaw et al. (1964, p. 513) in view of their statement that "accumulation of data at individual stations in older openings is not yet sufficient to determine with certainty if the closure rates are still decreasing after several decades". Their claim that "vertical closure rates have been shown to continue to decrease with time in openings up to 12 years old where the pillar stress is well below the ultimate strength", is not supported by measurement data; however, their comparison of assumed pillar stresses with ultimate strengths is indicative of the neglect of the fact that creep over decades proves plastic behavior. Plastic behavior makes it impossible to arrive at conclusive explanations based on elasticity theories.

The assumption of peak stresses close to the face, discussed at the beginning of this sub-section, is also based on elasticity theories. To summarize the above corrections, the statement by McClain and Bradshaw (1967) quoted on p. 254 may be changed as follows (Baar, 1971a, p. 137): "The results (of in-situ creep measurements) confirm the presence of a zone of partial stress-relief travelling ahead of advancing faces, and they suggest that the peak strain in the stress-relief creep zones is fairly close to the face even when it is advancing slowly. In a highly plastic material under triaxial stress, stress relief in one direction at an advancing face causes immediate reaction of the rock ahead of the face by creep into the opening."

At the Lyons mine as well as the Hutchinson mine, overburden load transfer over distances similar to those observed at other salt and potash mines took place immediately after extraction mining, resulting in pillar loading conditions which varied throughout the mined-out areas. Reloading of pillars located in umbrella zones along virgin ground and remnant pillars was not complete after time periods up to 70 years, but resumed when excavation resumed in abutment zones. This proves the difference in the time factors which govern creep deformations in salt rocks, and the deformation of overburden formations which may behave nearly as assumed in elasticity theories.

Recent studies dealing with the data accumulated in Project Salt Vault

(Starfield and McClain, 1973; Hardy et al., 1974) are also based on the assumption of elastic behavior of salt rocks; for this reason, such studies may be regarded as interesting mathematical exercises. However, practical application of the hypotheses derived from such theoretical studies is bound to result in unexpected events, as emphasized by Baar (1975) with reference to previous experience in potash mining.

5.5.3 Solution mining in Alpine salt diapirs, Austria

The difficulties encountered in mining Alpine salt diapirs are apparent from the short description of the geological conditions given in sub-section 4.2.3: the salt deposits contain considerable amounts of impurities which make selective mining necessary. For several thousand years, handpicked pieces of more or less clean salt made up the production from ancient mines.

The development of solution mining techniques for selective salt extraction began over thousand years ago (Kramm, 1973): the salt rock between development entries at different levels was dissolved, leaving the insolubles behind in the solution cavities with volumes of 2000—6000 m^3. This technique is still in use, as shown in Fig. 5-29; it requires long-term planning and development of entries at deeper levels. As the deposits contain large blocks of insoluble formations, the distribution of solution cavities cannot follow regular patterns over large areas; for details, the reader is referred to pertinent publications, e.g., Schauberger (1955).

The dimensions of the solution cavities dissolved in modern operations are considerably greater than those of early workings mentioned above. Some vertical dimensions are given in the schematic cross-section, Fig. 5-29. This figure also indicates minor changes in the location of the initial openings

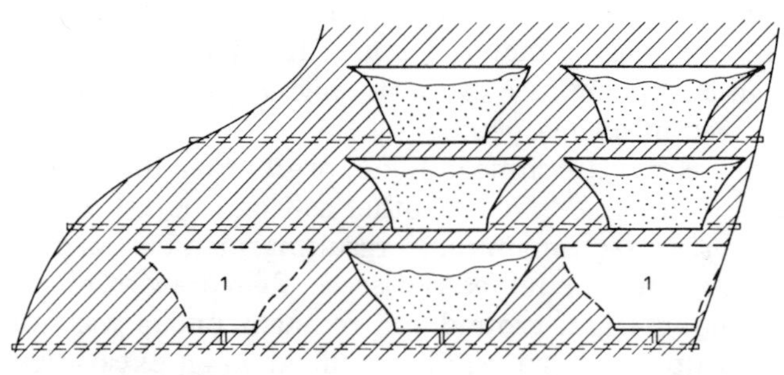

Fig. 5-29. Schematic cross-section through the Hallstatt salt diapir, Austria. (Modified after O. Schauberger, pers. comm., 1976.)

Approximately to scale. Vertical distance between development entries 30—40 m. Span of the ceilings of mined-out cavities up to 80 m. The mined-out cavities are almost completely filled with insoluble residue. *1* = projected cavities, the initial excavation by drilling and blasting completed.

relative to the development entries. The initial openings are excavated by drilling and blasting; they are connected to the upper entry system by inclines or by intermediate shafts. The horizontal dimensions of initial openings may vary within the following design limits: circles with diameters of 30—40 m, rectangles 20 × 30 m, and ellipses with corresponding lengths of axes.

The solutioning process is initiated by filling the initial opening with fresh water. The resulting salt solution is withdrawn at the bottom, and fresh water is added intermittently or continuously from the upper level. The insoluble residues accumulate at the bottom of the cavity, preventing salt dissolution. Solutioning takes place mainly at the ceiling, and the process is terminated when only approximately 10 m of salt are left between the ceiling and the bottom of a cavity at a higher level. The inclination of the walls of developing cavities is controlled by the rate of addition of fresh water.

The planned final dimension of the ceiling is a circle with the diameter of 70—90 m; however, the actual final dimensions often resemble an ellipse with the longer axis up to 100 m long, and the maximum width reaching only 50—70 m. As the planned distance between the final ceilings of two neighboring cavities is only 20 m, see Fig. 5-29, excess lengths of ceilings frequently resulted in the brining of cavities into adjacent existing cavities. The resulting total lengths may exceed 200 m, with widths in excess of 70 m. According to Schauberger (pers. comm., 1976), it depends on the composition of the 10-m ceiling plate rather than on time whether or not, in such cases, a cavity breaks through into an existing cavity above it; if the ceiling consists of relatively pure salt with 80—90% NaCl, ceiling failure does not occur. In such cases, up to 1.5 m of subsidence of 10-m ceilings were measured over time periods of 70 years; such homogeneous ceilings which exhibit no indications of fracturing span up to 100 m.

In contrast, ceilings which contain large amounts of brecciated shale and anhydrite frequently develop failure patterns as schematically shown in Fig. 5-30 (Schauberger, 1950). Such apparent shear failures are attributed to the combined effects of horizontal thrust and vertical deformation caused by the deadweight of roof strata, as indicated by arrows; the author emphasizes that the cross-section of the ceiling develops into the shape of a rather flat arch, implying "a certain degree of elastic behavior, i.e., the higher the salt content is, and the more regular the salt distribution is, the higher is the degree of elastic behavior".

This observation is considered extremely important as elastic behavior of homogeneous salt rocks requires the elimination of any stress differences which exceed the very low limits of elastic behavior of rock salt. The only possible way in which excess stress differences can be eliminated under the given conditions is by stress-relief creep into the large cavities during the brining process. In other words: under conditions as illustrated in Fig. 5-29, with only 2—3 cavities created in a cross-section that is flanked by compe-

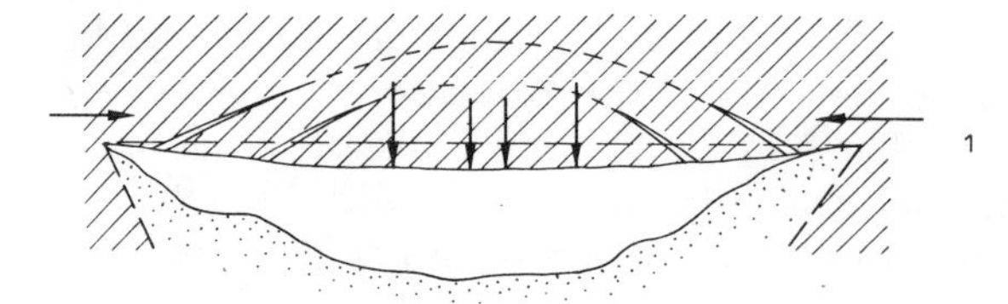

Fig. 5-30. Cross-section through the failing roof of a cavity, Hallstatt, Austria. (After Schauberger, 1955.)

The arrows indicate the directions of deformational forces. *1* = original level of ceiling.

tent limestone formations which exhibit elastic behavior, the remaining pillars must be considered stress-relieved unless and until overburden subsidence occurs.

Apparently, subsidence of the main overburden formations does not occur as these formations consist of thick, competent limestone and similar rocks (Schauberger, 1955, fig. 2); the total span of the salt diapir at the extraction levels shown in Fig. 5-29 is only 250—300 m, and the bridging, competent overburden formations are supported by competent abutments of limestone which prevent any horizontal deformations at the extraction levels. Under these conditions, detrimental horizontal creep deformations cannot occur—with the exception of stress-relief creep.

The relationships characterize the deformations around development entries — see Table 4-VII — as caused by stress-relief creep. The reason why similar deformations are not observed in and around the much larger solution-mined cavities appears obvious from the following differences: development entries are heading into salt rocks under relatively high stresses; measurements and other observations show the development of constant closure rates; according to Schauberger (1976), these entry closure rates are not significantly influenced by the development of large solution cavities.

Undoubtedly, the excavation of the huge solution cavities results in far more extensive stress-relief creep, and in dislocations of entries as examplified by Fig. 5-16. However, the decisive difference between entries and large cavities created by solutioning is that, in the latter case, the stress-relief deformations occur during the excavation process; it takes many years, even decades, until the solutioning of any individual cavity is terminated. Stress-relief creep is characterized by rapid initial rates which exhibit an exponential decay as shown in preceding chapters, and stress-relief creep zones extend to considerable distances from newly created openings—provided there are salt rocks under high stresses present.

The latter prerequisite is not fulfilled near the edge of a salt diapir — see Fig. 5-29 — and near large blocks of elastic rocks which are frequently encountered in the Alpine salt diapirs. Also, when a cavity is brined up towards an existing one, it approaches ground in which stress gradients corresponding to the creep limits of the respective salt rocks are already established. In any case, each individual solution cavity extends into its own stress-relieved zone; therefore, damages to the ceiling as shown in Fig. 5-30 are insignificant as the damaged rock is subsequently dissolved, or incorporated in the residue if local roof falls occur.

Summarizing these measurements and observations, it can be said that the surprising stability of large solution-mined cavities in Alpine salt diapirs can be attributed to the facts that: (1) the cavities are excavated in stress-relieved salt rocks; and (2) reloading of remaining pillars is prevented by elastic formations above the mined-out areas; these bridging formations are supported by flanking formations which also exhibit elastic behavior.

In recent years, the previously unexpected stability of large solution-mined cavities was utilized in the following way (Schauberger, pers. comm., 1976): instead of solutioning three individual cavities on top of each other as shown in Fig. 5-29, one single cavity is brined to the total height of up to 90 m; in order to prevent final ceiling spans over 80 m — this span is considered the maximum allowable span — the brining process had to be accelerated to prevent too much of solutioning at the walls.

The size of such cavities exceeds the size considered feasible for storage caverns in salt deposits according to various hypotheses which are based on laboratory experiments and theoretical expectations. It is emphasized that the above re-assessment of measurements and observations in Alpine salt diapirs is in line with the assessment of the comprehensive measurement data presented in other chapters of this Volume 1 of *Applied Salt-Rock Mechanics.*

The implications for the design of large cavities in close proximity to each other in salt deposits are dealt with in the second Volume. The effects of stress-relief creep must not be ignored, neither in conventional mining operations nor in solution mining, as such neglect has resulted in too many "unexpected" events, including mass accidents on account of gases released by stress-relief deformations, numerous mine floodings, and the largest rockbursts experienced in the mining industry as a whole. The implications of stress-relief creep are particularly significant when salt deposits are used for storage of oil and gas, and for the disposal of dangerous material such as radioactive waste. On the other hand, the application of stress-relief methods is necessary in deep salt and potash mines in order to keep openings for long-term use safe and operational.

REFERENCES

The following abbreviations are used for Proceedings frequently referred to:

Symp. Salt stands for: Symposium on Salt, Northern Ohio Geological Society, Inc., c/o Department of Geology, Case Western Reserve University, Cleveland, Ohio. The Proceedings of these Symposia are edited by:
Symp. Salt, 1963: A.C. Bersticker, K.E. Hoekstra, J.F. Hall
II Symp. Salt, 1966: J.L. Rau
III Symp. Salt, 1970: J.L. Rau and L.F. Dellwig
IV Symp. Salt, 1974: A.H. Coogan.

III Congr. ISRM stands for: Advances in Rock Mechanics, Proceedings of the Third Congress of the International Society for Rock Mechanics, Denver, Colorado, September 1—7, 1974. Publisher: Printing and Publishing Office, National Academy of Sciences, Washington, D.C.

Note: Sources referred to in quotations from publications are not necessarily included in the following list of references.

For comprehensive lists of reference papers, the reader is referred to the following publications: Geology: Lotze (1957); Mineralogy and Geochemistry: Kühn (1968), Braitsch (1971); Mechanical Properties of Salt: Odé (1968a); Theoretical Work: Gimm (1968), Dreyer (1972); In-Situ Measurements: Baar (1972a).

Abel, J.F., 1970. Rock mechanics—can it pay its way? *III Symp. Salt*, 2: 197—207.

Adachi, S., Serata, S. and Sakurai, S., 1968. Determination of underground stress field based on inelastic properties of rocks. *11th Symp. Rock Mech., Univ. California, Berkeley*, Preprint: 39 pp.

Ahlborn, O. and Richter-Bernburg, G., 1955. Exkursion zum Salzstock Benthe (Hannover) mit Befahrung der Kaliwerke Ronnenberg und Hansa. *Z. Dtsch. Geol. Ges.*, 105: 855—865.

Albrecht, H. and Langer, M., 1974. The rheological behavior of rock salt and related stability problems of storage caverns. *III Congr. ISRM.*, IIB: 967—974.

Allen, K., 1971. Eminence—natural gas storage in salt comes of age. *Soc. Pet. Eng., AIME, Texas*, Preprint SPE 3433: 8 pp.

Allen, K., 1974. Analysis of data from Eminence Dome storage caverns. (Internal report, Fenix and Scisson, Inc., Tulsa, Okla., 45 pp.)

Allen, M.H., 1966. Natural gas storage in caverns in Saskatchewan. *II Symp. Salt*, 1: 412—421.

Anthony, T.R. and Cline, H.E., 1974. Thermomigration of liquid droplets in salt. *IV Symp. Salt*, 1: 313—321.

Atwater, G.I., 1968. Gulf Coast salt dome field area. In: *Saline Deposits—Geol. Soc. Am. Spec. Pap.*, 88: 29—40.

Autenrieth, H., 1970. 50 Jahre deutsche Gemeinschaftsforschung auf dem Gebiet der Kalirohsalzverarbeitung. *Kali Steinsalz,* 5: 289—306.

Baar, A., 1952a. Entstehung und Gesetzmässigkeiten der Fazieswechsel im Kalilager am Südharz, 2. *Bergakademie,* 4: 138—150.

Baar, A., 1952b. Grubengase im Südharz-Kalibergbau. *Bergbautechnik,* 2: 469—473.

Baar, A., 1953a. Die gebirgsmechanischen Vorgänge beim Pfeilerrückbau in Bleicherode. *Bergbautechnik,* 3: 27—28; 35.

Baar, A., 1953b. Über Zusammenhänge zwischen Tektonik, Landschaftsform und Kalilagerstätte am Südharz. *Bergbautechnik,* 3: 120—124.

Baar, A., 1953c. Zur Gebirgsmechanik im Kalibergbau. *Freiberg. Forschungsh., A,* 12: 41—49.

Baar, A., 1954a. Ursachen und Mechanik der Gebirgsschläge und anderer Gebirgsdruckauswirkungen im mitteldeutschen Kalibergbau. *Bergbautechnik,* 4: 132—139.

Baar, A., 1954b. Untersuchungen des Bromgehalts im Zechsteinsalz. Auswertung des "Bromtests" bei Aufschlussarbeiten im Kalibergbau. *Bergbautechnik,* 4: 284—288.

Baar, A., 1954c. Zur Schlagwetterbekämpfung im Südharz-Kalibergbau. *Bergbautechnik,* 4: 339—343.

Baar, A., 1954d. Vorgeschichte, Auslösung and Vorschläge zur Verhütung von Gasausbrüchen im Kalibergbau. *Bergbautechnik,* 4: 441—446.

Baar, A., 1955. Zur stratigraphischen Orientierung im Kalibergbau durch den Bromtest. *Bergbautechnik,* 5: 40—47.

Baar, A., 1958. Über gleichartige Gebirgsverformungen durch bergmännischen Abbau von Kaliflözen bzw. durch chemische Umbildung von Kaliflözen in geologischer Vergangenheit. *Mineralsalze ozeanischen Ursprungs, Symp., 1958—Freiberg. Forschungsh., A,* 123: 137—159.

Baar, A., 1959a. Ist eine Reform der Abbauverfahren im Kalibergbau notwendig? *Bergbautechnik,* 9: 250—260.

Baar, A., 1959b. Discussion of papers presented at the "Internationale Gebirgsdrucktagung 1958, Leipzig" by K.H. Höfer, W. Buchheim, and G. Spackeler. In: *Internationale Gebirgsdrucktagung 1958, Diskussionen.* Akademie-Verlag, Berlin, pp. 51—55; 134—138; 170—173.

Baar, A., 1960. Über die fazielle Entwicklung der Kalilagerstätte des Stassfurtflözes. *Neues Jahrb. Geol. Paläontol., Abh.,* 111: 111—135.

Baar, C.A., 1961a. Unfallschutz bei Bohrungen in gasführendem Gebirge. Schlägel Eisen, 7: 468—469.

Baar, C.A., 1961b. Gebirgsdruckmessungen bei elasto-plastischer Verformung. *Bergbauwissenschaften,* 8: 461—468.

Baar, C.A., 1962. Gasausbrüche im Kupferschieferbergbau. Beitrag zur Deutung der gebirgsmechanischen Vorgänge bei Gasausbrüchen in verschiedenen Bergbauzweigen. *Erzmetall,* 15: 139—147.

Baar, C.A., 1963. Der Bromgehalt als stratigraphischer Indikator in Steinsalzlagerstätten. *Neues Jahrb. Mineral., Monatsh.,* 7: 145—153.

Baar, C.A., 1964a. Gesetzmässigkeiten bei Gas-Wasser-Einbrüchen in Salzbergwerken *Forschungsber. Dtsch. Forschungsgemeinsch.,* Bad Godesberg, 91 pp.

Baar, C.A., 1964b. Contributions to discussion of papers presented by Höfer, K.H., and by Gimm, W.A.R. and Pforr, H. *Proc. 4th Int. Conf. Strata Control Rock Mech., Columbia Univ., New York City,* pp. 68, 448—449.

Baar, C.A., 1966a. Measurements of rock pressure and pillar loads in deep potash mines. *II Symp. Salt.,* 2: 18—33.

Baar, C.A., 1966b. Bromine investigations on eastern Canada salt deposits. *II Symp. Salt,* 1: 276—292.

Baar, C.A., 1966c. Der Bromgehalt im Steinsalz als stratigraphischer und genetischer Indikator im norddeutschen Zechstein. *Z. Dtsch. Geol. Ges.,* 115: 572—608.

Baar, C.A., 1970a. Correct interpretation of data for better understanding of salt pillar behavior in mines. *III Symp. Salt*, 2: 280—289.

Baar, C.A., 1970b. Rock mechanics in deep potash mines in Canada. (Internal Report, International Minerals and Chemical Corporation (Canada) Ltd. (IMC), May 1970, 98 pp.)

Baar, C.A., 1971a. Applied rock mechanics in deep potash mines—creep and stress redistribution upon mining deep salt and potash beds. *SRC Rep.*, E 71-2, 1: 145 pp.; 2: 61 figs.

Baar, C.A., 1971b. Creep measured in deep potash mines vs. theoretical predictions. SRC Rep., E 71-6: 64 pp.

Baar, C.A., 1971c. Discussion to the paper by H. Habenicht (*Rock Mech.*, 3: 99—112) "Maschineller Streckenvortrieb im Bergbau—Entwicklung und Probleme". *Rock Mech.*, 3: 239—240.

Baar, C.A., 1972a. Creep measured in deep potash mines vs. theoretical predictions. *Proc. 7th Can. Rock Mech. Symp., Edmonton, March 1971*, M 37-1572: 23—77.

Baar, C.A., 1972b. Determination of stresses in salt and potash mines for application to mine design. *SRC Rep.*, E 72-8: 77 pp.

Baar, C.A., 1972c. Actual geological problems in Saskatchewan potash mining. *SRC Rep.*, E 72-18: 50 pp.

Baar, C.A., 1973. Discussion of papers by Winkel et al. (1972), and by Höfer and Knoll (1971). *Int. J. Rock Mech. Min. Sci. Geomech. Abstr.*, 10: 251—254.

Baar, C.A., 1974a. Geological problems in Saskatchewan potash mining due to peculiar conditions during deposition of potash beds. *IV Symp. Salt*, 2: 101—118.

Baar, C.A., 1974b. Discussion of the paper by King (1973). *Int. J. Rock Mech. Min. Sci. Geomech. Abstr.*, 11: 291—293.

Baar, C.A., 1975. The deformational behavior of salt rocks in situ: hypotheses vs. measurements. *IAEG/AIGI Bull.*, 12: 65—72.

Baar, C.A., 1976. Written contributions to the discussions on the following papers published in *III Congr. ISRM:* Vol. IA: Bernaix, J., 9—97; Everell, M.D., Herget, G., Sage, R. and Coates, D.F., 101—125; John, K.W., 173—186. Vol. IB: Salamon, M.D.G., 951—1395; Everling, G., 1441—1470; Lombardi, G., 1518—1554. Vol. IIA: Cvetković, M., 126—131; Mandžić, E., 186—191; Schuermann, F., 468—473; Langer, M., 619—624. Vol. IIB: Albrecht, H. and Langer, M., 967—974; Hardy, M.P., Crouch, S.L., Fairhurst, C. and Sinha, K.P., 1015—1021; Jašarević, I., 1022—1027; Wagner, H., 1076—1081; Bräuner, G., 1250—1255. (Discussions to be published in *III Congr. ISRM*, III.)

Baar, C.A. and Kühn, R., 1961. Der Werdegang der Kalisalzlagerstätten am Oberrhein. *Neues Jahrb. Mineral., Abh.*, 97: 289—336.

Baar, C.A., Von Hodenberg, R. and Kühn, R., 1971. Gelbes, lichtempfindliches Steinsalz von Esterhazy/Saskatchewan und gelber, lichtempfindlicher Boracit von Lehrte/Niedersachsen. *Kali Steinsalz*, 5: 460—472.

Barron, K. and Toews, N.A., 1964. Deformation around a mine shaft in salt. *Proc. Rock Mech. Symp., Queen's Univ., Kingston, December 1963*, pp. 115—136.

Baumert, B., 1955. Die Laugenspeicher in den Schichten des Zechsteins und ihre Gefahren für den Salzbergbau. *Z. Dtsch. Geol. Ges.*, 105: 729—733.

Boeke, H.E., 1908. Über das Kristallisationsschema der Chloride, Bromide, Jodide von Natrium, Kalium und Magnesium, sowie über das Vorkommen des Broms und das Fehlen von Jod in den Kalisalzlagerstätten. *Z. Kristall.*, 45: 346—391.

Borchert, H. and Muir, R.O., 1964. *Salt Deposits.* Van Nostrand, London, 338 pp.

Börger, H., 1954. Die Gebirgsdruckhypothesen für elastische Gesteine und ihre Anwendungsmöglichkeiten im Salzbergbau. *Kali Steinsalz*, 1(6): 10—20.

Bradshaw, R.L. and Lomenick, T.F., 1970. Rheology of salt in mine workings. *Proc. Symp. Geol. Technol. Gulf Coast Salt. Louisiana State Univ., Baton Rouge*, pp. 109—123.

Bradshaw, R.L. and McClain, W.C., 1970. Roof-bolt test and application in supporting a two-foot-thick roof bed. *III Symp. Salt*, 2: 466—470.

Bradshaw, R.L., Boegly, W.J. and Empson, F.M., 1964. Correlation of convergence measurements in salt mines with laboratory creep-test data. *6th Symp. Rock Mech., Univ. Missouri, Rolla*, pp. 501—514.

Bradshaw, R.L., Empson, F.M., Boegly, W.J., Kubota, H., Parker, F.L. and Struxness, E.G., 1968. Properties of salt important in radioactive waste disposal. In: *Saline Deposits—Geol. Soc. Am. Spec. Pap.*, 88: 643—658.

Braitsch, O., 1962. *Entstehung und Stoffbestand der Salzlagerstätten*. Springer, Berlin, 232 pp.

Braitsch, O., 1966. Bromine and rubidium as indicators of environment during sylvite and carnallite deposition of the Upper Rhine valley evaporites. *II Symp. Salt*, 1: 293—301.

Braitsch, O., 1971. *Salt Deposits—their Origin and Composition*. Springer, New York, N.Y., 297 pp.

Braitsch, O. and Herrmann, A.G., 1963. Zur Geochemie des Broms in salinaren Sedimenten, 1. *Geochim. Cosmochim. Acta*, 27: 361—391.

Brongersma-Sanders, M. and Groen, P., 1970. Wind and water depth and their bearing on the circulation in evaporite basins. *III Symp. Salt*, 1: 3—7.

Buchheim, W., 1958. Geophysikalische Methoden zur Erforschung des Spannungszustandes des Gebirges im Steinkohlen- und Kalisalzbergbau. In: *Internationale Gebirgsdrucktagung 1958, Vorträge*. Akademie-Verlag, Berlin, pp. 119—125.

Chao, R., 1974. Long-term creep closure of solution cavity system. *IV Symp. Salt*, 2: 119—127.

Clark, A.R. and Schwerdtner, W.M., 1966. Petrofabric analysis of potash rocks, Esterhazy, Saskatchewan. *II Symp. Salt*, 1: 102—121.

Coogan, A.H., 1974. Salt in the early '70s. *IV Symp. Salt*, 1: 3—6.

Coolbaugh, M.J., 1967. Special problems of mining in deep potash. *Min. Eng.*, May: 68—73.

Cornelius, C-D., 1954. Wie breiten sich Erdbebenwellen im Boden aus? Das Einsturzbeben im Werragebiet vom 22. Februar 1953. *Umschau*, 2: 48—50.

D'Ans, J., 1933. *Die Lösungsgleichgewichte der Systeme der Salze ozeanischer Salzablagerungen*. Kali-Forschungsanstalt, Berlin, 254 pp.

D'Ans, J., 1947. Über die Bildung und Umbildung der Kalilagerstätten. *Naturwissenschaften*, 34: 295—301.

D'Ans, J., 1969. Bemerkungen zu Problemen der Kalisalzlagerstätten, 2. *Kali Steinsalz*, 5: 152—157.

D'Ans, J. and Kühn, R., 1940. Über den Bromgehalt von Salzgesteinen der Kalisalzlagerstätten. *Kali*, 34: 42—46; 59—64; 77—83.

D'Ans, J. and Kühn, R., 1960. Bemerkungen zur Bildung und zu Umbildungen ozeanischer Salzlagerstätten. *Kali Steinsalz*, 3: 69—84.

Dellwig, L.F., 1953. Hopper crystals of halite in the Salina salt of Michigan. *Am. Mineral.*, 38: 730—731.

Dellwig, L.F., 1955. Origin of the Salina salt of Michigan. *J. Sediment. Petrol.*, 25: 83—110.

Dellwig, L.F. and Evans, R., 1969. Depositional processes in Salina salt of Michigan, Ohio, and New York. *Bull. Am. Assoc. Pet. Geol.*, 53: 949—956.

Dellwig, L.F. and Kühn, R., 1974. A re-evaluation of the Muschelkalk salt at Heilbronn. *IV Symp. Salt*, 1: 127.

Dreyer, W., 1955. Über das Festigkeitsverhalten sehr verschiedenartiger Gesteine. *Bergbauwissenschaften*, 2: 183—191.

Dreyer, W., 1967. *Die Festigkeitseigenschaften natürlicher Gesteine, insbesondere der Salz- und Karbongesteine*. Bornträger, Berlin-Nikolassee, 247 pp.

Dreyer, W., 1969. Planung und Inbetriebnahme der nordamerikanischen Kaligrube Cane Creek der Texas Gulf Sulphur Company in Staate Utah. *Bergbauwissenschaften*, 16: 441—446.
Dreyer, W., 1972. *The Science of Rock Mechanics, 1. The Strength Properties of Rocks.* Transl. Tech. Publ., 501 pp.
Dreyer, W., 1974a. *Gebirgsmechanik im Salz.* Enke, Stuttgart, 205 pp.
Dreyer, W., 1974b. Results of recent studies on the stability of crude oil and gas storage in salt caverns. *IV Symp. Salt*, 2: 65—92.
Dreyer, W. and Borchert, H., 1962. Kritische Betrachtung zur Prüfkörperformel von Gesteinen. *Bergbautechnik*, 12: 265—272.
Dreyer, W. and Borchert, H., 1963. Entwicklung eines neuartigen Messpatronentyps zur Bestimmung des absoluten Gebirgsdrucks in Salzgesteinspfeilern. *Kali Steinsalz*, 3: 330—337.
Duffield, A.H., 1972. Underground mining at Alwinsal Potash. *CIM Trans.*, 75: 223—229.
Empson, F.M., Boegly, W.J., Bradshaw, R.L., McClain, W.C., Parker, F.L. and Schaffer, W.F., 1966. Demonstration of disposal of high-level radioactive solids in salt. *II Symp. Salt*, 1: 432—443.
Empson, F.M., Bradshaw, R.L., McClain, W.C. and Houser, B.L., 1970. Results of the operation of Project Salt Vault: a demonstration of disposal of high-level radioactive solids in salt. *III Symp. Salt*, 1: 455—462.
Erasmus, T., 1965. Gebirgsbewegungen im Steinsalz durch Entwicklung eines Abbaufeldes. *Kali Steinsalz*, 4: 154—169.
Evans, R. and Linn, K.O., 1970. Fold relationships within evaporites of the Cane Creek anticline, Utah. *III Symp. Salt*, 1: 286—297.
Fine, J., Maurin, E., Michel, B., Sinou, P., Tincelin, E. and Vigier, G., 1964. Instability of mine workings: bumps, rockbursts in the floor and generalized collapses. *4th Int. Conf. Strata Control Rock Mech., Columbia Univ., New York City*, pp. 46—61.
Fulda, D., 1963. Auflöseerscheinungen durch Spülversatzlaugen in carnallitischen Feldesteilen des Südharz-Kalibergbaues. *Bergakademie*, 15: 448—455.
Garrett, D.E., 1970. The geochemistry and origin of potash deposits. *III Symp. Salt*, 1: 211—222.
Gimm, W., 1959. Abbauverfahren und Laugengefahr im Kalibergbau. *Freiberg. Forschungsh., A*, 114: 5—48.
Gimm. W., 1968. *Kali- und Steinsalzbergbau.* VEB Deutscher Verlag für Grundstoffindustrie, Leipzig, 600 pp.
Gimm, W. and Meyer, H., 1962. Die Besonderheiten der Laugengefahr im Südharz-Kalirevier. *Bergakademie*, 14: 403—407; 486—497.
Gimm, W. and Pforr, H., 1961. Gebirgsschläge im Kalibergbau unter Berücksichtigung von Erfahrungen des Kohlen- und Erzbergbaus. *Freiberg. Forschungsh., A*, 173: 190 pp.
Gimm, W.A.R. and Pforr, H., 1964. Breaking behavior of salt rock under rockbursts and gas outbursts. *4th Int. Conf. Strata Control Rock Mech., Columbia Univ., New York, N.Y.*, pp. 434—449.
Gimm, W., Duchrow, G. and Höfer, K.H., 1970. Neue wissenschaftliche Erkentnisse der Gebirgsmechanik im Salinar und ihre praktische Nutzanwendung in der modernen Technologie. *VI Int. Bergbaukongr., Madrid 1970, II-B*, 8: 11 pp.
Griggs, D., 1939. Creep of rocks. *J. Geol.*, 47: 225—251.
Griggs, D., 1940. Experimental flow of rocks under conditions favoring recrystallization. *Geol. Soc. Am. Bull.*, 51: 1001—1022.
Gropp, 1918. Gasvorkommen in Kalisalzbergwerken in den Jahren 1907—1917. *Z. Berg-Hütten- Salinenwesen*, 66: 238—257.
Gussow, W.C., 1970. Heat, the factor in salt rheology. *Proc. Symp. Geol. Technol. Gulf Coast Salt, Louisiana State Univ., Baton Rouge*, pp. 125—148.

Hardy, M.P., Crouch, S.L., Fairhurst, C. and Sinha, K.P., 1974. A hybrid computer system simulating inelastic seam behavior. *III Congr. ISRM*, IIB: 1015—1021.

Harrison, J.V., 1930. Geology of some salt plugs in Laristan (southern Persia). *Q. J. Geol. Soc. Lond.*, 86: 463—522.

Hedley, D.F.G., 1967. An appraisal of convergence measurements in salt mines. *4th Rock Mech. Symp., Ottawa, March 1967*, pp. 117—135.

Hedley, D.F.G., 1972. An evaluation of roof stability at a Canadian salt mine. *5th Int. Strata Control Conf., Lond., Pap.*, 30: 6 pp.

Hite, R.J., 1970. Shelf carbonate sedimentation controlled by salinity in the Paradox basin, southeast Utah. *III Symp. Salt*, 1: 48—66.

Hite, R.J., 1974. Evaporite deposits of the Khorat plateau, northeastern Thailand. *IV Symp. Salt*, 1: 135—146.

Höfer, K.H., 1958a. Beitrag zur Frage der Standfestigkeit von Bergfesten im Kalibergbau. *Freiberg. Forschungsh., A*, 100: 148 pp.

Höfer, K.H., 1958b. Die Gesetzmässigkeiten des Kriechens der Salzgesteine und deren allgemeine Bedeutung für den Bergbau. In: *Internationale Gebirgsdrucktagung, 1958, Vorträge.* Akademie-Verlag, Berlin, pp. 29—35.

Höfer, K.H., 1964. Results of rheological studies in potash mines. *4th Int. Conf. Strata Control Rock Mech., Columbia Univ., New York City*, pp. 62—69.

Höfer, K.H. and Menzel, W., 1964. Comparative study of pillar loads in potash mines established by calculation and by measurements below ground. *Int. J. Rock Mech. Min. Sci.*, 1: 181—198.

Höfer, K.H. and Knoll, P., 1971. Investigations into the mechanism of creep deformation in carnallitite, and practical applications. *Int. J. Rock Mech. Min. Sci.*, 8: 61—73.

Holdoway, K.A., 1974. Behavior of fluid inclusions in salt during heating and irradiation. *IV Symp. Salt*, 1: 303—312.

Holser, W.T., 1966. Bromide geochemistry of salt rocks. *II Symp. Salt*, 1: 248—275.

Hoppe, W., 1958. Die Bedeutung der geologischen Vorgänge bei der Metamorphose der Werra-Kalisalzlagerstätte. In: *Mineralsalze ozeanischen Ursprungs, Symp., 1958—Freiberg. Forschungsh., A*, 123: 41—60.

Hoppe, W., 1960. Die Kali- und Steinsalzlagerstätten des Zechsteins in der Deutschen Demokratischen Republik, 1: Das Werra-Gebiet. *Freiberg. Forschungsh., C*, 97/I: 166 pp.

Hsü, K.J., Cita, M.B. and Ryan, W.B.F., 1973. The origin of the Mediterranean evaporites. In: *Initial Reports of the Deep Sea Drilling Project, 13.* U.S. Gov. Print. Off., Washington, D.C., pp. 1203—1233.

Jaeger, J.C., 1962, 1969. *Elasticity, Fracture and Flow.* Methuen, London.

Junghans, R., 1953. Gebirgsschläge im Kalibergbau. *Bergakademie*, 4: 121—130.

Junghans, R. and Echtermeyer, H., 1953. Versuche zur Senkung der Abbauverluste im Kalibergbau des Südharzes. *Bergbautechnik*, 3: 21—26.

Kampf-Emden, G., 1956. Analyse abbaudynamischer Vorgänge durch Untersuchung der Verformungserscheinungen und der Spannungsverteilung im Gebirge um die Grubenbaue einer flach bis halbsteil gelagerten Kalisalzlagerstätte. *Bergbauwissenschaften*, 3: 189—199.

Kegel, K., 1957. Die Gebirgsschlaggefahr im Kalisalzbergbau. *Bergakademie*, 9: 480—488.

Keys, D.A. and Wright, J.Y., 1966. Geology of the IMC potash deposit, Esterhazy, Saskatchewan. *II Symp. Salt*, 1: 95—101.

King, M.S., 1973. Creep in model pillars of Saskatchewan potash. *Int. J. Rock Mech. Min. Sci.*, 10: 363—371.

King, M.S. and Acar, K.Z., 1970. Creep properties of Saskatchewan potash as a function of changes in temperature and stress. *III Symp. Salt*, 2: 226—235.

Kinsman, D.J.J., 1974. Evaporite deposits of continental margins. *IV Symp. Salt*, 1: 255—259.

Kozary, M.T., Dunlop, J.C. and Humphrey, W.E., 1968. Incidence of saline deposits in geologic time. In: *Saline Deposits—Geol. Soc. Am. Spec. Pap.*, 88: 43—57.

Kramm, E., 1973. Kurzbericht über die Montangeschichtliche Tagung 1973 in Hallstatt. *Kali Steinsalz*, 6: 184—185.

Kühn, R., 1955a. Tiefenberechnung des Zechsteinmeeres nach dem Bromgehalt der Salze. *Z. Dtsch. Geol. Ges.*, 105: 646—663.

Kühn, R., 1955b. Uber den Bromgehalt von Salzgesteinen, insbesondere die quantitative Ableitung des Bromgehalts nichtprimärer Hartsalze oder Sylvinite aus Carnallitit. *Kali Steinsalz*, 1(9): 3—16.

Kühn, R., 1968. Geochemistry of the German potash deposits. In: *Saline Deposits—Geol. Soc. Am. Spec. Pap.*, 88: 427—504.

Kühn, R. and Baar, A., 1955. Ein ungewöhnliches Vorkommen von Danburit. *Kali Steinsalz*, 1(10): 17—21.

Kühn, R. and Dellwig, L.F., 1971. Salt deposits of Permian, Triassic and Tertiary age in W-Germany. In: *Sedimentology of Parts of Central Europe, Guidebook, 8th Int. Sediment. Congr.*, pp. 303—326.

Kupfer, D.H., 1970. Conflicting strain patterns in the salt of Gulf Coast salt domes and their genetic implications. *III Symp. Salt*, 1: 271—282.

Kupfer, D.H., 1974. Boundary shear zones in salt rocks. *IV Symp. Salt*, 1: 215—225.

Langer, M., 1974. Rheologische Grundlagen felsmechanischer Modellversuche. *III Congr. ISRM*, IIA: 619—624.

Langer, M. and Hofrichter, E., 1969. Standsicherheit und Konvergenz von Salzkavernen. *Proc. Int. Symp. Large Permanent Underground Openings, Oslo*, pp. 147—156.

Lees, G.M., 1927. Salzgletscher in Persien. *Mitt. Geol. Ges. Wien*, 20: 29—34.

Lepeschkow, I.N., 1958. Untersuchungen der Schule N.S. Kurnakows zur physikalischen Chemie der natürlichen Salze und Salzsysteme. In: *Mineralsalze ozeanischen Ursprungs, Symp., 1958—Freiberg. Forschungsh., A*, 123: 105—118.

Linn, K. and Adams, S.S., 1966. Barren halite zones in potash deposits, Carlsbad, New Mexico. *II Symp. Salt*, 1: 59—69.

Löffler, J., 1962. Die Kali- und Steinsalzlagerstätten des Zechsteins in der Deutschen Demokratischen Republik, Sachsen-Anhalt. *Freiberg. Forschungsh., C*, 97(3): 347 pp.

Lomenick, T.F. and Bradshaw, R.L., 1969. Deformation of rock salt in openings mined for the disposal of radioactive wastes. *Rock Mech.*, 1: 5—30.

Lotze, F., 1957. *Steinsalz und Kalisalze.* Bornträger, Berlin-Nikolassee, 465 pp.

Mackintosh, A.D., 1975. The development of safe travel ways at the Cominco potash mine. *Proc. 10th Can. Rock Mech. Symp., Kingston, September 1975, Queen's Univ.*, 1: (preprint) 29 pp.

Mǎndzić, E., 1974. Effective strength of rock salt. *III Congr. ISRM*, IIA: 186—191.

Matthews, D.R. and Egleson, G.C., 1974. Origin and implications of a mid-basin potash facies in the Salina salt of Michigan. *IV Symp. Salt*, 1: 15—34.

McClain, W.C., 1964. Time-dependent behavior of pillars in the Alsace potash mines. *6th Symp. Rock Mech., Univ. Missouri, Rolla*, pp. 489—500.

McClain, W.C. and Bradshaw, R.L., 1967. Stress redistribution in room and pillar salt mines. *Int. J. Rock Mech. Min. Sci.*, 4: 245—255.

McKinlay, D.W., 1972. Practical rock mechanics at Allen Potash Mines. (Paper presented at the Annual Western CIM Meeting, Saskatoon, Saskatchewan, 19 pp.)

Mohr, F., 1954. Planung von Gebirgsdruckmessungen in Kaligruben. *Bergbauwissenschaften*, 1: 99—104.

Mohr, F., 1955. Gebirgsdruckmessungen in einem Kalibergwerk und ihre Aussagen über das Verhalten des Hangenden über Abbauräumen. *Glückauf*, 91: 1209—1217.

Mohr, F., 1960. Contribution to discussion of the paper presented by Spackeler et al., 1960. *Int. Kongr. Gebirgsdruckforschung, Paris, 1960*, p. 599.

Mraz, Z.F., 1972. The theory of flow and its practical application for pillar design in deep potash mines. *Western Miner*, April 1972: 22—26.

Mraz, D.Z.F., 1973. Behavior of rooms and pillars in deep potash mines. *CIM Trans.*, 76: 138—143.

Nair, K. and Deere, D.U., 1970. Creep behavior of salt in triaxial extension tests. *III Symp. Salt*, 2: 208—215.

Nair, K., Chang, C-Y, Singh, R.D. and Abdullah, A.M., 1974. Time-dependent analysis to predict closure in salt cavities. *IV Symp. Salt*, 2: 129—139.

Neuwirth, G., 1960. Bewegungsvorgänge im Kalibergbau des Werra-Fulda-Gebietes. *Kali Steinsalz*, 3: 37—54.

Obert, L., 1964. Deformational behavior of model pillars made of salt, trona, and potash. *6th Symp. Rock Mech., Univ. Missouri, Rolla*, pp. 539—560.

O'Brien, C.A.E., 1955. Salztektonik in Südpersien. *Z. Dtsch. Geol. Ges.*, 105: 803—813.

Odé, H., 1968a. Review of mechanical properties of salt relating to salt dome genesis. In: *Saline Deposits—Geol. Soc. Am. Spec. Pap.*, 88: 543—595.

Odé, H., 1968b. Physical properties work sessions. In: *Saline Deposits—Geol. Soc. Am. Spec. Pap.*, 88: 683—701.

Pawlick, B., 1943. Über die Untersuchungsarbeiten im Gebirgsschlagfeld des Kaliwerks Krügershall. *Kali*, 37(8—11): 130 ff.

Pearson, W.J., 1963. Salt deposits of Canada. *Symp. Salt*, pp. 197—239.

Peyfuss, K.F. and Jacoby, C.H., 1966. Measurement of salt pillar movement. *II Symp. Salt*, 2: 46—56.

Piper, T., 1974. Rock mechanics applied to conventional mining—panel discussion. *IV Symp. Salt*, 2: 177.

Potts, E.L.J., Potts, W.H. and Szeki, A., 1972. Entwicklung von Verfahren zur Beherrschung des Gebirges und Kenngrössen für die Abbauplanung im Steinsalzbergbau. *5th Int. Strata Control Conf., Lond., Pap.* 26: 10 pp.

Raup, O.B., 1966. Bromine distribution in some halite rocks of the Paradox Member, Hermosa Formation, in Utah. *II Symp. Salt*, 1: 236—247.

Raup, O.B., 1970. Brine mixing: an additional mechanism for formation of basin evaporites. *Bull. Am. Assoc. Pet. Geol.*, 54: 2246—2259.

Raup, O.B., Hite, R.J. and Groves, H.L., 1970. Bromine distribution and paleosalinities from well cuttings, Paradox basin, Utah and Colorado. *III Symp. Salt*, 1: 40—47.

Richter-Bernburg, G., 1955. Über salinare Sedimentation. *Z. Dtsch. Geol. Ges.*, 105: 593—645.

Richter-Bernburg, G., 1970. Post-depositional structures in salt. *Proc. Symp. Geol. Technol. Gulf Coast Salt, Louisiana State Univ., Baton Rouge*, pp. 161—166.

Rininsland, H., 1972. Der Bergwerksbetrieb der Alwinsal Potash of Canada Ltd. in Lanigan (Provinz Saskatchewan, Canada). *Kali Steinsalz*, 6: 33—39.

Röhr, H.U., 1974. Mechanical behavior of a gas storage cavern in evaporitic rocks. *IV Symp. Salt*, 2: 93—100.

Salamon, M.D.G., 1974. Rock mechanics of underground excavations. *III Congr. ISRM*, IB: 951—1395.

Salustowicz, A., 1958. Neue Anschauungen über den Spannungs- und Formänderungszustand im Gebirge in der Nachbarschaft bergmännischer Hohlräume. In: *Internationale Gebirgsdrucktagung 1958, Vorträge.* Akademie-Verlag, Berlin, pp. 5—11.

Sannemann, D., 1968. Salt-stock families in northwestern Germany. In: *Diapirism and Diapirs—Am. Assoc. Pet. Geol., Mem.*, 8: 261—270.

Schauberger, O., 1950. Gebirgsdruckerscheinungen im alpinen Haselgebirge. In: *Internationale Gebirgsdrucktagung, 1950, Leoben*, pp. 138—140.

Schauberger, O., 1955. Zur Genese des alpinen Haselgebirges. *Z. Dtsch. Geol. Ges.*, 105: 736—751.

Schmidt, W., 1943. Gesteinsfestigkeit im Bergbau. *Kali,* 37: 98—105.
Schmidt, W., 1944. Zur Versatzfrage im Kalibergbau. Kali, 38: 135—138.
Schultz, W.G., 1971. Planning the Sylvite project. *Western Miner,* 44(11): 37—45.
Schultz, W.G., 1973. Automated continuous mining at Sylvite. *CIM Trans.,* 76: 110—119.
Schulze, G., 1958. Beitrag zur Stratigraphie und Genese der Steinsalzserien I—IV des mitteldeutschen Zechsteins unter besonderer Berücksichtigung der Bromverteilung. In: *Mineralsalze ozeanischen Ursprungs, Symp., 1958—Freiberg. Forschungsh., A,* 123: 175—196.
Schwerdtner, W.M., 1974. Schistosity in deformed anhydrite—a reinterpretation. *IV Symp. Salt,* 1: 235—240.
Schwerdtner, W.M. and Morrison, M.J., 1974. Internal-flow mechanism of salt and sylvinite in the Anagance diapiric anticline near Sussex, New Brunswick. *IV Symp. Salt,* 1: 241—248.
Schwerdtner, W.M. and Wardlaw, N.C., 1963. Geochemistry of bromine in some salt rocks of the Prairie Evaporite Formation of Saskatchewan. *Symp. Salt,* pp. 240—246.
Sellers, J.B., 1970. Rock mechanics instrumentation for salt mining. *III Symp. Salt,* 2: 236—248.
Serata, S., 1964. Theory and model of underground opening and support system. *6th Symp. Rock Mech., Univ. Missouri, Rolla,* pp. 260—292.
Serata, S., 1966. Continuum theory and model of rock salt structures. *II Symp. Salt,* 2: 1—17.
Serata, S., 1968. Application of continuum mechanics to design of deep potash mines in Canada. *Int. J. Rock Mech. Min. Sci.,* 5: 293—314.
Serata, S., 1970. Prerequisites for application of finite-element method to solution cavities and conventional mines. *III Symp. Salt,* 2: 249—279.
Serata, S., 1972. The Serata stress control method of stabilizing underground openings. *Proc. 7th Can. Rock Mech. Symp., Edmonton, March 1971,* pp. 99—118.
Serata, S., 1973. Rock mechanics problems of solution cavities used for storage of gaseous and solid matters. *Soc. Min. Eng. AIME,* Preprint 73-AM-79: 26 pp.
Serata, S., 1974. Utilization of stress envelopes in design of solution cavities. *IV Symp. Salt,* 2: 51—64.
Serata, S., 1976. Stress control technique—an alternative to roof bolting. *Min. Eng.,* 28(5): 51—56.
Serata, S. and Schultz, W.G., 1972. Application of stress control in deep potash mines. *Min. Congr. J.,* 58(11): 36—42.
Serowy, F., 1958. Spezielle Probleme der Mineralsalzforschung und -auswertung. In: *Mineralsalze ozeanischen Ursprungs, Symp., 1958—Freiberg. Forschungsh., A,* 123: 287—302.
Shearman, D.J. and Fuller, J.G., 1969. Anhydrite diagenesis, calcitization, and organic laminites, Winnipegosis Formation, Middle Devonian, Saskatchewan. *Can. Pet. Geol. Bull.,* 17: 182—193.
Shitkov, E.F. and Permyakov, R.S., 1972. Room and pillar mining with controlled continuous subsidence of the overburden at the Soligorsk potash mines *Gorn. Zh.,* 5: 33—35 (in Russian). (Summary in German by H. Börger, 1972, *Kali Steinsalz,* 6: 115—116. Translation and discussion by C.A. Baar, 1973, *SRC Rep.,* E 73-3; 41 pp.)
Shlichta, P.J., 1968. Growth, deformation, and defect structure of salt crystals. In: *Saline Deposits—Geol. Soc. Am. Spec. Pap.,* 88: 597—617.
Simon, P. and Haltenhof, M., 1970. Feinstratigraphie, Fazies und Bromgehalt des Stassfurt Steinsalzes (Zechstein 2) im Kali- und Steinsalz-Bergwerk "Asse" (Schacht II) bei Braunschweig. *Geol. Jahrb.,* 88: 159—202.
Singhal, R.S., 1971. Determination of stress in evaporites. *Can. Min. J.,* August 1971: 35—47.

Smith, J.D., 1966. The use of rock mechanics in mine design at the Fairport mine, Morton Salt Company. *II Symp. Salt*, 2: 34—45.

Sorensen, H.O. and Segall, R.T., 1974. Natural brines of the Detroit River Group, Michigan basin. *IV Symp. Salt*, 1: 91—99.

Spackeler, G., 1956. Abbauverfahren und Abbauwirkung im Salzbergbau. *Bergakademie*, 8: 569—580.

Spackeler, G., 1958. Beobachtungen über die Spannungsverteilung um Steinkohlen- und Salzabbaue. In: *Internationale Gebirgsdrucktagung, 1958, Vorträge*. Akademie-Verlag, Berlin, pp. 162—168.

Spackeler, G. and Sieben, 1944. Zusammenstellung und Auswertung zahlenmàssiger Angaben über die physikalischen Eigenschaften der Salzgesteine. *Kali*, 38: 101—104; 115—127.

Spackeler, G., Gimm, W., Höfer, K.H. and Duchrow, G., 1960. Neue Erkenntnisse über Gebirgsschläge im Kalibergbau. *Int. Kongr. Gebirgsdruckforschung, Paris, 1960*, Vortrag G 5: 583—603.

Stewart, F.H., 1963. Data of geochemistry—marine evaporites. *Am. Geol. Surv., Prof. Pap.*, 440-Y: 53 pp.

Stöcke, K. and Borchert, H., 1936. Fliessgrenzen von Salzgesteinen und Salztektonik. *Kali*, 30: 191—195; 204—207; 214—217.

Stöcklin, J., 1968. Salt deposits of the Middle East. In: *Saline Deposits—Geol. Soc. Am. Spec. Pap.*, 88: 157—181.

Storck, U., 1952. Das Wiederherstellen der Förderfähigkeit des ersoffenen Kalibergwerks Königshall-Hindenburg. *Kali Steinsalz*, 1(1): 3—24.

Storck, U., 1954. Die Entstehung der Vertaubungen und des Hartsalzes im Flöz Stassfurt im Zusammenhang mit regelmässigen Begleiterscheinungen auf dem Kaliwerk Königshall-Hindenburg. *Kali Steinsalz*, 1(6): 21—31.

Storck, U., 1962. Auswertung jüngster Erkenntnisse der Gebirgsdruckforschung für die Praxis im Kalibergbau. *Bergbauwissenschaften*, 9: 341—351.

Sturmfels, E., 1943. Das Kalilager von Buggingen (Südbaden). *Neues Jahrb. Mineral., Abh.*, 78: 131—216.

Tincelin, E. and Sinou, P., 1960. Der Einsturz des Hangenden bei Örterbau. Praktische Schlussfolgerungen und Versuch einer mathematischen Erfassung der bei Gebirgsschlägen beobachteten Vorgänge. *Int. Kongr. Gebirgsdruckforsch., Paris, 1960*, Vortrag G 6: 21 pp.

Trusheim, F., 1960. Mechanism of salt migration in northern Germany. *Bull. Am. Assoc. Pet. Geol.*, 44: 1519—1540.

Uhlenbecker, F.W., 1968. *Verformungsmessungen in der Grube und ergänzende Laboruntersuchungen auf dem Kaliwerk Hattorf (Werra-Revier) im Hinblick auf eine optimale Festlegung des Abbauverlustes bei grösstmöglicher Sicherheit der Grubenbaue.* Thesis, Tech. Univ., Clausthal, 180 pp.

Uhlenbecker, F.W., 1971. Gebirgsmechanische Untersuchungen auf dem Kaliwerk Hattorf (Werra-Revier). *Kali Steinsalz*, 5: 345—359.

Uhlenbecker, F.W., 1971. Gebirgsmechanische Untersuchungen auf dem Kaliwerk Hattorf Salzbergbau. *Kali Steinsalz*, 6: 308—314.

Valyashko, M.G., 1956. Geochemistry of bromine in the processes of salt deposition and the use of the bromine content as a genetic and prospecting criterion. *Geochemistry*, 6: 570—589.

Valyashko (Waljaschko), M.G., 1958. Die wichtigsten geochemischen Parameter für die Bildung der Kalisalzlagerstätten. In: *Mineralsalze ozeanischen Ursprungs, Symp. 1958—Freiberg. Forschungsh., A*, 123: 197—235.

Wagner, W., 1955. Die tertiären Salzlagerstätten im Oberrheintal-Graben. *Z. Dtsch. Geol. Ges.*, 105: 706—728.

Wardell, K., 1958. Probleme der Analysierung und Erklärung beobachteter Bodenbewegungen. In: *Internationale Gebirgsdrucktagung, 1958*, Vorträge, Akademie-Verlag Berlin, pp. 110—118.

Wardell, K., 1968. Design of partial extraction systems in mining. *4th Rock Mech. Symp., Ottawa, March 1967*, pp. 271—284.

Wardlaw, N.C., 1963. Bromine in some Middle Devonian salt rocks of Alberta and Saskatchewan. *3rd Int. Williston Basin Symp., Sask. Geol. Soc.*, pp. 270—273.

Wardlaw, N.C., 1968. Carnallite—sylvite relationships in the Middle Devonian Prairie Evaporite formation, Saskatchewan. *Geol. Soc. Am. Bull.*, 79: 1273—1294.

Wardlaw, N.C. and Reinson, G.E., 1971. Carbonate and evaporite deposition and diagenesis, Middle Devonian Winnipegosis and Prairie Evaporite formations of south-central Saskatchewan. *Am. Assoc. Pet. Geol. Bull.*, 55: 1759—1786.

Wardlaw, N.C. and Schwerdtner, W.M., 1966. Halite—anhydrite seasonal layers in the Middle Devonian Prairie Evaporite Formation, Saskatchewan, Canada. *Geol. Soc. Am. Bull.*, 77: 331—342.

Wardlaw, N.C. and Watson, D.W., 1966. Middle Devonian salt formations and their bromide content, Elk Point area, Alberta. *Can. J. Earth Sci.*, 3: 263—275.

Weaver, P., 1963. Some early geological and production problems in the industry beginning 100 years ago. *Symp. Salt*, pp. 282—287.

Wieselmann, E.A., 1968. Closure measurements, an important tool in mine design at the Cane Creek potash mine. (Paper presented at 10th Rock Mech. Symp., Austin, Texas, May 1968; Preprint: 14 pp.)

Wilkening, W., 1958. Die Verformung der Gebirgsschichten im Hangenden und Liegenden von Abbaufeldern im Kalisalzbergbau und die daraus zu folgernde Deutung der abbaudynamischen Vorgänge. *Bergbauwissenschaften*, 5: 1—15.

Wilkening, W., 1959. Discussion of papers presented at the "Internationale Gebirgsdrucktagung 1958, Leipzig" by K.H. Höfer, and G. Spackeler. In: *Internationale Gebirgsdrucktagung, 1958, Diskussionen.* Akademie-Verlag, Berlin, pp. 50—51; 174.

Winkel, B.V., Gerstle, K.H. and Ko, H.Y., 1972. Analysis of time-dependent deformations of openings in salt media. *Int. J. Rock Mech. Min. Sci.*, 8: 249—260.

Wjasowow, W.W., 1958. Die Arbeiten des Allunions-Forschungsinstituts für Halurgie auf dem Gebiet der Verarbeitung von Kalisalzen. In: *Mineralsalze ozeanischen Ursprungs, Symp. 1958—Freiberg. Forschungsh., A*, 123: 376—388.

Zahary, G., 1965. Rock mechanics at International Minerals and Chemical Corporation (Canada) Limited. *3rd Can. Symp. Rock Mech. Univ. Toronto, January 1965*, pp. 1—17.

Zak, I., 1974. Sedimentology and bromine geochemistry of marine and continental evaporites in the Dead Sea basin. *IV Symp. Salt*, 1: 349—361.

SUBJECT INDEX